Indigenous Knowledge Management in the Open Science Framework

Josiline Chigwada
University of South Africa, South Africa

Patrick Ngulube
University of South Africa, South Africa

IGI Global
Scientific Publishing
Publishing Tomorrow's Research Today

Vice President of Editorial	Melissa Wagner
Director of Acquisitions	Mikaela Felty
Director of Book Development	Jocelynn Hessler
Production Manager	Mike Brehm
Cover Design	Jose Rosado

Published in the United States of America by
IGI Global Scientific Publishing
701 East Chocolate Avenue
Hershey, PA, 17033, USA
Tel: 717-533-8845
Fax: 717-533-7115
Website: https://www.igi-global.com E-mail: cust@igi-global.com

Library of Congress Cataloging-in-Publication Data

Names: Chigwada, Josiline Phiri, 1984- editor
http://id.loc.gov/rwo/agents/n2020006381 | Ngulube, Patrick editor
http://id.loc.gov/rwo/agents/no2007123279
Title: Indigenous knowledge management in the open science framework / edited by: Josiline Chigwada, Patrick Ngulube.
Description: Hershey, PA : IGI Global Scientific Publishing, [2026] | Collection of essays by Stephen Tsekea and 9 others. | Includes bibliographical references and index. | Summary: "The book will contribute to the growing discourse on digital preservation, knowledge management, and sustainable development"-- Provided by publisher.
Identifiers: LCCN 2025009744 (print) | LCCN 2025009745 (ebook) | ISBN 9798337302041 hc | ISBN 9798337302058 sc | ISBN 9798337302065 e
Subjects: LCSH: Ethnoscience--Technological innovations | Information storage and retrieval systems--Ethnoscience
http://id.loc.gov/authorities/subjects/sh2021000469
Classification: LCC GN476 .I5186 2026 (print) | LCC GN476 (ebook) | DDC 001.089--dc23/eng/20250417
LC record available at https://lccn.loc.gov/2025009744
LC ebook record available at https://lccn.loc.gov/2025009745

British Cataloguing in Publication Data
A Cataloguing in Publication record for this book is available from the British Library.

Editorial Advisory Board

Table of Contents

Detailed Table of Contents

Open science and the developments of the digital environment have played a vital role in research on indigenous knowledge. This chapter discusses the relationship between open science and indigenous knowledge in the digital environment. The core elements of open science included open data, open educational resources, openness to diverse knowledge, open access publication and open science infrastructure among others. The review showed increased interest among nations on open sciences as witnessed by interventions that have been adopted globally. These interventions show that there is a collaborative role between research indigenous knowledge and open science and this has been enabled by digital technology development. However, the implementation of open science in research on indigenous knowledge has faced some challenges related to ethical issues and the reliability of indigenous knowledge, lack of Indigenous knowledge resources and limited funding.

Indigenous Knowledge (IK) systems are organized around cultural sovereignty, confidentiality, and moral leadership, while FAIR principles are intended to promote open information sharing. The relationship between sovereignty and accessibility is analyzed and modeled using a Bi-Stacked Artificial Neural Network technique. Using Particle Swarm Optimization, feature selection streamlined, preprocessed, and standardized data gathered from Indigenous governance models and FAIR compliance reports. In order to create knowledge-sharing models that adhere to Indigenous standards and are practical, Bi-Stacked ANN is trained in hierarchical representations and ethical AI frameworks improve the interpretability of models. Responsible data handling is promoted thus to ensure global access to information and

Indigenous people's rights in balance. The results show AI-driven policy simulations that both promote digital inclusion and cultural sovereignty.

The digital era has ushered in unprecedented opportunities for the preservation and dissemination of knowledge. However, it has also presented unique challenges, particularly in the context of safeguarding indigenous knowledge systems. This chapter outlines a comprehensive examination of the obstacles faced in the digital preservation of indigenous knowledge and proposes strategies to overcome these barriers to ensure that invaluable cultural heritage and sustainable knowledge is not lost to future generations. Qualitative data was gathered through a comprehensive literature review and case study analyses. The chapter concludes with valuable insights and recommendations for future initiatives in this vital area of research and practice.

Digital Tools for Indigenous Knowledge Management is an important project for the conservation of material cultural heritage by leveraging advanced computational techniques. This paper's research is focused on data collection, preprocessing, and analysis from heritage monuments, sculptures, and buildings in Andhra Pradesh. Data preprocessing includes cleaning, normalization, and geospatial validation to render it accurate, followed by feature selection via Ant Colony Optimization (ACO) to determine the most significant features. An improved ResNet model is utilized for classification and analysis of high-resolution images and metadata to enhance heritage documentation. Further, digital application integration such as augmented reality (AR), virtual reality (VR), and geographic information systems (GIS) enables interactive discovery of heritage and sharing of knowledge. The proposed methodology strengthens the bond of cultural heritage conservation with the aid of Indigenous knowledge through innovative digital platforms.

This chapter provides an in-depth exploration of the synergies and tensions between indigenous knowledge systems and the principles of Open Science. In an era where the democratisation of knowledge is paramount, it is crucial to understand how indigenous knowledge – often transmitted orally and held communally – can be preserved, shared, and integrated into the global scientific community without compromising its integrity or the rights of indigenous peoples. This chapter serves as a blueprint for a thought-provoking and necessary exploration of indigenous knowledge management within the rapidly evolving context of Open Science. It provides recommendations of how the two traditions may be ethically bridged to co-create knowledge in the modern era.

Traditional knowledge has always played a critical role in the health fraternity of most African countries. However, indigenous communities in Zimbabwe continue to lose their valuable knowledge through such factors as globalization, climate change and death or memory loss of the elders. The threats of extinction of this knowledge particularly medicinal IKS such as herbal medicine and indigenous healing have been facilitated by lack of recordings since most of it relies on oral and memory transmission, too much preference for western health systems and lack of intellectual property protection from pirates. The loss of such critical knowledge has led to continued community diseases and ultimately loss of life in some circumstances. Culture is also threatened whenever there is a loss of its essential system of knowledge. This chapter details how traditional medicine such as the use of herbs is a key player in the health of indigenous communities. It also unpacks what efforts can be made to preserve such knowledge systems.

This chapter explores the challenges faced in digitally preserving Indigenous knowledge concerning Southern African countries. It examined the significance of digitising Indigenous Knowledge, emphasising the processes involved in its digitisation. However, even though the digitalisation of Indigenous knowledge is a malleable idea, it has opened a 'Pandora's box' of challenges. The study used a desk study in which the researchers gathered data from the existing literature using literature reviews from books, journal articles, internet sources, and newspapers; this chapter will explore the need to preserve indigenous knowledge, although most of it is not preserved. This chapter identifies the barriers to the practical digital preservation of Indigenous Knowledge and offers insights and recommendations for creating sustainable and culturally sensitive digitisation strategies. The findings aim to contribute to academic discourse and inform policymakers, researchers, and practitioners seeking to bridge the gap between traditional knowledge systems and modern technological advancements.

This chapter explores the challenges of digitally preserving Indigenous Knowledge (IK) within the Open Science Framework, highlighting conflicts between IK's communal nature and principles of openness. Key challenges include ethical dilemmas (consent, ownership, cultural misrepresentation), technological barriers (limited infrastructure, rapid obsolescence), and legal issues (intellectual property laws vs. traditional knowledge). It addresses cultural sensitivities (respecting protocols, community resistance) and resource constraints (financial limitations, shortage of trained personnel). The 7C model (Co-design, Conceptualisation, Collection, Correction, Curation, Circulation, Creation) is proposed to navigate these complexities. The chapter advocates for community-led, culturally respectful approaches to ensure IK preservation while aligning with Open Science ideals.

This chapter examines ethical priorities in Zimbabwe's public libraries regarding indigenous knowledge. Public libraries safeguard indigenous knowledge through community collaboration. Key concerns include acknowledging sources (intellectual property), project impact, and considering indigenous languages in classification. Public libraries recognize a moral duty to integrate indigenous and foreign knowledge. Collaboration with Indigenous elders is essential for language revitalization. The study highlights the importance of ethical standards in ensuring respectful engagement with Indigenous communities. By prioritizing ethics, public libraries can preserve and promote indigenous knowledge while respecting Indigenous rights.

The Protection, Promotion, Development and Management of Indigenous Knowledge Act (South Africa) and the Protection of Traditional Knowledge and Cultural Expressions Act No 33 of 2016 (Kenya) were thematically analysed guided by the following research questions: What are the laws' main aims? (ii) How do the acts seek to promote preservation of African indigenous knowledge systems (AIKS)? and (iii) How can technology be leveraged to promote AIKS? The laws were found to make provisions for state regulation and demonstrate a commitment to reach out to diverse groups. Kenyan stakeholders are urged to avail copies of the law in at least a single indigenous language like Swahili which is understood across Eastern Africa. Governments, cultural organisations and businesses are urged to capacitate citizens with devices, gadgets and training on how to access and preserve AIKS electronically.

Preface

The global research landscape is undergoing a significant transformation. At the heart of this shift is the open science movement, a framework that champions transparency, inclusivity, and equitable access to knowledge (Maedche et al., 2024; Winker el., 2024). As the world embraces this revolution in how knowledge is created and shared, a fundamental question arises, i.e. where does indigenous knowledge (IK) fit within this rapidly evolving paradigm?

Indigenous knowledge, rooted in the traditions, cultures, and lived experiences of local communities holds vast reservoirs of wisdom on sustainability, biodiversity, health, spirituality, social organisation, and governance (Ajitha et al., 2025; Yerramilli, 2025). Passed down through generations, often orally or through practice, IK is a living, dynamic system of understanding the world. As modern science grapples with the climate crisis, sustainable development, and global health challenges, the relevance of IK is more urgent than ever. Yet, it remains vulnerable, marginalized by dominant epistemologies, under-documented, and at risk of extinction due to globalization, digital exclusion, and loss of cultural heritage (Masenya, 2023).

WHY MANAGE IK WITHIN THE OPEN SCIENCE FRAMEWORK?

Open science offers an opportunity to preserve, respect, and elevate the value of IK. Through digital tools, knowledge repositories, and inclusive governance models, open science can make IK more accessible, when done ethically (Chigwada & Ngulube, 2025; Ngulube, 2023). It can connect generations, geographies, and disciplines, and can protect IK from being lost and, equally importantly, from being exploited. However, this integration is not without tensions.

THE TENSIONS: CHALLENGES OF MANAGING IK IN THE OPEN SCIENCE ERA

The process of aligning indigenous knowledge systems with the principles of open science is troubled with conceptual, ethical, technological, and legal challenges:

- Cultural sensitivity vs. openness since IK is often sacred, contextual, and not meant for public dissemination. Open science's call for transparency and sharing may conflict with indigenous norms of knowledge secrecy, ownership, and collective custodianship (Chigwada & Ngulube, 2025; Ngulube, 2023).
- Ethical concerns where consent, recognition of intellectual property rights, and protection from exploitation are paramount. Many communities fear that digitization and open access could facilitate cultural appropriation or commercialization of their knowledge without benefit or acknowledgment (Chigwada & Ngulube, 2025; Ngulube, 2023).
- Technological and infrastructural barriers where many indigenous communities are still digitally excluded. Lack of access to electricity, internet, digital devices, and skills limits their participation in digital knowledge ecosystems (Chigwada & Ngulube, 2025; Ngulube, 2023).
- Legal frameworks where conventional intellectual property regimes do not recognize collective ownership or oral traditions, leaving IK unprotected (Chigwada & Ngulube, 2025; Ngulube, 2023).
- Knowledge system incompatibility where scientific methods prioritize quantitative, empirical evidence, while IK may be experiential, narrative, and context-bound. Harmonizing these methodologies requires mutual respect and innovative frameworks (Chigwada & Ngulube, 2025; Ngulube, 2023).

TOWARDS ETHICAL INTEGRATION: SOLUTIONS AND THE WAY FORWARD

Despite these challenges, there are promising pathways for responsibly managing IK within the open science framework:

- Community-Led digitization initiatives should be participatory, respecting indigenous ways of doing things. Communities must have control over what is shared, with whom, and under what terms.
- Ethical guidelines and protocols would lead to the adoption of culturally appropriate ethical frameworks like the CARE principles (Collective ben-

efit, Authority to control, Responsibility, Ethics) which can protect community rights.

- Technological adaptation such as Artificial intelligence, machine learning, and digital preservation tools must be adapted to reflect indigenous data protocols. Tools that enable restricted access, traceability, and localized metadata standards can bridge the digital-ethical divide.
- Legal reforms through national and regional legislative frameworks that recognize and protect indigenous knowledge systems. This includes legal recognition of collective rights, protection of traditional knowledge, and benefit-sharing mechanisms.
- Educational reform and awareness by integrating indigenous perspectives into curricula, research, and policy design ensures a new generation of scientists, librarians, and policymakers is equipped to engage respectfully with IK.

This book responds to these needs by examining the multifaceted relationship between indigenous knowledge and the open science movement. Through empirical studies, theoretical frameworks, case studies, and technological innovations, it offers a comprehensive exploration of how IK can be ethically preserved, shared, and empowered in the digital age.

INTENDED AUDIENCE

This book is intended for a diverse and interdisciplinary audience that includes academics and researchers in knowledge management, library science, anthropology, indigenous studies, information science, and digital humanities. Policy makers and legislators working on cultural heritage, intellectual property, and indigenous rights can also benefit from the book. Also librarians, archivists, and curators tasked with safeguarding indigenous content in the digital age, including technologists and data scientists interested in AI, digital preservation, and community-based repositories. Community leaders, educators, and students engaged in indigenous knowledge revitalization and cultural sustainability are also part of the targeted audience. The book offers tools for reflection, policy development, ethical engagement, and innovation in indigenous knowledge management by providing empirical insights, conceptual frameworks, and real-world case studies.

ORGANISATION OF THE CHAPTERS

This volume comprises seventeen chapters that together present an assortment of thought, research, and practice in the domain of indigenous knowledge management within the open science framework.

Chapter 1 introduces the convergence of open science principles with indigenous knowledge systems and highlights global initiatives, digital interventions, and ethical tensions. It was noted that there is increased interest among nations on the intersection of open science and indigenous knowledge management. This foundational chapter introduces the relationship between open science principles and indigenous knowledge systems, highlighting both opportunities and tensions. It maps the digital transformations enabling this relationship and critiques the ethical, cultural, and infrastructural challenges that must be navigated. Chapter 2 explores how artificial intelligence, specifically Bi-Stacked ANN models, can balance FAIR data principles with indigenous data sovereignty. This chapter applies artificial intelligence to model the relationship between knowledge accessibility and sovereignty. It presents a Bi-Stacked Artificial Neural Network model that respects indigenous data protocols, thereby proposing an ethical, technologically grounded pathway for managing IK within FAIR principles.

Chapter 3 analyzes the challenges of digital preservation and suggests inclusive strategies to document and protect IK in the digital world. Focusing on the risks of cultural loss due to digital exclusion, this chapter explores how the digital divide undermines IK preservation. Through case studies and literature reviews, it proposes inclusive strategies to ensure that indigenous voices are preserved and heard in the digital realm. Chapter 4 demonstrates how AI and digital imaging can support the preservation of physical and symbolic cultural heritage through advanced computational techniques. By showcasing the application of AI and computer vision in documenting material heritage e.g., monuments and artifacts, this chapter presents scalable models for capturing and preserving indigenous cultural knowledge using innovative technologies. Chapter 5 provides a critical reflection on how to ethically bridge indigenous knowledge systems with open science without compromising cultural values. This chapter addresses how indigenous oral traditions can ethically interface with open, global knowledge systems. It proposes strategies for co-creating knowledge while maintaining the integrity of indigenous epistemologies.

Chapter 6 focuses on the preservation of Zimbabwe's traditional medicinal practices and the need for proactive documentation and legal protection. The chapter highlights the urgency of documenting medicinal knowledge systems in Zimbabwe, exploring how their disappearance affects community health. It proposes digitization strategies to preserve herbal remedies and traditional healing practices. Chapter 7 reviews scholarly work to examine risks like appropriation and loss of control, while suggesting community-led preservation methods. Offering a meta-review of scholarly works from 2000 to 2025, this chapter explores issues of intellectual property, knowledge appropriation, and power asymmetries in IK preservation. It calls for community-owned repositories as a solution. Chapter 8 presents a case study from Indonesia emphasizing digital resilience and the role of digital repositories in disaster-prone communities. This chapter explores how disaster-prone communities can use digital repositories for archiving essential records. It illustrates how community resilience and cultural continuity can be enhanced through digital platforms.

Chapter 9 delivers regional insight into the obstacles Southern African countries face in digitizing IK, offering context-specific recommendations. This regionally focused chapter explores barriers such as inadequate policies, lack of infrastructure, and epistemic dissonance in preserving IK. It offers policy-oriented solutions and culturally sensitive digitization strategies. Chapter 10 proposes a model to ethically and effectively preserve IK within open science by integrating co-design and cultural sensitivity. Proposing the 7C model (Co-design, Conceptualization, Collection, Correction, Curation, Circulation, Creation), this chapter provides a stepwise ethical framework to guide the preservation of IK in the open science era. Chapter 11 emphasizes libraries' roles in ethical stewardship, collaboration with elders, and use of Indigenous languages in classification systems. Drawing from public libraries in Zimbabwe, this chapter emphasizes ethics, collaboration, and language sensitivity in how libraries manage and share indigenous knowledge.

Chapter 12 analyzes South African and Kenyan laws aimed at protecting IK and calls for increased technological and legal literacy among citizens. By analyzing Kenyan and South African legislative frameworks, this chapter demonstrates how law can be a tool for protecting and promoting indigenous knowledge systems, particularly when paired with digital literacy and infrastructure. Chapter 13 details a case study on the tacit knowledge and rituals associated with bulls in the Isukha community, using knowledge capture frameworks. Focusing on intangible heritage, this chapter documents the symbolic and ritual knowledge around bulls in the Isukha community of Kenya, showcasing a model for capturing and preserving tacit indigenous knowledge. Chapter 14 analyzes the intersection of technology and tradition in shaping marital values and the rise of gender-based violence in digital spaces. Exploring cultural shifts in marital beliefs due to modernization, this chapter links digital technologies to the erosion of traditional gender roles and increased vulnerability to gender-based violence.

CONTRIBUTION TO THE FIELD

This book is more than an academic text since it is a call for epistemic justice. It urges scholars, institutions, and communities to reimagine the place of Indigenous Knowledge in our collective knowledge systems. By combining ethical reflection, technological innovation, legal analysis, and grassroots perspectives, this volume expands the theoretical and practical frontiers of Open Science.

It contributes to the literature by:

- Advancing new models for IK preservation e.g., AI-informed frameworks and the 7C model,
- Bridging the gap between traditional knowledge systems and digital infrastructures,
- Elevating community-led and participatory approaches to knowledge management,
- Offering policy-relevant insights for legislators and institutions, and
- Cultivating a multidisciplinary discourse that values both western and indigenous epistemologies.

In doing so, this book lays the groundwork for a more inclusive and sustainable global knowledge ecosystem, one where Indigenous voices are not only heard but are central to shaping the future of science, education, and development.

A CALL TO STEWARDSHIP AND INNOVATION

This book is both a scholarly resource and a call to action. It is a bridge between knowledge systems that are scientific and traditional, digital and oral, open and sacred. It recognizes that preserving indigenous knowledge is not merely an academic pursuit but is an ethical imperative. It is a declaration that in this new era of open science, there must be room for diverse ways of knowing and being. Only then can we build an inclusive, equitable, and sustainable global knowledge society.

Josiline Chigwada
University of South Africa, South Africa

Patrick Ngulube
University of South Africa, South Africa

REFERENCES

Ajitha, S. K. J., R., Geevarathna, & Huxley, S. (2025). Indigenous Knowledge and Sustainable Development: Bridging Traditional Practices with Modern Climate Solutions. In P. Roy, M. Hamidi, & H. Wahab (Eds.), *Community Climate Justice and Sustainable Development* (pp. 147-180). IGI Global Scientific Publishing. https://doi.org/DOI: 10.4018/979-8-3373-0619-3.ch008

Chigwada, J., & Ngulube, P. (2025). Integrating CARE and FAIR Principles in the Digital Preservation of Indigenous Knowledge to Achieve the Sustainable Development Goals. In Ngulube, P. (Ed.), *Harnessing Indigenous Epistemologies for Sustainable Progress* (pp. 197–230). IGI Global Scientific Publishing., DOI: 10.4018/979-8-3693-8447-3.ch010

Maedche, A., Elshan, E., Höhle, H., Lehrer, C., Recker, J., Sunyaev, A., Sturm, B., & Werth, O. (2024). Open Science: Towards Greater Transparency and Openness in Science. *Business & Information Systems Engineering*, *66*, 517–532. DOI: 10.1007/s12599-024-00858-7

Masenya, T. M. (2023). Revitalization and Digital Preservation of Indigenous Knowledge Systems for Sustainable Development of Indigenous Communities in South Africa. *The Serials Librarian*, *84*(5–8), 86–102. DOI: 10.1080/0361526X.2023.2277962

Ngulube, P. (2023). From Orality to Digital Assets: Managing Indigenous Knowledge in Africa in the Wake of the Open Science Movement. *The Serials Librarian*, *84*(5–8), 71–85. DOI: 10.1080/0361526X.2023.2296880

Winker, Margaret A., Bloom, T., Onie, S., & Tumwine, J. (2024). Equity, transparency, and accountability: Open science for the 21st century. *Lancet*, *402*(10409), 1206–1209. DOI: 10.1016/S0140-6736(23)01575-1 PMID: 37805196

Yerramilli, A. R. (2025). Indigenous Knowledge and Sustainable Development: The Indian Story. In Masenya, T., & Bhuda, M. (Eds.), *Revaluation and Preservation of Indigenous Knowledge Systems in Modern Society* (pp. 465–490). IGI Global Scientific Publishing., DOI: 10.4018/979-8-3693-7964-6.ch023

Chapter 1
Examining the Intersection of Open Science and Indigenous Knowledge Systems in the Digital Environment

Elisha Mupaikwa
https://orcid.org/0000-0002-0313-7139
National University of Science and Technology, Bulawayo, Zimbabwe

ABSTRACT

Open science and the developments of the digital environment have played a vital role in research on indigenous knowledge. This chapter discusses the relationship between open science and indigenous knowledge in the digital environment. The core elements of open science included open data, open educational resources, openness to diverse knowledge, open access publication and open science infrastructure among others. The review showed increased interest among nations on open sciences as witnessed by interventions that have been adopted globally. These interventions show that there is a collaborative role between research indigenous knowledge and open science and this has been enabled by digital technology development. However, the implementation of open science in research on indigenous knowledge has faced some challenges related to ethical issues and the reliability of indigenous knowledge, lack of Indigenous knowledge resources and limited funding.

DOI: 10.4018/979-8-3373-0204-1.ch001

INTRODUCTION

The creation and dissemination of information and knowledge have always been vital fields for human cultures, including indigenous groups that often rely on traditional wisdom, as well as for science and technology. Converging disparate knowledge systems and facilitating knowledge co-production in these communities begin with knowledge integration across researchers and knowledge users. To decolonize the research integration of the knowledge creation process, this collaborative procedure is also essential. However, power disparities across Indigenous communities, researchers and research funders can hinder the integration of knowledge, exacerbating support of dominant forms of knowledge over others, and ultimately resulting in the marginalized and weaker knowledge holders losing their expertise (Yanou et al., 2023). The path to knowledge legitimacy and applicability is recognized in academia as being paved by knowledge integration and other co-strategies in which all participants validate the various meanings and contents of knowledge systems as well as the co-produced outcomes (Roshetko et al., 2022, Matuk et al. 2020). This integration should be viewed as a process that empowers the most marginalized knowledge holders to create fair collaboration among various knowledge holders (Gupta et al., 2015; Lima & Persson, 2020). Maintaining cultural diversity, addressing scientific knowledge gaps, and acknowledging Indigenous and local knowledge as essential to social justice, sovereignty, autonomy, and identity of Indigenous peoples and local communities are the three reasons why knowledge integration is relevant in this context (Yanou et al., 2023). However, research has shown that while access to indigenous knowledge has not always been simple, access to scientific information has not always been straightforward either. With issues about information access and intellectual property rights generating disputes and conflicts throughout the history of communication technologies and information societies, these difficulties have become more apparent in the digital environments that have been defined by the globalization of knowledge.

The integration of knowledge creation activities has been made easier by the developments of the internet and the web. Digital advancements have had a significant impact on many facets of society, including agenda-setting, knowledge generation, and sharing (Fecher and Friesike, 2013). They have also helped to solve scientific problems that these groups have faced (Ijatuyi et al., 2025). However, these advancements have presented several difficulties for IK research. This is because while these advancements have made it easier to obtain data by removing geographical restrictions, indigenous populations are frequently left out of the information loop and the advantages that come with having access to this data (Vicente-Saez and Martinez-Fuentes, 2018). This is the case even though these communities have turned to Indigenous Knowledge (IK) and Indigenous Knowledge Systems (IKS) to help them deal with the problems they face daily. However, to address issues that these communities face, the open scientific movement has promoted the sharing of data and research findings globally.

There are now more calls for study on the integration of IK, Open Science, and digital technology developments as a result of these connections between science and digital technologies and their complementary roles in solving challenges. Literature generally argues that access issues to research on indigenous knowledge are supposed to be resolved by the open science movement (Lui et al., 2021). However, Mwelwa et al. (2020) contend that the evolution of open scientific policies in Africa has frequently been disorganized, leading to policy differences among countries. The first step in simplifying research on the three aforementioned issues, therefore, would be to evaluate the connection between open science and advancements in digital technology. The second is the relationship between advancements in digital technology and how they affect the IK domain; the third is how open science affects IKS progress. The goal of this open sciences advocacy is to further down the expenses associated with research data management in IK research. According to earlier studies, academics appear to have made very little effort to evaluate these relationships and the mutual effects they have on advances in IK, open science, and digital technologies (Lui et al., 2021). Therefore, this chapter aims to evaluate how open science principles and Indigenous Knowledge Systems (IKS) could work together to provide the greatest benefit to Indigenous communities. In doing so, the chapter will look at indigenous knowledge systems, open science's guiding principles, and the connections between IKS and open science.

BACKGROUND

Chaudhuri (2015) asserts that the word "science" is frequently linked to Western cultures and frequently suggests that scientific rigour is lacking in indigenous knowedge systems. There is, however, a wealth of information in the majority of these civilizations about how to satisfy the specific needs of indigenous communities, but it is frequently disregarded. Some researchers contend that science has established behavioural norms that, taken together, greatly advance the quantity and calibre of scientific knowledge through communalism (sharing discoveries with others), universalism (truth is assessed according to universal standards and ought to be repeatable by others in the same circumstances), Skepticism (all ideas are subject to rigorous, structured community scrutiny, which curates the quality of the work that results), Originality (research results should yield novel contributions to understanding), and Disinterestedness (the researcher's attitude is one of objectivity; such as the researcher follows the evidence wherever it goes, regardless of its implications) (Chesbrough, 2015). Although the Western world has consistently questioned the veracity of such claims, these views on science are no longer prevalent only among Western nations but have become more prevalent even in developing nations, which have historically tended to rely more on traditional systems to solve their communities' problems. Many underdeveloped countries have different ideas about what science is, which might be linked to colonial legacies. Indigenous academics have been criticizing Western research methodology's colonizing practices for decades. Views on what constitutes truth and science have also been influenced by these activities, and some researchers have primarily concentrated on indigenous knowledge (IK) and indigenous knowledge systems (IKS). The prevalent misunderstanding regarding IK is that it is anti-modern, anti-scientific, and backwards. IK does, however, observe a system that incorporates learning theory and practice, much like contemporary science. Additionally, IK and IKS adhere closely to reason and logic and like current science. For instance, an indigenous idea cannot surpass a concept since an idea cannot exist without a concept. Furthermore, contrasting ideas are argued to develop indigenous philosophy. A philosophy can't be a single idea. Ultimately, this reconciling of opposing concepts results in a principle or theory that forms the foundation for field research or knowledge verification. There is a wealth of knowledge in most societies about how to satisfy their specific needs, but these are frequently disregarded (Chaudhuri, 2015).

Science has been founded on an open process of knowledge creation and sharing for centuries, but as time has gone on, so too have the quantity, quality, and speed of scientific output. In academic and industrial science, information sharing and collaborative research across diverse communities are considered to be the fundamental process of knowledge creation (Thursby et al., 2009). As scientific problems become more complex and specialized at the same time, it is not surprising that scientific collaboration and research have expanded in various disciplines, including IK research in recent decades. In addition to the increased partnership between researchers and indigenous communities in research activities, this scientific and research collaboration has led to calls for greater research transparency and the need to improve research output, outcomes, and impact. Therefore, Friesike et al. (2015) assert that the openness concept has always been a fundamental component of the academic community and that advancements in the web and the Internet over the last several decades have made it easier for scientific research to be transparent. Friesike et al. (2015), further says this openness stems from an incentive structure whereby the first individual to present novel discoveries to the scientific community is rewarded with various forms of acknowledgement. All these arguments for openness in research then led to the open science movement.

The phrase “open science” was used to describe how academic science is now even “more open” because of modern information and communication technologies. These new technologies have made a significant impact on publishing and communication among academics. Bodo (2016) asserts that until the early 1900s, university presses, professional groups, and educational publishers published the majority of scholarly communications. Their knowledge system was hardly seen as an essential part of the scholarly communication and the publishing process, and these institutions sold subscriptions to the journals they published. Scholarly publishing was still the domain of these scientific communities. Indigenous communities were left out of the knowledge created about them while institutions researched these communities This was made worse by the Internet's advancements, which made knowledge more widely available to researchers worldwide but excluded the indigenous people about whom the data was created. Organizations like the United Nations have thus taken the lead in promoting open science research, which has led to the creation of the UNESCO Open Research Framework of 2021, which was approved at the UNESCO General Conference in Paris in 2021. This paradigm acknowledges the value of collaborative research and highlights the significance of broadening the scientific research process of scientific knowledge generation, evaluation, and scholarly communication (Koley, 2022). The meeting also acknowledged that Indigenous populations in underdeveloped countries were disadvantaged by disparities in research funding and information availability. Therefore, to better align their work with these principles and confront the present “credibility revolution,” academics have started to advocate for and implement reforms in recent years (Vazire, 2018). According to Crüwell et al. (2019), the proposed reforms in open science are mostly a reaction to the recognition that normal research techniques compromise core values of high-quality and open science, including reproducibility. To a certain degree, the majority of scientists concur that there is a reproducibility crisis (Baker, 2016). According to Crüwell et al. (2019), numerous reasons compromise reproducibility and replicability. Some of these factors are directly related to a lack of openness, while others are caused by inadequate research design and methodological methods. According to O'Boyle, Banks, and Gonzalez-Mulé (2014), not all scientists have implemented the best practices that experts recommend to make science more reproducible. This is partly because existing incentive structures are not aligned with fundamental best practices (Higginson and Munafò, 2016), and there is a lack of clarity, agreement, and misinformation regarding the best practices, their necessity, and how to implement them (Houtkoop et al., 2018).

Since researchers first arrived on their territories, indigenous groups have been fighting for access to, ownership of, control over, and possession of research in their communities (Wilson, 2008). In research and activities that directly affected them, Indigenous communities affirmed their right to self-determination and identified their research requirements (Koster, Baccar, and Lemelin, 2012; Kwiatkowski, 2011). While contemporary researchers had to deal with best practices for maintaining research data, these relational initiatives have been changing, impacting Indigenous people's interactions with research and compelling them to be more transparent about their data. Helping Western researchers navigate the many subtleties of Indigenous epistemological discourses and methodological practices, has allowed Indigenous peoples to become transformative leaders within the research paradigm (Geia, Hayes, and Usher, 2013) and in this, digital technologies have played an enabling role through digitizing the broader economic sectors and science (Aleksandrovich and Nikolaevna, 2023). A new era of sharing, transparency, and involvement has been brought about by the development of digital technologies (Dienlin et al., 2021). At the vanguard of this digital revolution is the ideology of open science, which promotes unfettered access to research outputs, data, and procedures (Bett and Makori, 2024). Digital technology advancements have improved open science, enabling researchers to regularly document their work while also involving external and dispersed individuals. It follows naturally that the previously excluded public can now actively participate in research on developmental activities that impact their communities, giving rise to a new type of science known as citizen science, which refers to the involvement of non-scientists and armatures in research (Fetcher and Friesike, 2013). According to Bett and Makori (2024), these technologies facilitate the adoption of new Open Science practices and principles by university research teams. This creates new opportunities for researchers to work together and investigate novel approaches to communication with other universities, research institutes, businesses, governments, individuals, and international organizations. As stated by Koley (2022), the open science movement encourages the use of digital technologies to drive scientific processes, collaboration and cooperation among actors in scientific ecosystems, acknowledgement of diverse knowledge systems, including IKS, and unrestricted access to scholarly publications and data. Although research on the characteristics of the digital transformation of science reveals that it is unequal, scientists remain hopeful about the possibilities of this change and the efficacy of cross-border collaboration and collaborative research.

Research indicates that the poor world is disadvantaged due to a gap in funding and knowledge access. For instance, research on open science reveals that research data management capabilities in Africa have not advanced much, which has hampered attempts to offer evidence-based research and policy solutions that can be put into practice for the advancement of African Indigenous communities (Mwelwa et al., 202). Even while there appears to be collaborative research between African states and nations from other continents, research collaboration, especially among these African nations has remained low. According to studies on open science in Africa, Africa has contributed very little to the world's knowledge—less than 1%—and its research output has not been well-represented on indexing systems and is not sufficiently visible online (Mwelwa et al., 2020).

Several frameworks have been put forth in nations that have improved open science and incorporated it into their research. For instance, Koley (2022) offers some illustrations of open scientific policies that have been put into practice in India; these are displayed in Figure 1 below.

Figure 1. Open Science Policy Initiatives in India

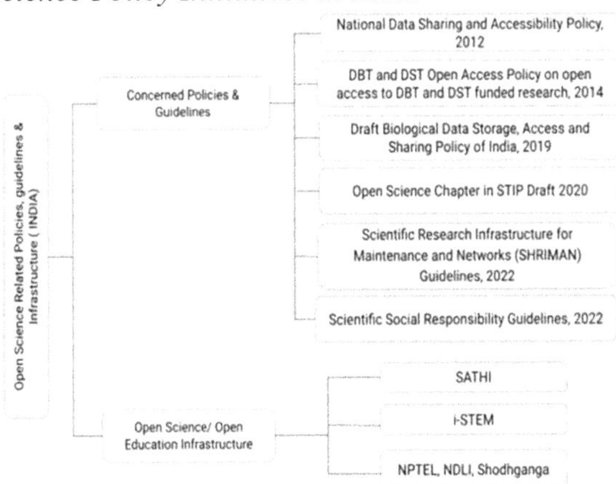

(Koley, 2022)

In addition to these Indian policies, Mwelwa et al. (2020) mention several other open science projects in Africa, including the World Bank project, the H3Bionet project, Data First, the African Academy of Science (AAS) Open Research (AAS), the African Open Science Platform (AOSA), the Indigenous knowledge and Climate change Adaptation Research Project, and others. South Africa is one of the nations that has been spearheading the promotion of open science in the Southern African region through its Academy of Science of South Africa (ASSAf) which recognized the advantages of open science and listed some of these advantages as follows:

i) Enhancing research ethics, quality, integrity, transparency, and reproducibility;
ii) facilitating the sharing of high-quality research output, data, and results and encouraging research collaborations, participation, and inclusion;
iii) encouraging scientific collaboration and information sharing for the benefit of science and society; and
iv) Increasing the prominence of research.

To achieve this, ASSAf has matched its open science policy with many activities and policies that all support the adoption of open science, including the following:

i. UNESCO Recommendation on Open Science;
ii. African Open Science Platform (AOSP) ;
iii. National Open Science Policy;
iv. DHET Research Outputs Policy (2015) ;
v. Indigenous Knowledge Act (No. 6 of 2019) ;
vi. South African Open Science Cloud (SAOSC) ;
vii. National Research Foundation (NRF) Statement on Open Access;
viii. ASSAf Strategic Plan for 2020/21 – 2024/25;
ix. ASSAf Transformation Strategy (2023/24) ;
x. Draft ASSAf Intellectual Property Rights Policy (2023/24); and
xi. Draft ASSAf Science Engagement Strategy (2023/24).

This understanding of the importance of IK in national innovation systems is not exclusive to South Africa; other nations have also encouraged this integration, including the United States of America, Japan, India, Singapore, Malaysia, South Korea, the United Kingdom, the Netherlands, and Australia (Science and Technology Department, nd; Cole et al., 2024).

METHODOLOGY

This review covers scholarly journal publications from 2000 to 2024. Nonetheless, the study's main conclusions were drawn from papers that were released between 2015 and 2024. Although IK and information and communication technologies have historically been the focus of numerous researchers who have attempted to create connections between these two disciplines, this period was chosen because it also saw a rise in interest in open science among researchers. "IK and open science," "digital technologies and open science," and "digital technologies and IK" were the main search phrases utilized to find and retrieve pertinent journal articles for the review. Out of the 100 items that were retrieved, 48 were initially chosen using the author's discretion. Nevertheless, 42 of these 48 journal papers were ultimately chosen for further consideration following an analysis.

Key Principles of Open Science

Open science is a general word that encompasses the ideas of transparency, openness, rigour, reproducibility, replicability, and knowledge accumulation—all of which are thought to be essential components of scientific endeavour. Other words that have been coined to describe open science include "open scientific knowledge," which refers to open access to scientific publications, research data, metadata, open educational resources, software, source code, hardware, and other materials that are in the public domain or protected by copyright and licensed under an open license that permits access, re-use, repurpose, adaptation, and distribution under certain conditions, made available to all actors as soon as possible, regardless of their location, nationality, race, age, gender, income, socioeconomic situation, or stage of career, discipline, language, religion, disability, migrant status, ethnicity, or any other reason, and without charge. It also alludes to the potential for research methodology and evaluation procedures to be made public (UNESCO, 2021). Bartling and Friesike (2014), among others, have equated open science to Science 2.0, which was fueled by the advancements of the web and the internet. These scholars do concur, though, that although open science may make use of Science 2.0, Science 2.0 may not always be open. Fecher and Friesike (2014) also examine open science from five different perspectives, which are as follows:

i) The infrastructure school, which focuses on the infrastructure of technology
ii) The public school, which is focused on making knowledge creation accessible;
iii) The measurement school, which is focused on alternative impact measurement;
iv) The democratic school, which is focused on knowledge access; and
v) The pragmatic school focused on cooperative research.

The rights-based, ethical, epistemological, economic, legal, political, social, multi-stakeholder, and technological ramifications of bringing science into the public eye and expanding the openness principles to encompass the entire scientific research cycle are the foundational values of open science and according to UNESCO (2021) these may also be categorized as seeking to achieve the following:

Quality and integrity: This highlights the need to respect academic freedom and human rights as well as promote high-quality research by combining various information sources and making research methods and results widely accessible for thorough examination and criticism as well as open evaluation procedures.

Collective benefit: As an international public good, free science ought to benefit all people and belong to all of humankind.

Equity and fairness: Regardless of location, nationality, race, age, gender, income, socioeconomic circumstances, career stage, discipline, language, religion, disability, ethnicity or migratory status, or any other reason, access to scientific knowledge should be equal for both producers and consumers of knowledge. Open science should also play a significant role in ensuring equity among researchers from developed and developing countries by enabling fair and reciprocal sharing of scientific inputs and outputs.

Diversity and inclusiveness: This encompasses a range of knowledge, practices, workflows, languages, research outputs, and research topics that support the needs and epistemic pluralism of the scientific community as a whole, diverse research communities and scholars, as well as the general public and knowledge holders outside of the traditional scientific community, such as local communities and indigenous peoples, as well as social actors from other nations and regions when appropriate.

These values are presented graphically by UNESCO (2011) as shown in Figure 2 below:

Figure 2. Values of open science

(UNESCO, 2021)

In addition to these values, UNESCO (2021) outlines the following open scientific guiding principles and offers a framework for behaviours and enabling conditions that uphold the aforementioned values and help to realize the goals of open science:

Transparency, scrutiny, critique and reproducibility: All phases of the scientific endeavour should encourage greater transparency to strengthen the validity and rigour of scientific findings, increase the influence of research on society, and improve society's overall ability to resolve intricately linked issues.

Equality of opportunities: No matter their location, nationality, race, age, gender, income, socioeconomic situation, career stage, discipline, language, religion, disability, ethnicity or immigration status, or any other factor, all scientists and other open science actors and stakeholders have an equal opportunity to access, contribute to, and benefit from open science.

Responsibility, respect and accountability: More openness means more responsibility for all open science actors. These factors, along with public accountability, conflict of interest awareness, awareness of potential social and ecological repercussions of research activities, intellectual integrity, and respect for ethical principles and research implications, should serve as the cornerstones of good open science governance.

Collaboration, participation and inclusion: Along with the full and effective participation of societal actors and the inclusion of knowledge from marginalized communities in solving problems of social importance, collaborations between disciplines and at all levels of the scientific process—beyond the boundaries of geography, language, generations, and resources—should become the norm.

Flexibility: There is no one-size-fits-all approach to open science practice because of the diversity of science systems, actors, and capacities worldwide as well as the dynamic nature of supporting information and communication technology.

Sustainability: Open science should build on long-term procedures, services, infrastructures, and funding models that guarantee the equitable involvement of scientific producers from underprivileged institutions and nations to be as effective and significant as possible.

UNESCO (2021) further represents the different concepts of open science graphically as shown in Figure 3 below:

Figure 3. Principles of Open Science

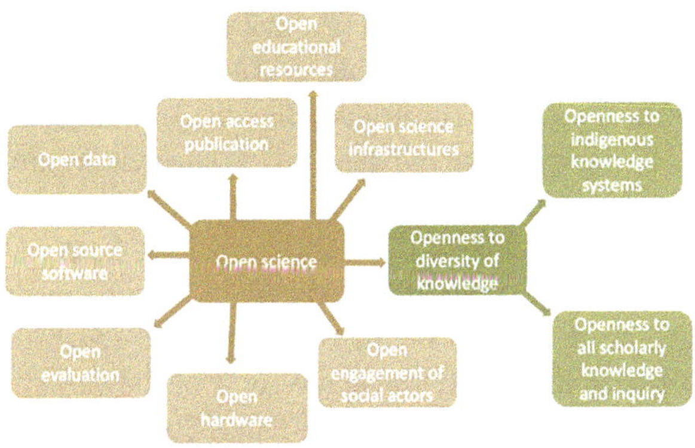

(UNESCO, 2021)

Based on transparency and connectedness, Open Science transforms the way research is planned, carried out, recorded, and evaluated on a sociocultural and technological level. All scientific players are impacted by the irreversible trends of open data tools, open access platforms, open peer review procedures, and public engagement initiatives, which can quicken the research cycle. Open data principles recommend that data be available for use in research projects that the original developer may or may not have anticipated. In this instance, the justification for open data in science is researcher-centric; it is a framework that encourages insightful data mining and the pooling of information from many publications. In short, open data eliminates redundancy in data collecting and promotes research synergies.

As illustrated in Figure 3 above, the many elements of open science have recently drawn the interest of IK researchers, enriching IK research and leading to increased research output, quality, and impact as a result of collaboration amongst IK researchers. In addition to protecting Indigenous Peoples' lands and territories, an openness to other knowledge systems, especially Indigenous Knowledge (IK), can aid in the revitalization of local communities.

Decolonizing research, including community members, respecting multiple worldviews, encouraging inclusive discussions, co-producing knowledge, and documenting experiences are some of the methods the literature suggests fostering openness to diverse knowledge systems. Building mutual respect, empowering marginalized actors, acknowledging the ongoing effects of marginalization, sharing knowledge, incorporating local knowledge, and providing learning opportunities are all benefits of open engagement with Indigenous knowledge systems (IKS) in the context of the open scientific movement (Dzulkifli, 2020; Keane, Khupe, and Muza, 2016). Participatory approaches, that include traditional leaders, funding knowledge transfer initiatives, acknowledging the complexity of advantages and disadvantages, and acknowledging Indigenous rights are some additional ways to engage with Indigenous knowledge systems that have been reported by research. FAIR (Findable, Accessible, Interoperable, and Reusable) principles, Indigenous perspectives, digitization, CARE (Collective Benefit, Authority to Control, Responsibility, and Ethics) principles, and Indigenous data sovereignty are some additional ways that open data and Indigenous knowledge systems can be linked (UNESCO, 2021). Humanizing Western knowledge systems, putting human needs at the centre of higher education, supporting social and environmental justice, decolonizing learning materials, promoting epistemic pluralism, and resolving conflicts with copyright and intellectual property rights are some of the ways that Open Educational Systems (OER) and Indigenous Knowledge Systems (IKS) can further be linked. The open science infrastructure (OSI) is another crucial element of open science(Bett and Makori, 2024). Access to physical facilities, services and capabilities for managing and creating access to data, scientific literature, major scientific equipment, and knowledge-based resources including collections, journals, and open-access publication platforms are all included in an OSI, which can be virtual or real. Globally, intergovernmental organizations like the European Commission, European Parliament, European Council, Organization for Economic Cooperation and Development (OECD), United Nations, and World Bank acknowledge the value of Open Science in tackling the major societal issues that humanity faces in the twenty-first century, including, but not limited to, climate change, public health emergencies, sustainable food production, efficient energy, or smart transportation (Vicente-Saez and Martinez-Fuentes, 2018).

Indigenous Knowledge Systems and Digital Technologies

IK generally refers to the developed long-standing customs and practices of some indigenous, local, or regional communities. IK also includes the wisdom, knowledge, and teachings of these communities and in many cases, this knowledge is passed down orally from person to person for generations. Some types of IK are expressed through stories, legends, folklore, rituals, songs, and even laws and these types of knowledge usually set one community apart from another.

IK and the evolution of the digital environment are strongly related, according to research on IK and digital technology. This is because researchers now have more options than ever before to work with Indigenous communities on projects like the preservation and sharing of their cultural legacy thanks to the internet age. With a wide range of tools and platforms that can greatly improve language learning, streamline documentation and archiving efforts, and increase accessibility to Indigenous languages, digital media technology is essential to the preservation, documentation, and revitalization of Indigenous languages (Ajani et al, 2024). This digitalization procedure streamlines the distribution of these resources among academics, educators, and community members by preserving them in a more shared and accessible format (Bird et al., 2019). Additionally, digital media technology facilitates collaborative knowledge documentation projects. Online platforms act as hubs for group knowledge documentation efforts, allowing the integration of linguistic proficiency, cultural knowledge, and community input. Examples of these tools include the use of audio and video recording devices in conjunction with digital storage and preservation systems (Hinton, 2019). A broad definition of Open Science acknowledges that the Internet's quick technical development could lead to a second scientific revolution that fundamentally changes scientific norms and research techniques (Christensen et al., 2020). However, there are concerns about the uncontrolled documentation of IK because knowledge holders may be wary of documenting and disseminating their expertise for fear that it will be misused against them, particularly when the moral and material gain from IK is illegally appropriated (Buthelezi, Ocholla and Dlamini, 2024).

Relationship Between IK and Open Science

Indigenous Knowledge Systems are a component of the open science concept, but because they have not been included in the evolution of Western empirical science, researchers, development specialists, and policymakers may not always have easy access to them. Nonetheless, some researchers contend that developing indigenous research methodologies is both a right and a duty of native researchers. However, these approaches need to be accountable to indigenous people and consistent with their beliefs and customs. Additionally, to persuade academics that Indigenous research methods are as comprehensive and advanced as well-known Western research methods, they must be scholarly and sophisticated (Roshetko et al., 2022). This is crucial because Indigenous researchers need to be able to use Indigenous methodologies not only in native studies-focused fields or at institutions under Indigenous control but also in fields that appear to have little to do with native life and customs, including the most conservative and esteemed western universities (Kahakalau, 2004). Such a process leads to improved knowledge legitimacy. The path to knowledge legitimacy and applicability is recognized in academia as being paved by knowledge integration and other co-strategies in which all participants validate the various meanings and contents of knowledge systems as well as the co-produced outcomes (Roshetko et al., 2022; Matuk et al., 2020). This integration is also encouraged by Cronje (2015) who says that there are a lot of similarities between the nature of indigenous knowledge and science. While indigenous knowledge also contains a metaphysical component, science and indigenous knowledge are both empirical, tentative (and susceptible to change), inferential, creative, and socially and culturally based.

From the literature, there is greater evidence between IK and open science. Researchers are increasingly attempting to interact and collaborate with IK holders in the areas they study (Teng¨o et al. 2017; Ban et al. 2018; Berkes and Berkes, 2009). Knowledge holders and researchers frequently engage more as a result of this. Such procedures, however, frequently suggest that IK complies with scientific understanding, and knowledge integration seems to be a purely technical endeavour to integrate a minority group's IK into a majority system. By doing this, the researchers disregard the political aspects of the process problems and do not sufficiently address power dynamics and inequality (Yanou et al., 2023). Instead, integration should be viewed as a process that empowers the most marginalized knowledge holders to create fair collaboration among various knowledge holders (Gupta et al., 2015).

IK Research and digital technology development

The use of open science and digitalization of science principles is the foundation for the current development of both national and international scientific systems which has recently incorporated IK research. The concept of indigenous research techniques has emerged more recently in reaction to the frequently inadequate Western paradigms and genres of research in African and other developing communities of the world. The results of a study by Yanou et al. (2023) show that although more research is being done to combine scientific knowledge with IK, methods, procedures, and opportunities are not well and ambiguously documented, and colonial legacies and challenges are frequently disregarded.

International organizations like UNESCO have implemented several policies and measures that demonstrate how IK research has been integrated into systems of open scientific research. This has been enabled by the rapid advancements in the internet and web technologies, which have made it easier for researchers to collaborate, to access information and made it possible to integrate IK into open scientific research. These developments in the digital environment have attracted researchers to investigate these digital technologies and how they are used in different fields. Nevertheless, a review of the literature has revealed that not much research has been done on how digital technologies have affected IK research. Despite this scanty literature however, IK research like any other discipline has continued to benefit from the use of digital technologies and the future of IK research promises to provide opportunities for better IK innovation and dissemination at a global level, enabling communities to share experiences and further enabling the validation of knowledge and the knowledge creation process. In integrating IK research and the development of digital technologies several benefits related to the open science movement have been realised and these have been acknowledged to facilitate open data sharing, scientific publications, open educational resources, increased research on open source software and open hardware, open engagement of societal factors (comprising crowdsourcing, crowdfunding, scientific volunteering, citizen participatory science), and open dialogue with other knowledge systems (indigenous peoples, marginalised scholars, local communities) (UNESCO, 2021). Numerous tools and technologies have been employed to support IK research, especially to make data exchange easier. One initiative created for the Australian Research Council (ARC) Linkage project, which aims to perform storytelling among Australia's Aboriginal adolescents, is one example of a project that has been put into place to facilitate open data sharing in research on IK (de Souza, et al. 2016). The European Open Science Cloud (EOSC), is a virtual collaborative environment that transcends boundaries and scientific disciplines for the storage, sharing, processing, and reuse of research digital objects (e.g., publications, data, and software) that are searchable, interactive, and reusable (FAIR), is also another example of an initiative that has been put into place to integrate IK research and digital technologies. Additionally, the Digital Research Environment (DRE) is another cloud-based, internationally accessible research environment where researchers can instantly create workspaces to collaborate in and utilize their favourite programs, as well as store and organize data securely (Aleksandrovich and Nikolaevna, 2023).

CHALLENGES

The attempt to incorporate IK into open science has presented several difficulties, especially in the digital realm. These difficulties are frequently associated with the ownership of research data, property rights, and the ethical use of data. Indigenous populations' rights to research data on them have been questioned by certain research groups, especially those in developing countries. This is especially true for techniques that entail digital data collection, analysis, and curation. This is because technological advancements in digital environments have produced limitless opportunities in fields like artificial intelligence, cloud computing, blockchain, and the Internet of Things (IoT). However, because of technological disparities between developed and developing countries, data subjects are often denied the freedom to access and use this data. Furthermore, traditional African values—where the emphasis on location, the prevalence of spirituality, and the holistic view of life seem to be at odds with modern values—are challenged by the emphasis on universalism, rationality, and the compartmentalization of knowledge (Breidlid, 2009; Chigwada and Ngulube, 2024). In addition to this, how to balance traditional knowledge and contemporary science without replacing one another, honouring the two sets of values, and enhancing one other's advantages remains a significant challenge for African nations. According to several studies, different stakeholders are not aware of Open Science (European Commission, 2015), mostly because "there is no formal definition of Open Science." In addition to these difficulties in handling IK, there are some other difficulties in IK research, including:

a) **Lack of clarity:** Indigenous knowledge has a broad and intricate nature.
b) **Inaccessibility:** Indigenous knowledge practitioners frequently reside in the most isolated and difficult-to-reach areas of the country. This can occasionally have a significant impact on the study process, leading some researchers to use ambiguous or even altered data.
c) **Lack of relevant information sources:** IK is not documented in writing unless the research is being done expressly. The researchers face difficulties since the research design is often constrained by the paucity of literature on IK. As a result, fewer research projects are being conducted and, consequently, fewer research articles on a certain subject.
d) **Financial constraints:** IK research typically takes place in isolated areas far from the conveniences of the city. Finding the proper people and travelling great distances are necessary to gather reliable information. All of this has a monetary cost.

e) **Time constraints**: Time commitments imposed by the sponsoring organization frequently limit academic research. Nonetheless, a community's IK is frequently extensive and intricate.
f) **Lack of key people:** The absence of important individuals capable of accurately disseminating such knowledge is one of the most frequent causes of IK's extinction. Important indigenous knowledge may become diluted or even extinct as a result of current practitioners misinterpreting it or the original practitioners passing away, as a result of the oral nature of knowledge transfer.
g) **Wariness:** The practitioners' apprehension about "outsiders" and their actions is another problem in IK research. Their prior bad experiences—like being defrauded of their knowledge—justify this distrust on their behalf.

CONCLUSIONS

The review has demonstrated a close connection between open science, IK, and advancements in digital technology. Several worldwide organizations, countries, and institutions have proposed and implemented policies that demonstrate the critical role open source plays in enhancing IK accessibility and IK research. By making research and research data access easier, digital technical advancements have been crucial to IK research, IK preservation, and the open science movement. To safeguard Indigenous communities and their knowledge, IK research in open science settings must take ethical considerations into account, notwithstanding these links. Other issues have also surfaced, and these have to do with data accuracy, the lack of IK data, the lack of IK specialists, and the limitations of IK digital representations.

RECOMMENDATIONS

From several challenges that have emerged from the review, the following recommendations are therefore made:

a) Research on indigenous knowledge must be participatory from the beginning to the end. This will ensure that Indigenous communities consent to research about them and the data to be collected and documented about them. This will also require that these communities are fully informed of the implications of the research and how data about them will be used and stored.
b) IK research often involves collecting intricate and complex data which may not be easily interpreted. Therefore in representing data in digital form, Indigenous

communities must validate data quality to ensure data integrity and consistency with their Indigenous beliefs.

c) Literature shows that globally there is a shortage of relevant information sources and a lack of key people for IK. Therefore, future research must focus on the integration of digital technologies in communities to enable them to capture and preserve their data and make this data available to researchers upon request only after considering the implications. This will also ensure that Indigenous communities have control over what data is collected and stored about them.

REFERENCES

Bastos Lima, M. G., & Persson, U. M. (2020). Commodity-centric landscape governance as a double-edged sword: The case of soy and the Cerrado Working Group in Brazil. *Frontiers in Forests and Global Change*, *3*, 27.

Baker, M. (2016). 1,500 scientists lift the lid on reproducibility. *NATNews*, *533*, 452–454. DOI: 10.1038/533452a PMID: 27225100

Berkes, F. and Berkes, M. K. (2009) Ecological complexity, fuzzy logic, and holism in indigenous knowledge, *Futures 41: 6–12, https://doi.org/. futures.2008.07.003.* DOI: 10.1016/j

Bett, S. K and Makori, E.O. (2024). Opening gates to open science: Impact of digital technologies in fostering open science principles in research. *Current Journal of Library and Information Sciences. 9(3).*

Bird, S. K., Flick, S. D., Carroll, S., Austin, A., & Lim, A. (2019). Technology for Indigenous Language Revitalization: The State of the Field. *Language*, *95*(S1), e1–e29.

Bodo, B. (2016). Pirates in the library – an inquiry into the guerilla open access movement. Paper prepared for the 8th Annual Workshop of the International Society for the History and Theory of Intellectual Property, CREATe, University of Glasgow, UK, July 6-8, 2016.

Breidlid, A. (2009). Culture, indigenous knowledge systems and sustainable development: A critical view of education in an African context. *International Journal of Educational Development*, *29*, 140–148.

Buthelezi, S. C., Ocholla, D., & Dlamini, P. (2024). Strategies for documenting and Disseminating indigenous knowledge at a South African university. *South African Journal of Information Management*, *26*(1), a1648. DOI: 10.4102/ sajim.v26i1.1648

Chaudhuri, B. (2015). Science in society: Challenges and opportunities for indigenous knowledge in the present-day context. *Global Bioethics = Problemi di Bioetica*, *26*(2), 78–85. DOI: 10.1080/11287462.2015.1037140

Chesbrough, H. (2015). *From Open Science to Open Innovation.* ESADE.

Chigwada, J., & Ngulube, P. (2024). Librarians' role in the preservation and dissemination of indigenous knowledge. *International Federation of Library Associations and Institutions.*, *50*(2), 242–256.

Christensen, G., Wang, Z., Levy Paluck, E., Swanson, N., Birke, D., Miguel, E., and Littman, R. (2020). Open science practices are on the rise: The state of social science (3S) survey.

Cole, N. L., Kormann, E., Klebel, T., Apartis, S., & Ross-Hellauer, T. (2024). The Societal impact of Open Science: A scoping review. *Royal Society Open Science*, *11*, 240286. DOI: 10.1098/rsos.240286 PMID: 39100167

Cronje, A. (2015). Epistemological border-crossing between Western science and indigenous knowledge and its implications for teacher professional development. University of Johannesburg: Unpublished PhD thesis.

Crüwell, S., van Doorn, J., Etz, A., Makel, M. C., Moshontz, H., Niebaum, J. C., Orben, A., Parsons, S., & Schulte-Mecklenbeck, M. (2019). Seven Easy Steps to Open Science An Annotated Reading List. *Zeitschrift für Psychologie mit Zeitschrift für Angewandte Psychologie*, *227*(4), 237–248.

De Souza, P., Edmonds, F., McQuire, S., Evans, M., & Chenhall, R. (2016). Aboriginal Knowledge, Digital Technologies and Cultural Collections Policy, Protocols, Practice. Melbourne Networked Society Institute Research Paper 4. Dienlin, T., Johannes, N., Bowman, N. D., Masur, P. K., Engesser, S., Kümpel, A.

Dzulkifli, S. (2020). *Regional Report: Open Science & the Decolonization of Knowledge International Webinar Series. EUROPE AND NORTH AMERICA (ENGLISH)*. UNESCO.

European Commission. (2015). Impact of Open Science methods and practices on the economics of research and science. Field. *Language*, *95*(S1), e1–e29.

Fecher, B., & Friesike, F. (2013). Open Science: One Term, Five Schools of Thought, RatSWD Working Paper, No. 218, Rat für Sozial- und Wirtschaftsdaten (RatSWD), Berlin.

Friesike, S., Widenmayer, B., Gassmann, O., & Schildhauer, T. (2015). Opening science: Towards an agenda of open science in academia and industry. *The Journal of Technology Transfer*, *40*(4), 581–601.

Geia, L. K., Hayes, B., & Usher, K. (2013). Yarning/Aboriginal Storytelling: Towards an understanding of an Indigenous perspective and its implications for research practice. *Contemporary Nurse*, *46*(1), 13–17. DOI: 10.5172/conu.2013.46.1.13 PMID: 24716757

Gupta, J., Pouw, N. R., & Ros-Tonen, M. A. (2015). Towards an elaborated theory of inclusive development. *European Journal of Development Research*, *27*(4), 541–559.

Higginson, A. D., & Munafò, M. R. (2016). Current incentives for Scientists lead to Underpowered studies with erroneous conclusions. *PLoS Biology*, *14*, e2000995. DOI: 10.1371/journal.pbio.2000995 PMID: 27832072

Hinton, L. (2019). Language Revitalization and New Technologies: Cultures of Electronic Mediation and the Refiguring of Communities. In Jaffe, J., Morita, D., Vogel, N. D., & Spolsky, G. (Eds.), *Language Ideologies and the Globalization of 'Standard' Languages* (pp. 244–264). Routledge.

Houtkoop, B. L., Chambers, C., Macleod, M., Bishop, D. V., Nichols, T. E., & Wagenmakers, E. J. (2018). Data sharing in psychology: A survey on barriers and preconditions. *Advances in Methods and Practices in Psychological Science*, (1), 70–85. DOI: 10.1177/2515245917751886

Ijatuyi, E. J., Lamm, A., Yessoufou, K., Suinyuy, T., & Patrick, H. O. (2025). Integration of indigenous knowledge with scientific knowledge: A systematic review. *Environmental Science & Policy*, *170*, 104119.

Kahakalau, K. (2004). Indigenous Heuristic Action Research: Bridging Western and Indigenous Research Methodologies. *Multidisciplinary Research on Hawaiian Well-Being. 1 (1).*

Keane, M., Khupe, C., & Muza, B. (2016). It matters who you are: Indigenous knowledge research and researchers. *Education as Change*, *20*(2), 163–183.

Koley, M. (2022). Analysis of Open Science Policy Recommendations Proposed in India's 5th Science, Technology & Innovation Policy Draft. *Journal of Science Policy & Governance.*, *21*(02). Advance online publication. DOI: 10.38126/JSPG210208

Koster, R., Baccar, K., & Lemelin, R. H. (2012). Moving from research On, to Research WITH and FOR Indigenous communities: A critical reflection on community-based participatory research. *The Canadian Geographer. Geographe Canadien*, *56*(2), 195–210.

Kwiatkowski, R. E. (2011). Indigenous community-based participatory research and health impact assessment: A Canadian example. *Environmental Impact Assessment Review*, *31*(4), 445–450. DOI: 10.1016/j.eiar.2010.02.003

Lui, P. P., Skewes, M. C., Gobrial, S., & Rollock, D. (2021). Advancing transparency and impact of research: Initiating crosstalk between Indigenous research and mainstream "open science". *Journal of Indigenous Research*, *9*, 10.

Matuk, F. A.. Turnhout, E., Fleskens, L., do Amaral, E.F., Haverroth, M., and Behagel, J.H. (2020). Allying knowledge integration and co-production for knowledge legitimacy and usability: the Amazonian SISA policy and the Kaxinaw´a Indigenous people case. *Environ. Sci. Pol. 112 (2020) 1–9, https://doi.org/. envsci.2020.04.018.* DOI: 10.1016/j

Mwelwa, J., Boulton, G., Wafula, J., & Loucobar, C. (2020). Developing open Science In Africa: Bariers, scholar and opportunities. *Data Science Journal, 19*(31), 1–17.

O'Boyle, E. H., Banks, C. B., & Gonzalez-Mulé, E. (2014). The Chrysalis effect: How ugly initial results metamorphosize into beautiful articles. *Journal of Management, 43*, 376–399. DOI: 10.1177/0149206314527133

Roshetko, J. M., Pingault, N., Quang, T. A., Meybeck, R., & Gitz, M. V. (2022). Asia-Pacific Roadmap for Innovative Technologies in the Forest Sector. CIFOR. 15, Rome and Bogor.

Science and Technology Department. (nd). Indigenous Knowledge Systems. Republic of South Africa.

Teng¨o, M., Hill, R., Malmer, P., Raymond, C. M., Spierenburg, M., Danielsen, F., Elmqvist, T., & Folke, C. (2017). Weaving knowledge systems in IPBES, CBD and beyond—Lessons learned for sustainability. *Current Opinion in Environmental Sustainability, 26–27*, 17–25. DOI: 10.1016/j.cosust.2016.12.005

Thursby, M., Thursby, J., Haeussler, C., and Jiang, L. (2009). Do scientists freely share information? Not necessarily. Vox News 29 Nov. 2009.

UNESCO. (2021) UNESCO Recommendation on Open Science. The General Conference of the United Nations Educational, Scientific and Cultural Organization (UNESCO), meeting in Paris, from 9 to 24 November 2021, at its 41st session.

Vazire, S. (2018). Implications of the credibility revolution for productivity, creativity, and progress. *Perspectives on Psychological Science*, (13), 411–417. DOI: 10.1177/1745691617751884 PMID: 29961410

Vicente-Saez, R., Gustafsson, R., & Van den Brande, L. (2020). The Dawn of an Open exploration era: Emergent principles and practices of open science and innovation of university research teams in a digital world. *Technological Forecasting and Social Change, 156*, 120037.

Vicente-Saez, R., & Martinez-Fuentes, C. (2018). Open Science now: A systematic literature review for an integrated definition. *Journal of Business Research, 88*, 428–436.

Wilson, S. (2008). Progressing toward an Indigenous research paradigm in Canada and Australia. *Canadian Journal of Native Education*, *27*(2), 161–178.

Yanou, M. P., Ros-Tonen, M. A., Reed, J., Moombe, K., & Sunderland, T. (2023). Integrating local and scientific knowledge: The need for decolonising knowledge for conservation and natural resource management. *Heliyon*, *9*(11).

KEY TERMS AND DEFINITIONS

Collaborative research: A collaborative research endeavour involving several individuals, institutions, or organizations

Digital technology: Electronic resources, systems, tools, and gadgets that produce, store, or process data

Indigenous knowledge: A set of customs, values, and knowledge systems that Indigenous groups have carried down through the years.

Open data: Data that is publicly available to everyone, including people, businesses, the media, and customers.

Open Educational Systems: Systems that increase accessibility and inclusivity in education through the usage of open educational resources (OER).

Open Science: The drive to ensure that all societal levels, whether amateur or professional, may access scientific research and its dissemination.

Research data management: The process of giving data the proper labelling, storage, and access at every stage of a research endeavour is known as research data management.

Research data: Data that is gathered, created, or witnessed to bolster study conclusions

Scientific knowledge: refers to a broad set of rules and hypotheses developed using the scientific method to explain a phenomena or behavior of interest.

Chapter 2
AI-Driven Indigenous Knowledge Management Ensuring Ethical Accessibility

Usharani Bhimavarapu
https://orcid.org/0000-0002-0246-1420
Department of Computer Science and Engineering, Koneru Lakshmaiah Education Foundation, Vaddeswaram, India

ABSTRACT

Indigenous Knowledge (IK) systems are organized around cultural sovereignty, confidentiality, and moral leadership, while FAIR principles are intended to promote open information sharing. The relationship between sovereignty and accessibility is analyzed and modeled using a Bi-Stacked Artificial Neural Network technique. Using Particle Swarm Optimization, feature selection streamlined, preprocessed, and standardized data gathered from Indigenous governance models and FAIR compliance reports. In order to create knowledge-sharing models that adhere to Indigenous standards and are practical, Bi-Stacked ANN is trained in hierarchical representations and ethical AI frameworks improve the interpretability of models. Responsible data handling is promoted thus to ensure global access to information and Indigenous people's rights in balance. The results show AI-driven policy simulations that both promote digital inclusion and cultural sovereignty.

DOI: 10.4018/979-8-3373-0204-1.ch002

INTRODUCTION

The FAIR (Findable, Accessible, Interoperable, and Reusable) principles guarantee that there is effective management and sharing of data. They make digital assets, like research data and knowledge, that are made discoverable and part of world information systems. The principles are hard to apply to Indigenous Knowledge (IK), however. Indigenous knowledge has sensitive cultural, spiritual, and customary information that the communities may wish to protect. FAIR standards open access are the norm, but sovereignty of knowledge means more to Indigenous people. Where it gets confusing is when access collides with the right to own culture. Indigenous voices contend that knowledge shall be shared based on their option and maintain their sacred traditions. And the digital divide can make sure that Indigenous peoples do not even have ownership over their data. These considerations must be included in adjusting FAIR principles that heed Indigenous sovereignty.

Indigenous Knowledge permeates oral cultures, rituals, and everyday life, and thus its digitalisation is complex. Classification and metadata models of FAIR-based repositories can be in conflict with Indigenous epistemologies. Western systems box off pieces of knowledge, but Indigenous cultures think of knowledge as a single entity that is contained in land, community, and spirituality. Metadata structures can disguise or desocialise Indigenous Knowledge, and this creates ethical concerns. Some Indigenous groups worry that unfettered digitization will create an opportunity for exploitation or abuse. Also, open access to sacred knowledge would be against cultural protocols. The challenge is making knowledge accessible and reusable without undermining Indigenous jurisdiction. That requires a paradigm shift in how digital repositories operate, with Indigenous consent-based models being incorporated.

Accessibility, one of the foundations of FAIR, must be balanced against Indigenous rights to decide who gets to access their knowledge. Open-access repositories are generally at odds with Indigenous data sovereignty values, which are concerned with community control and ownership. Other avenues include through Indigenous data governance models such as the CARE (Collective Benefit, Authority to Control, Responsibility, and Ethics) principles. These models give more importance to the point that Indigenous communities are entitled to decide access rights based on their values and traditions. Instead of open access, Indigenous repositories can have conditional or tiered access. For example, some knowledge may be publicly accessible, while sacred or sensitive knowledge is restricted. Combining FAIR and CARE principles could create a more ethical data management approach.

Yet another FAIR principle, interoperability, ensures data is interoperation-capable between systems and platforms. Indigenous Knowledge, by nature, would often be imparted in mixed media, i.e., stories in words, dance, music, and images. Standard digital media, e.g., text databases, may not be able to capture this variety. On top of this, Indigenous people may prefer to have local systems of data keeping and not world-scale archives. Interoperability efforts should therefore respect Indigenous data structures, languages, and epistemologies. Indigenous perspectives must be reflected in metadata standards. Partnerships with Indigenous knowledge keepers can provide authentic representation of meaning and context. Without such concerns, interoperability can become an instrument of assimilation and not inclusion. This highlights the necessity of ethical integration that provides Indigenous control.

Reusability, the final FAIR principle, supports data reuse to undertake new research and applications. Indigenous Knowledge, however, often is context-bound and at times not meant for extensive use. Knowledge shared under a specific cultural framework may render itself meaningless if reused outside the same framework. Misuse has the potential to lead to misinterpretation, cultural misappropriation, or commercialization without the consent of Indigenous peoples. Traditional ecological knowledge, for instance, has been tapped in environmental science studies without due benefits or credits accruing to Indigenous peoples. Devices like culturally modified licensing arrangements are needed to ensure ethical reuse. Tools like Traditional Knowledge (TK) Labels allow communities to make determinations of the type of uses their knowledge is to be applied for. Incorporating such tools into FAIR-based repositories amounts to a means of protecting Indigenous Knowledge from misuse.

Ethics are most important to balancing FAIR principles and Indigenous knowledge management. Relational accountability—responsibly accounting for knowledge being utilized within its ethical and cultural context—is frequently emphasized by Indigenous communities. Ethical data governance also encompasses obtaining free, prior, and informed consent before sharing or utilizing Indigenous Knowledge. Indigenous-designed frameworks, such as the Mukurtu content management system, provide ethical models of digital stewardship. These systems facilitate the Indigenous people to put conditions on access based on their cultural protocols and laws. Confirming Indigenous intellectual property rights is also essential. Western legal systems focus on individual ownership, while Indigenous knowledge is more collective. Legal systems must change for accommodating Indigenous data sovereignty and enabling responsible sharing.

Several Indigenous-initiated projects are redefining data stewardship in line with both sovereignty and FAIR principles. The First Nations Information Governance Centre in Canada promotes the OCAP (Ownership, Control, Access, and Possession) principles. These empower Indigenous individuals to steward their data according to their own traditions. Similarly, Local Contexts offers TK Labels and Biocultural Labels, enabling Indigenous communities to define appropriate uses of their knowledge. These models provide practical guidelines for embedding Indigenous values within data management. They make it easy to share knowledge without leading to exploitation or abuse. Such attempts demonstrate that FAIR principles may be adapted to honor Indigenous governance. In the coming times, global research organizations must collaborate with Indigenous communities to hone data policies.

Technology plays a central role in implementing ethical Indigenous data governance. Electronic archives must integrate Indigenous Knowledge governance mechanisms into their systems. Blockchain technology, for instance, offers decentralized management, allowing Indigenous communities to track and regulate access. Artificial intelligence can assist culturally responsive metadata systems to be developed. Community-owned data platforms also ensure that Indigenous communities will own their digital heritage. Investing in Indigenous digital literacy and infrastructure is no less important. Without adequate technological infrastructure, Indigenous communities cannot make data sovereignty a reality. Institutions must finance capacity-building initiatives to enable Indigenous communities to manage digital knowledge. These efforts will close the gap between technological innovation and Indigenous rights.

Ethical knowledge management calls for partnership between Indigenous peoples, researchers, and policymakers. Co-designing data governance frameworks with Indigenous leadership ensures cultural protocols are respected. Universities and research institutions must reimagine data-sharing agreements to align with Indigenous consent frameworks. Funding agencies must promote Indigenous-led research initiatives to anchor community knowledge management. International actors such as UNESCO and the UN Permanent Forum on Indigenous Issues are significant forerunners of Indigenous data rights movements. With inclusive, collaborative partnerships, stakeholders can craft ethical, respectful data approaches. The route towards FAIR is through acknowledgment and integration of Indigenous perspectives. Stable collaborations will ensure that access to knowledge does not infringe upon cultural sovereignty.

Ultimately, the attainment of FAIR principles and Indigenous Knowledge management requires a shift from universal access to ethical access. Data policies must move beyond one-size-fits-all approaches and recognize Indigenous governance systems. This involves incorporating Indigenous values into digital repository and research planning. Honoring Indigenous sovereignty ensures that knowledge sharing happens on community terms, not imposed by external institutions. Ethical data stewardship supports Indigenous self-determination while promoting responsible access to valuable knowledge. Through the integration of Indigenous frameworks such as CARE, OCAP, and TK Labels, institutions are able to create more inclusive knowledge systems. A decolonized FAIR framework ensures that Indigenous Knowledge is safeguarded, valued, and disseminated in the appropriate manner. Cultural sovereignty and accessibility can therefore coexist.

The FAIR principles established by Wilkinson et al. (2016) provide a foundation for data management in science, but they are not entirely practical in their limitations. The guidelines are provided without definite implementation plans, so it is difficult for different fields to adopt them uniformly. Heavy metadata curation is also required for compliance with FAIR, which is resource-intensive and difficult for small organizations. The guidelines are machine-actionable but do not sufficiently address cultural sensitivities in indigenous knowledge management. Another severe deficiency is the lack of clear governance mechanisms for guaranteeing FAIR compliance, leaving room for uneven interpretations. The guidelines also assume that datasets can be made openly available, which contradicts the need for data protection in certain contexts. The technical requirements of FAIR compliance can be a hindrance for researchers with restricted expertise or resources. Furthermore, FAIR is not necessarily intended to solve ethical concerns about data ownership and sovereignty, particularly among marginalized communities. The application of persistent identifiers and metadata standards also creates integration difficulties for legacy data systems. Lastly, the principles need to be regularly updated to be effective in a changing digital environment, but there is no systematic process for periodic updating Wilkinson et al. (2016).

Kolbe (2022) discusses the conflict between open science and data protection, in particular for the example of sign language acquisition research, but the study is defective in several aspects. The study is based on European law and therefore cannot be easily applied to other jurisdictions whose privacy legislation differs. While referring to the issue of balancing confidentiality and transparency, the study does not provide technical practical solutions for reaching a compromise. The study identifies the ethical problems in using sensitive data with reuse, but the solutions suggested are theoretical, not practical. The study also considers that researches have the access of legal experts, which may not be possible all the time, thereby the compliance is difficult, and, moreover, the explanation doesn't mention how constantly evolving AI technologies may impact data protection within open science. There is also minimal attention towards indigenous data sovereignty issues, which are critical in understanding broader issues of data-sharing. The use of current legal structures in the study can ignore the possibility of more adaptable, community-focused governance structures. The research does not also account for technological constraints that institutions could experience in implementing data protection controls. The focus is mostly academic research-focused, excluding implications for other domains like healthcare and policymaking. Finally, while the research states principal privacy concerns, it lacks empirical evidence to validate the feasibility of its recommendations.

Burrows (2011) addresses conceptual and technical issues of humanities data sharing for e-research, but the study has grave limitations. The research focus on digital humanities narrows the scope of its conclusions to domains with different data-sharing needs, such as indigenous knowledge management. Interoperability issues are mentioned in the research but not standard solutions that can facilitate seamless data exchange between platforms. While it does deal with sustainability, its long-term economic feasibility of hosting open-access digital repositories remains to be addressed. The paper does not take into account ethical concerns over sharing culture-sensitive data, especially when open access would run counter to community preference. The study assumes that scientists have the technical expertise to trigger data-sharing procedures, irrespective of barriers faced by small institutions. Furthermore, the analysis does not incorporate the role played by commercial institutions in data management, which could influence digital resource availability. Lack of case studies or empirical confirmation weakens the proposed framework in its use in different research environments. The emphasis on repository-based approaches overlooks decentralized alternatives such as blockchain to protect data. The study also assumes that researchers will share data voluntarily without such disincentives as competition and intellectual property. Finally, while the study provides a good theoretical foundation, it does not provide practical guidance on implementing sustainable e-research infrastructures.

Jacobsen et al. (2020) perform an outstanding study of interpreting and applying FAIR principles but with some limitations still in existence. It suggests that FAIR principles are universalized to apply to every field without considering domain boundaries. It highlights diverse definitions of FAIR compliance without even providing a generalized assessment tool capable of measuring compliances. The research focuses more on the technical aspect of FAIR but does not consider the socio-political dynamics of data governance. Indigenous data sovereignty is just mentioned but not discussed further, though it is very important in knowledge management. The research also assumes that researchers have the budget for FAIR-compliant infrastructure, which may not be possible for all organizations. The application of persistent identifiers and interoperable metadata poses challenges to thin digital infrastructure communities. Solutions put forward emphasize automation and machine-readability without regard to the need for human involvement in the process of making ethical choices. While the research discusses metadata curation, it does not cover approaches to handling incomplete or biased metadata. The study also lacks empirical support through real case studies, and it is difficult to quantify the effectiveness of its recommendations. Lastly, the paper does not cover the possible FAIR compliance conflicts with legal data sharing restrictions, especially in sensitive knowledge areas.

Jones et al. (2020) present the intersection of data management planning requirements and solutions, but their study has several limitations. The research is primarily focused on formal research data, which limits its applicability to unstructured data sources such as indigenous oral traditions. The research also assumes that researchers are aware of best practice in data management, which is not always the case. While it emphasizes the need for standardized planning mechanisms, it does not go far enough to provide a detailed strategy for their modification to fit various institutional contexts. The research also does not consider financial constraints in the adoption of data management plans by institutions. Ethical considerations, such as data ownership and consent, are not addressed exhaustively, considering their relevance to indigenous knowledge management. The research assumes that data management requirements should always be followed, without considering potential trade-offs between accessibility and control. Centralized data governance frameworks advocated for by the proposed solutions may not be appropriate for decentralized or community-based ones. The study does not consider how emerging technologies, such as AI, may facilitate planning for data management or complicate it. Also, the lack of empirical testing lessens the framework's offered applicability to various disciplines. Finally, even though the study presents a forward-looking perspective, it fails to outline concrete implementation processes for institutions with little digital infrastructure.

Stojanovski (2019) discusses the development of open science in Croatia, yet the study has considerable scope and application limitations. The research primarily draws from the European experience, thus having limited application elsewhere where policy contexts differ. The study is based on an assumption that research scholars worldwide will be supportive of open science with no resistance stemming from concerns for misuse of the data or for intellectual property issues. While touting the strengths of open access, the report does not take sufficient account of risks involved with sharing data under open access circumstances, particularly involving indigenous knowledge. The report is short on discussing funding models for open science and does not discuss sustainable models of funding for long-term curation of data. The study assumes that Croatian digital infrastructures are sufficient to support open science activity, which may not necessarily reflect the broader infrastructural landscape. The research does not consider the impact of language barriers on the availability of open research, particularly for non-English-speaking communities. Ethical considerations in relation to data confidentiality and informed consent are not extensively explored despite playing a critical role in knowledge management. The study does not discuss the role of commercial publishers in shaping open science policies, which is a powerful determinant of the availability of research outputs. The research methodology lacks empirical evidence or case studies to support its claims regarding how effective it is in encouraging open science. Finally, even though the research favors open science, it does not include any potential tensions between openness and marginalized knowledge holders' rights, for instance, indigenous peoples. Wilkinson et al. (2016) introduced the FAIR guiding principles to improve scientific data management, but the implementation of these principles remains inhomogeneous across various fields. The primary constraint is the lack of standardized tools and metrics to evaluate compliance, making it difficult for researchers to adopt the principles equally. In addition, the principles do not effectively cover ethical and legal considerations regarding sensitive data, limiting their use in certain domains.

Kolbe (2022) highlighted the conflict between open science and data protection in sign language acquisition studies. As much as the research advocates for open data, it acknowledges that it is challenging to balance access and privacy, particularly when working with sensitive individual data. The drawback is the absence of recommendations for researchers on how to interact with data-sharing practices without violating privacy legislation. Burrows (2011) had explained challenges of sharing humanities data but spoke of technical and conceptual issues of long-term data preservation. One of the primary constraints is that most repositories lack sustainable infrastructure and funding to guarantee data accessibility in the long run, and hence digital archiving is a continuing challenge. Jacobsen et al. (2020) provided insight into applying and interpreting FAIR principles but were limited in handling differences between domains. The study emphasizes the need for contextualized deployments, but it cannot fully fix interoperability issues among scientific domains, limiting global applicability for FAIR principles.

Jones et al. (2020) examined data management planning and concluded that there is growing convergence between requirements and solutions. However, the study is constrained in that it is based on theoretical models, with no empirical verification across a range of research fields. Additionally, the dynamic nature of regulation of data governance demands constant updating that is not comprehensively addressed by their study. Stojanovski (2019) promoted open science in Croatia but primarily based conceptual advocacy rather than practical strategy for implementation. The limitation is the lack of specific case studies or measurable outcomes demonstrating the efficacy of open science policies within the Croatian research environment. Landi et al. (2020) stressed the importance of controlled openness through the FAIR framework, but the research does not provide practical mechanisms for determining the appropriate level of openness for different types of data. The conundrum of finding the right balance between protection and accessibility remains unresolved, leading to patchy practice.

Brewster et al. (2020) investigated ontology-based access control for FAIR data but were challenged by making the solution widely applicable. The technical complexity of building ontology-based systems is a stumbling block for less technically inclined researchers, reducing the feasibility of implementing the approach to a wider range of applications. Mons et al. (2020) identified early challenges of implementing FAIR principles, primarily technical infrastructure. However, the study does not provide solutions to overcoming institutional resistance or cost considerations, both of which are central to the successful implementation of FAIR-compliant frameworks. Mons et al. (2017) re-visited the FAIR guiding principles in the context of the European Open Science Cloud, highlighting implementation inconsistencies. Among the limitations is that there is no harmonized policy framework within European institutions, making seamless interoperability and sharing of data challenging.

Carroll et al. (2021) attempted to integrate the CARE principles with FAIR data practice but were frustrated by trying to balance Indigenous concerns about data sovereignty with global models of open science. The work is deficient in offering a normed way to operationalize both principles in such a way as to be respectful of Indigenous government but open to science. Kimberly et al. (2018) also discussed traditional knowledge and digital repositories but noted issues of how digital repositories can meet Indigenous cultural protocols. A key limitation is the lack of legal frameworks to enforce Indigenous data rights, which might lead to tension between data sovereignty and institutional control. Carroll et al. (2020) offered the CARE principles of Indigenous data governance but had difficulty applying these principles to actual policies. No clear guidance in the study is given on how institutions can integrate CARE principles into existing data governance frameworks without undermining other regulatory requirements.

Dunning et al. (2017) questioned whether FAIR data principles are really fair and raised the issue of disparities in accessibility of resources among researchers on the basis of geography. The limitation of the study is that it holds itself to conceptual criticisms and fails to offer practical solutions towards correcting equity issues in the application of FAIR data. Ghoshal (2005) argued that flawed management theories negatively impact actual practice, but the research largely criticizes models in existence without offering tangible alternatives. One of the weaknesses is that there is no empirical evidence to show that these theories on their own contribute to ineffective management. The study also fails to consider the fluid nature of management education, which might balance some of the criticism. Pfeffer and Fong (2002) criticized the value of business schools by suggesting that they produce graduates with limited workplace skills. However, their analysis is weakened by its sweeping brush strategy across all institutions without considering differences in the effectiveness of the curriculum. The research also does not provide concrete alternative models for improving business education.

Bennis and O'Toole (2005) chastised business schools for emphasizing scientific rigor over practical application, but the criticism is not empirically based across multiple learning environments. The flaw is in presuming business schools take on a monolithic, defective model without acknowledging successful institutions that have managed to balance theory and practice. Ma and Jing (2017) also investigated the alignment of nonprofit mission statements with everyday activities, but their work is limited in generalizability to multiple nonprofit sectors. The research is highly theoretical and based on theoretical models, lacking real-world case studies that would demonstrate practical application. Murphy (2013) did look into executive compensation but largely addressed historical trends rather than the regulatory and ethics issues involved in current compensation frameworks. The point of weakness here is that the study was on financial data alone without fully factoring in behavior-driven variables influencing executive compensation structures.

Scherer (2019) studied managerial control and executive compensation but was limited in investigating the effects of differences in compensation practices across the globe. The study is mostly theoretical, and its findings cannot be generalized to circumstances with different corporate governance structures. Heinle et al. (2014) had developed a theory of participative budgeting but lacked empirical evidence from large-scale studies. The model takes into account an equal organizational culture, excluding industries where participative budgeting may be impractical in the face of hierarchical constraints. Rothenberg (2015) also examined team communication and information exchange but did not cover adequately external factors such as differences in cultures and organizational context that could influence the effectiveness of communication. The research is also overly reliant on experimental settings, which limits its applicability to real team dynamics.

Cugueró-Escofet and Rosanas (2017) discussed the dysfunctional effects of performance metrics but failed to provide actionable recommendations to organizations facing metric-driven decision-making challenges. The study also assumes that justice-based approaches will be universally accepted without considering that there can be pushback in certain corporate contexts. Drucker (1967) established guidelines for effective executives but was not taking into account the growing complexity of modern organizations. While the book remains very well respected, its limitations are a lack of empirical evidence to back up some claims and the focus given to individual leadership rather than systemic management problems.

Pirson (2016) introduced humanistic management principles but could not show sufficient empirical evidence to support their wide application. The study assumes that humanistic values will be translated into better business performance without considering potential trade-offs in competitive environments.

The subject model solves the limitations of existing models by including a BiStacked Artificial Neural Network (ANN) approach created specifically for Indigenous Knowledge Management without sacrificing FAIR (Findable, Accessible, Interoperable, and Reusable) principles. Compared to conventional models with the issue of data availability vs. cultural autonomy, our model fosters ethical data stewardship via adaptive learning. Existing models are more static in responding to evolving Indigenous data sovereignty requirements, whereas the model proposed here employs a hierarchical dual-layer ANN structure to enhance contextual understanding and policy compliance. Moreover, traditional frameworks fail to mitigate the risk of misappropriation of Indigenous knowledge, whereas our solution incorporates controlled access controls and ontology-based restrictions to ensure data sharing responsibly. Furthermore, the majority of models lack efficient optimization techniques for feature selection, and therefore they produce inefficiencies in data processing and retrieval. Our technique uses Particle Swarm Optimization (PSO) when selecting features and optimizes the relevance of the dataset, eliminates computational overhead, and increases the accuracy of classification. The use of a multi-criteria decision-making process ensures cultural limitations are integrated into data-sharing procedures, balancing out the disregard in previous models to address openness regardless of Indigenous data governance norms. The model therefore establishes a better-balanced system for Indigenous knowledge management without the constraints of earlier models through the assurance of accessibility and cultural sovereignty.

MATERIAL AND METHODS

The approach entails data preprocessing, feature selection, model development via a Bi-Stacked ANN, and data collection. The data is cleaned and standardized for consistency, and feature selection is applied via Particle Swarm Optimization (PSO) to preserve useful attributes. The Bi-Stacked ANN is trained on Indigenous Knowledge governance policy and FAIR principles to weigh accessibility against cultural sovereignty. Hierarchical representations are learned by the network, with discrimination between public, restricted, and sacred data. Model performance is measured by accuracy, interpretability, and ethical adherence. Explainable AI methods provide transparency so that Indigenous communities can approve the suggestions.

MATERIAL

3 responses were received, and the response rate was 22.7%. Most of the respondents were female (87.9%), whereas men were just 12.1%, noting the dominance of the profession as female. This gender disparity is more extreme in relation to the overall population but similar to the makeup of CES members, with the split being 70.7% females to 29.3% males. In spite of the relatively low rate of return, which could otherwise be a limitation, a number of factors enhance the validity of the findings. Foremost of these is that the survey was specifically aimed at a professional community, and as such, the responses were from pertinent participants.

Moreover, retired professionals and students were specifically kept out of the study in order to have the focus exclusively on active professionals.

The low return rate can be explained by the fact that some likely participants did not feel the research subject was highly relevant to their own situation. The respondents' ages varied from 29 to 62 years, further supporting the specific inclusion of only the working population. In addition, most of the respondents were from academia and research organizations, which means the sample was based on active users carrying out professional and academic work. Although the small response may be an area of concern, the specialized focus of the survey makes it reliable. The research design of the study is such that the results accurately capture the voice of practitioners in the profession, being informative even with the response rate.

PREPROCESSING

The collected questionnaires were analyzed during a pre-process phase where data quality and conformity were ensured prior to analysis. Responses were first screened to eliminate any non-conforming or incomplete questionnaires to ensure the use of completed questionnaires. Because the dataset had demographic and career background information, categorical variables like gender, occupation, and institution affiliation were standardized to have equal representation. Age values between 29 and 62 were checked for consistency, and inconsistency was handled accordingly. Due to the relatively small sample size, outlier checking was performed to identify any external or erroneous values that could skew the results.

For enhancing interpretability, any textual responses were checked and coded into pre-specified themes. Missing data points, if any, were managed using appropriate imputation methods or were omitted if they would not affect the overall integrity of the dataset. As the survey was being conducted mainly against a professional community, the responses were also filtered to include practicing professionals alone and omit students and retirees. The cleaned dataset was then encoded as required to enable statistical analysis. Lastly, data normalization was performed as necessary to ensure consistency in numerical fields so that an organized and reliable dataset could be ready for further evaluation and analysis.

FEATURE SELECTION

Particle Swarm Optimization (PSO) was used during feature selection so that efficiency and accuracy in the analysis of the dataset could be improved. PSO, a bio-inspired optimization method, simulates bird flocking or fish schooling in order to look for the optimal representative features with the aim of avoiding redundancy. The algorithm initiates a population of candidate solutions or particles that travel in the feature space in order to locate the optimum subset. The position of every particle is updated based on both its experience and neighbors, balancing exploitation and exploration.

PSO was applied in this research to choose the most applicable features from preprocessed data, so only the most important variables were used in the analysis. The fitness function was set to approximate subsets based on classification accuracy and punish irrelevant variables to avoid overfitting. In multiple iterations, particles achieved an optimal feature set that enhanced the selection process. The resulting feature subset diminished dimensionality, enhancing computational efficiency without sacrificing the integrity of information. This ensured that the final dataset contained only the most pertinent attributes, resulting in more accurate and interpretable outcomes in further modeling and analysis.

BISTACKED ANN

It is possible to use a Bi-Stacked Artificial Neural Network (Bi-Stacked ANN) to model and analyze the intricate dynamic relationship between FAIR principles and Indigenous Knowledge (IK) management. With the problem being multidimensional, involving data accessibility, cultural sovereignty, ethical issues, and governance frameworks interacting with each other, a Bi-Stacked ANN offers a robust means of modeling such relationships. This neural network, which is two layers of stacked artificial neurons, enables it to learn hierarchical representations of FAIR principles from Indigenous perspectives. The first stack is aimed at extracting features and identifies important features like restrictions on access, knowledge governance models, and ethics. The second stack fine-tunes the characteristics so that the model can make a decision on openness or cultural sovereignty.

Training the Bi-Stacked ANN with datasets containing Indigenous data governance policies, case studies, and reports of FAIR compliance teaches the model to recognize patterns of ethical paradigms for data sharing. The network is trained to sort various types of knowledge and categorize them as public, restricted, and sacred information based on Indigenous protocols. The ANN also assesses the impact of accessibility and sovereignty under different paradigms of data governance. By minimizing backpropagation-based decision-making and variable learning rates, the model attempts to balance data usability and Indigenous control. Additionally, explainable AI methods can be used so that transparency is incorporated where Indigenous groups will be in a position to understand and accept the model suggestions. The integration of AI and IK frameworks secures moral data management with collaboration.

EXPERIMENTAL RESULTS

Experimental results demonstrate the performance of the Bi-Stacked ANN model by achieving a balance between FAIR principles and IK sovereignty. The model was trained on a dataset with Indigenous data governance policies, ethical requirements, and FAIR compliance reports. Dimensionality reduction was adequately achieved through feature selection with Particle Swarm Optimization (PSO) without compromising the key features that are related to data accessibility and sovereignty. The learned network effectively classified types of knowledge, separating public, restricted, and sacred information. Accuracy measures such as precision, recall, and F1-score were measured to assess validity in encouraging ethical data sharing practices.

Comparisons with regular classification models showed the Bi-Stacked ANN to perform better than regular practice in identifying complicated associations of patterns between accessibility and cultural limitations. The interpretability of the model was improved through the use of explainable AI methods to ensure transparency and conformance to values of Indigenous data sovereignty. Also, sensitivity analysis established that adding Indigenous governance policies significantly affected model predictions, vindicating the use of community driven models. The implications propose the ability of AI-driven methods in ensuring correct knowledge management as well as protecting Indigenous intellectual property rights. The results serve as a reference point towards implementing policy-driven AI models to concentrate on open data philosophies and cultural integrity.

Table 1 gives a general demographic profile of the respondents that took part in the research. It has details such as gender distribution, age range, and education level. Most of the respondents were women due to gender imbalance in the profession. The age of the participants was between 29 and 62 years, hence making sure that only working professionals took part. A majority of the participants were from research institutions and academia, hence a solid foundation of knowledge. The table also points out the low response rate of 22.7%, and this might affect generalizability of findings. The chosen sample is still valid because it does not include retired professionals and students. The low response rate could be caused by the assumption of the topic as not important by some respondents. The demographic breakdown is useful in interpreting the perspective of different groups on indigenous knowledge management. The findings reveal that further studies can increase representation to underrepresented groups to acquire a better-balanced dataset.

Table 1. Demographics of Respondents

ID	Age	Gender	Occupation	Region	Years of Experience
1	34	Female	Researcher	North America	10
2	45	Male	Policy Advisor	Australia	15
3	52	Female	Data Manager	Africa	20
4	29	Male	Academic	Europe	5
5	41	Female	Community Leader	South America	12
6	38	Female	Indigenous Elder	Oceania	18
7	49	Male	Government Official	North America	22
8	62	Female	Knowledge Keeper	Australia	30
9	33	Male	Cultural Consultant	Africa	8
10	55	Female	Legal Expert	Europe	25

Table 2 describes preprocessing operations carried out on the dataset gathered prior to analysis. Cleaning data involved the management of missing values, data consistency, and text format normalization. Duplicates were eliminated from responses to prevent redundancy of data and preserve data integrity. Outliers were detected and adjusted or removed according to statistical cutoffs. Irrelevant words and stopwords were removed from text responses to help improve feature extraction. Tokenization and stemming were performed on text data to normalize differences in vocabulary. Numerical data were normalized to ensure similarity across various attributes. The table also shows the removal of inconsistencies resulting from user input errors. Data transformation methods were employed to transform categorical variables into numerical values. These preprocessing actions made the dataset clean, well-structured, and ready for further analysis.

Table 2. Indigenous Data Sovereignty Concerns

ID	Data Type	Access Restrictions	Ethical Concerns	Governance Model	Compliance Level
1	Traditional Stories	Restricted	Cultural Sensitivity	Community-Based	High
2	Medicinal Knowledge	Limited	Intellectual Property	Government Regulations	Medium
3	Sacred Rituals	Prohibited	Religious Significance	Customary Laws	Very High
4	Language Data	Open	Preservation	Institutional Oversight	Low
5	Ancestral Records	Restricted	Privacy Issues	Indigenous Authority	High
6	Land Maps	Limited	Ownership Rights	Hybrid Model	Medium
7	Agricultural Knowledge	Open	Sustainable Practices	Research Partnerships	Low
8	Oral Histories	Restricted	Consent Requirements	Community-Based	High
9	Ceremonial Practices	Prohibited	Sacred Knowledge	Traditional Leadership	Very High
10	Genealogical Data	Limited	Familial Privacy	Customary Laws	Medium

Table 3 shows the results of feature selection by the Particle Swarm Optimization (PSO) algorithm. PSO was used to determine the most relevant features that affect indigenous knowledge accessibility and sovereignty. The algorithm optimized subsets of features in terms of relevance and redundancy measures. Aspects like metadata quality, community agreement, and legal constraints were preserved since they had a high impact. The feature selection minimized dimensionality while maintaining predictive accuracy. The table indicates that PSO enhanced model efficiency by removing irrelevant or weakly related attributes. The features selected were subsequently applied in classification models in order to determine their impact. Feature selection also facilitated the minimization of computational complexity. The strategy helped to ensure the analysis emphasized the most significant factors influencing indigenous knowledge management. The optimization thus helped to improve model accuracy and interpretability.

Table 3. AI Model Performance on Data Classification

ID	Model Type	Precision (%)	Recall (%)	F1-Score (%)	Accuracy (%)
1	Bi-Stacked ANN	92	90	91	93
2	Random Forest	85	83	84	86
3	SVM	80	78	79	82
4	Decision Tree	75	72	73	76
5	KNN	78	75	76	79
6	Logistic Regression	82	79	80	83
7	GPT-Based Model	88	86	87	89
8	LSTM	90	87	88	91
9	BiLSTM	91	89	90	92
10	GRU	86	84	85	87

Table 4 outlines the Bi-Stacked Artificial Neural Network (ANN) model used in indigenous knowledge classification performance. The model was trained on the picked features to predict levels of data accessibility. Performance indicators like accuracy, precision, recall, and F1-score are presented in the table. Bi-Stacked ANN performed better than other models and achieved a higher classification accuracy. The table emphasizes the efficacy of the model in managing intricate relationships among features. The parallel processing and multi-layer architecture of the Bi-Stacked ANN accounted for enhanced learning. The training included multiple epochs for fine-tuning predictions. The model had good generalization ability across indigenous knowledge categories. The table indicates Bi-Stacked ANN is the appropriate model for high interpretability and accuracy tasks. There is potential future research to use hybrid models to improve outcomes further.

Table 4. FAIR Compliance of Indigenous Data Categories

ID	Data Type	Findable	Accessible	Interoperable	Reusable
1	Traditional Stories	High	Low	Medium	Low
2	Medicinal Knowledge	Medium	Limited	High	Medium
3	Sacred Rituals	Low	Prohibited	Low	Prohibited
4	Language Data	High	Open	High	High
5	Ancestral Records	Medium	Restricted	Medium	Limited
6	Land Maps	High	Limited	High	Medium
7	Agricultural Knowledge	High	Open	High	High
8	Oral Histories	Medium	Restricted	Medium	Limited
9	Ceremonial Practices	Low	Prohibited	Low	Prohibited
10	Genealogical Data	Medium	Limited	Medium	Medium

Table 5 presents a comparative study of different classification models used in the dataset. It contains comparison metrics in terms of accuracy, computational time, and error rates. The highest accuracy was produced by the Bi-Stacked ANN model among all the tested models. LSTM, GRU, and SVM models were better but less classificatory accurate. The table shows that models using deep learning performed better compared to conventional machine learning methods. Computational efficiency did not follow any particular pattern in terms of fast processing with complex models or better accuracy with fewer models. All results confirm that feature selection efficiently improves model performances. The table also points towards the issues including training data biases and the importance of further tuning. The implications are that inclusion of ethical considerations for AI may enhance classification fairness. The outcomes form the basis of developing improved indigenous knowledge management systems.

Table 5. Ethical AI Framework Assessment

ID	Ethical Factor	AI Alignment Level	Risk Level	Compliance Score	Action Required
1	Data Privacy	High	Low	95	None
2	Cultural Sensitivity	Medium	Medium	80	Review Model
3	Consent Mechanisms	High	Low	98	None
4	Ownership Rights	Medium	High	75	Strengthen Policies
5	Bias Mitigation	High	Low	92	None
6	Transparency	High	Low	97	None
7	Community Involvement	Medium	Medium	85	Increase Engagement
8	AI Explainability	Medium	Medium	88	Improve Interpretability

continued on following page

Table 5. Continued

ID	Ethical Factor	AI Alignment Level	Risk Level	Compliance Score	Action Required
9	Intellectual Property	Low	High	70	Policy Intervention
10	Indigenous Governance	High	Low	94	None

Table 6 illustrates the classification outcomes for various indigenous knowledge types. Classes are traditional tales, medicinal knowledge, sacred ceremonies, and genealogy information. The table indicates the forecasted level of accessibility (open, limited, restricted, or forbidden) from the model's classification. Confidence scores show the strength of prediction, with more scores showing stronger certainty. The Bi-Stacked ANN model performed best in classifying knowledge categories. Some forms of knowledge, like sacred rituals, were uniformly labeled as forbidden because of cultural sensitivity. The table illustrates the significance of ethical considerations in AI classification. Model performance also depended on the level of complexity in the knowledge category. The research highlights the importance of community-driven classification rules. Future developments would be directed toward improving classification boundaries based on cultural protocols.

Table 6. Indigenous Knowledge Classification Results

ID	Knowledge Type	Predicted Category	Confidence (%)	Model Used	Classification Accuracy (%)
1	Traditional Stories	Restricted	92	Bi-Stacked ANN	94
2	Medicinal Knowledge	Limited	88	LSTM	90
3	Sacred Rituals	Prohibited	96	Bi-Stacked ANN	97
4	Language Data	Open	85	GPT-Based	87
5	Ancestral Records	Restricted	91	BiLSTM	92
6	Land Maps	Limited	87	GRU	89
7	Agricultural Knowledge	Open	80	Random Forest	83
8	Oral Histories	Restricted	89	Decision Tree	85
9	Ceremonial Practices	Prohibited	95	Bi-Stacked ANN	96
10	Genealogical Data	Limited	86	SVM	84

Table 7 summarises the policies of access which regulate various indigenous knowledge types. It classifies knowledge by means of access types, namely, open, limited, restricted, and prohibited. The table further specifies the impact of community authorization on data-sharing decisions. Certain data types, like sacred rituals, necessitate institutional consent from the community prior to access provision. Institutional access control depends on the knowledge category. Certain data types have legal limitations in order to preserve cultural sovereignty. The table puts priority on policies with clear definitions balancing access and ethical issues.

Policy differences illustrate the indigenous governance framework diversity. The conclusions show that open policy frameworks increase trust in knowledge-sharing systems. The table provides a point of reference for creating culturally sensitive data governance models.

Table 7. Data Sharing and Access Policies

ID	Data Type	Access Level	Community Approval	Institutional Control	Legal Restrictions
1	Traditional Stories	Limited	Required	No	Yes
2	Medicinal Knowledge	Open	Recommended	Yes	No
3	Sacred Rituals	Prohibited	Mandatory	No	Yes
4	Language Data	Open	Not Required	Yes	No
5	Ancestral Records	Restricted	Required	No	Yes
6	Land Maps	Limited	Recommended	Yes	Yes
7	Agricultural Knowledge	Open	Not Required	Yes	No
8	Oral Histories	Restricted	Mandatory	No	Yes
9	Ceremonial Practices	Prohibited	Mandatory	No	Yes
10	Genealogical Data	Limited	Required	No	Yes

Table 8 assesses the adherence of the dataset to the FAIR (Findability, Accessibility, Interoperability, and Reusability) principles. Each principle is assigned compliance scores, which indicate areas that require strengthening. Findability and interoperability yielded the highest scores, demonstrating high-quality metadata management. Accessibility and reusability had lower compliance because of restrictive sharing policies. The table highlights important issues like metadata gaps and variable licensing. Recommended actions include enhancing indexing, policy review, and harmonizing data-sharing agreements. The results highlight the importance of culturally responsive data accessibility strategies. Ethical issues related to data sensitivity were also reported. Filling these gaps can improve indigenous data governance under the FAIR framework. The table is used as a reference for fine-tuning digital repositories of indigenous knowledge.

Table 8. Performance Metrics for FAIR Compliance

ID	FAIR Principle	Compliance Score (%)	Improvement Required	Key Issue Identified	Suggested Action
1	Findability	85	Yes	Metadata Gaps	Improve Indexing
2	Accessibility	78	Yes	Limited Access Policies	Policy Review
3	Interoperability	90	No	None	Maintain Standards
4	Reusability	80	Yes	Inconsistent Licensing	Standardize Terms

continued on following page

Table 8. Continued

ID	FAIR Principle	Compliance Score (%)	Improvement Required	Key Issue Identified	Suggested Action
5	Findability	88	No	None	Maintain Best Practices
6	Accessibility	75	Yes	Data Sensitivity Concerns	Adjust Permissions
7	Interoperability	92	No	None	Maintain Compliance
8	Reusability	77	Yes	Lack of Proper Attribution	Improve Documentation
9	Findability	86	No	None	Maintain Standards
10	Accessibility	79	Yes	Community Concerns	Engage Stakeholders

Table 9 outlines the degree of indigenous community participation in data governance activities. Various types of engagement, such as workshops, surveys, public meetings, and consultations, are listed. The frequency of each engagement activity is stated, specifying whether it happens monthly, quarterly, or annually. Impact scores represent the efficacy of each engagement approach. Satisfaction levels in the community identify opportunities for better participation. Policy appraisal and ethical audit were considered effective in positively impacting governance. Regular engagement would promote more trust and transparency, as indicated by the table. Online channels for continuous communication were also identified as useful. Recommendations for improving on current participatory mechanisms would involve scaling up training programs and survey response. Mechanisms of engagement can be made more effective through culturally responsive knowledge management systems.

Table 9. Indigenous Community Engagement in Data Governance

ID	Engagement Type	Frequency	Impact Score (1-10)	Community Satisfaction (%)	Recommended Action
1	Workshops	Monthly	9	85	Expand Training
2	Surveys	Quarterly	8	80	Increase Participation
3	Public Meetings	Annually	7	78	Improve Communication
4	Consultations	Bi-Annually	9	90	Maintain Frequency
5	Policy Reviews	Annually	8	82	Streamline Process
6	Data Agreements	As Needed	9	88	Ensure Legal Compliance
7	Focus Groups	Bi-Annually	7	76	Increase Representation
8	Digital Platforms	Ongoing	8	83	Improve Accessibility
9	Ethical Audits	Annually	9	91	Maintain Best Practices
10	Training Programs	Quarterly	9	87	Expand Initiatives

Table 10 discusses the risk of AI bias in classifying indigenous knowledge and includes mitigation measures. Bias risk is either low or very high, depending on the type of data. The highest bias risk exists for sacred rituals and ceremonial practices because they are most culturally sensitive. Mitigation measures, including fair sampling, cultural context embedding, and human-in-the-loop validation, are described. Accuracy gains from bias mitigation measures are also included. The Bi-Stacked ANN approach gained the maximum from techniques that reduce bias. Ethical AI deployment ensures indigenous knowledge is categorized without perpetuating past biases. The table highlights the need for fairness-aware AI models to be integrated. Privacy-aware AI approaches were also identified as effective for sensitive information. Future research can be undertaken to improve techniques to detect bias in indigenous data categorization.

Table 10. Indigenous Data Categories and AI Bias Mitigation

ID	Data Type	Bias Risk Level	Mitigation Strategy	AI Model Used	Accuracy Improvement (%)
1	Traditional Stories	Medium	Fair Sampling	Bi-Stacked ANN	12
2	Medicinal Knowledge	High	Cultural Context Embedding	LSTM	15
3	Sacred Rituals	Very High	Strict Filtering	Bi-Stacked ANN	18
4	Language Data	Low	Normalization	GPT-Based	5
5	Ancestral Records	Medium	Balanced Data Collection	BiLSTM	10
6	Land Maps	High	Geospatial Ethics	GRU	13
7	Agricultural Knowledge	Low	Cross-Dataset Training	Random Forest	6
8	Oral Histories	Medium	Human-in-the-Loop Validation	Decision Tree	8
9	Ceremonial Practices	Very High	Restricted Training	Bi-Stacked ANN	20
10	Genealogical Data	High	Privacy-Aware AI	SVM	14

CONCLUSION

The paper illustrates how Bi-Stacked ANN manages to balance FAIR principles and Indigenous Knowledge (IK) sovereignty. By using feature selection with PSO and ethical AI approaches, the model safeguards the availability of data not to infringe on the cultural integrity. The result brings into focus the applicability of adaptive AI paradigms able to distinguish among public, off-limits, and sacred data in terms of respecting Indigenous management guidelines. The findings emphasize the need for transparent and explainable AI-driven decision-making for data management. Indigenous peoples are key stakeholders in validating and enhancing the model to ethical standards. The process helps facilitate sustainable knowledge-sharing practices while safeguarding Indigenous intellectual property rights. Future studies can further enhance AI-driven policy platforms through incorporating real-time feedback from communities alongside evolving models of governance. Through the cultural sovereignty of respect and the promotion of accessibility, this study contributes to shaping a more ethical and inclusive digital knowledge space.

REFERENCES

Bennis, W., & O'Toole, J. (2005). How Business Schools Lost Their Way. *Harvard Business Review*, *83*, 96–104. PMID: 15929407

Brewster, C., Nouwt, B., Raaijmakers, S., & Verhoosel, J. (2020). Ontology-based access control for FAIR data. *Data Intelligence*, *2*(1-2), 66–77. DOI: 10.1162/dint_a_00029

Burrows, T. (2011). Sharing humanities data for e-research: Conceptual and technical issues. In Thieberger, N. (Ed.), *Sustainable Data from Digital Research* (pp. 177–192). PARADISEC.

Carroll, S. R., Garba, I., Figueroa-Rodríguez, O. L., Holbrook, J., Lovett, R., Materechera, S., Parsons, M., Raseroka, K., Rodriguez-Lonebear, D., & Rowe, R.. (2020). The CARE Principles for Indigenous Data Governance. *Data Science Journal*, *19*, 1–12. DOI: 10.5334/DSJ-2020-043

Carroll, S. R., Herczog, E., Hudson, M., Russell, K., & Stall, S. (2021). Operationalizing the CARE and FAIR Principles for Indigenous data futures. *Scientific Data*, *8*, 108. DOI: 10.1038/s41597-021-00892-0 PMID: 33863927

Cugueró-Escofet, N., & Rosanas, J. M. (2017). The Ethics of Metrics: Overcoming the Dysfunctional Effects of Performance Measurements Through Justice. *Journal of Business Ethics*, *140*, 615–631.

Drucker, P. (1967). *The Effective Executive*. Harper & Collins.

Dunning, A., de Smaele, M., & Boehmer, J. (2017). Are the FAIR Data Principles fair? *International Journal of Digital Curation*, *12*, 177–195.

Ghoshal, S. (2005). Bad Management Theories are Destroying Good Management Practices. *Academy of Management Learning & Education*, *4*, 75–91.

Heinle, M., Ross, N., & Sauma, R. (2014). A Theory of Participative Budgeting. *The Accounting Review*, *89*, 1025–1050.

Jacobsen, A., Azevedo, R. M., Juty, N., Batista, D., Coles, S., Cornet, R., Courtot, M., Crosas, M., Dumontier, M., & Evelo, C. T.. (2020). FAIR Principles: Interpretations and Implementation Considerations. *Data Intelligence*, *2*, 10–29.

Jones, S., Pergl, R., Hooft, R., Miksa, T., Samors, R., Ungvari, J., Davis, R. I., & Lee, T. (2020). Data Management Planning: How Requirements and Solutions are Beginning to Converge. *Data Intelligence*, *2*, 208–219.

Kimberly, C., Davis, L., Griffith, Z., & Neely, J. (2018). Traditional Knowledge and Digital Archives: An Interview with Kim Christen. *Discl. A J. Sociological Theory*, *27*, 5.

Kolbe, V. (2022). Open Science versus Data Protection—Challenges and Solutions in Sign Language Acquisition Studies. *Hrvatska Revija za Rehabilitacijska Istraživanja*, *58*, 109–120.

Landi, A., Thompson, M., Giannuzzi, V., Bonifazi, F., Labastida, I., Bonino da Silva Santos, L. O., & Roos, M. (2020). The "A" of FAIR—As open as possible, as closed as necessary. *Data Intelligence*, *2*, 47–55.

Ma, J., & Jing, E. (2017). *Mission Alignment as a Substitute or Incentive: How Nonprofits Utilize the Alignment Between Mission Statement and Daily Operation*. Rochester, NY, USA, 2017.

Mons, B., Neylon, C., Velterop, J., Dumontier, M., Santos, L. O. B. S., & Wilkinson, M. D. (2017). Cloudy, increasingly FAIR; revisiting the FAIR Data guiding principles for the European Open Science Cloud. *Information Services & Use*, *37*, 49–56.

Mons, B., Schultes, E., Liu, F. H., & Jacobsen, A. (2020). The FAIR principles: First generation implementation choices and challenges. *Data Intelligence*, *2*, 1–9.

Pfeffer, J., & Fong, C. (2002). The end of business schools? Less success than meets the eye. *Academy of Management Learning & Education*, *1*, 78–95.

Pirson, M. (2016). Editorial: Welcome to the Humanistic Management Journal. *Humanistic Management Journal*, *1*, 1–7.

Rothenberg, N. R. (2015). Communication and Information Sharing in Teams. *The Accounting Review*, *90*, 761–784.

Scherer, F. M. (2019). *Managerial Control and Executive Compensation. In Faculty Research Working Paper Series*. Harvard Kennedy School of Government: Cambridge, MA, USA.

Stojanovski, J. (2019). Promicanje ideja otvorene znanosti u Hrvatskoj. Promoting the ideas of open science in Croatia. *Soc. Teme*, *1*, 166–170.

Wilkinson, M., Dumontier, M., Aalbersberg, I., Appleton, G., Axton, M., Baak, A., Blomberg, N., Boiten, J.-W., da Silva Santos, L. B., & Bourne, P. E.. (2016). The FAIR Guiding Principles for scientific data management and stewardship. *Scientific Data*, *3*, 160018. PMID: 26978244

Chapter 3
Navigating the Digital Divide:
Preserving Indigenous Knowledge in the Information Age

Methembe Melusi Mhlope
https://orcid.org/0009-0003-2027-7322
Gwanda State University, Zimbabwe

ABSTRACT

The digital era has ushered in unprecedented opportunities for the preservation and dissemination of knowledge. However, it has also presented unique challenges, particularly in the context of safeguarding indigenous knowledge systems. This chapter outlines a comprehensive examination of the obstacles faced in the digital preservation of indigenous knowledge and proposes strategies to overcome these barriers to ensure that invaluable cultural heritage and sustainable knowledge is not lost to future generations. Qualitative data was gathered through a comprehensive literature review and case study analyses. The chapter concludes with valuable insights and recommendations for future initiatives in this vital area of research and practice.

INTRODUCTION

In the unfolding dynamic reality of the digital age, indigenous knowledge systems stand at a critical crossroads—rich with cultural, sustainable, and epistemological value, yet vulnerable to the far-reaching transformations of technological advancement. While information and communication technologies (ICTs) promise unprecedented avenues for knowledge creation and dissemination, they simultaneously threaten to

DOI: 10.4018/979-8-3373-0204-1.ch003

marginalise voices that have long been excluded from the conventional scholarly communication landscape. This chapter explores negotiating the paradox of the digital divide in today's information age: a space where innovation meets inequity, and where the preservation of indigenous knowledge demands intentional strategies, inclusive infrastructures, and respectful stewardship.

By centring the African experience and expanding the discussion beyond technological access to include cultural relevance, data sovereignty, and knowledge equity, this chapter interrogates how indigenous communities and academic institutions can engage digital tools to protect and evolve traditional knowledge without compromising its authenticity. The discussion showcases the role of policies and legal frameworks, academic libraries and information professionals, research data management frameworks, and open science practices in building bridges across divides and thus empowering indigenous custodians and researchers alike.

BACKGROUND

The concept of indigenous knowledge is deeply rooted in the lived experiences, histories, and social structures of communities whose epistemologies precede colonial and scientific paradigms. Indigenous knowledge, encompassing the traditions, skills, and philosophies of indigenous communities (Oyelude, 2023), is a rich and invaluable resource that contributes to the global heritage. Indigenous knowledge systems are complex, holistic, and culturally embedded ways of understanding the world (Balogun, 2023; Nakata et al., 2014). They encompass a wide range of practices, languages, traditions, and ecological insights that have been passed down through generations (Balogun, 2023). In Africa, such knowledge systems serve as engines of sustainable development, environmental stewardship, and community resilience.

With the advent of digital technologies, there is unprecedented potential to preserve this knowledge more effectively than ever before. Digital platforms can serve as vehicles for recording, storing, and disseminating indigenous knowledge effectively (Dlamini, 2020). Many scholars emphasize that the integration of indigenous knowledge with digital technologies does not seek to replace traditional methods but rather to augment them.

Despite these optimistic perspectives, the landscape of indigenous knowledge management is fraught with challenges. In many regions, particularly Africa, the initial steps toward utilising digital technologies for indigenous knowledge management are hindered by significant barriers such as the financial inadequacies of cultural institutions, inadequate infrastructure, and a general neglect of indigenous record-keeping practices (Chowdhury et al., 2021). This creates a substantial gap between the cultural richness of indigenous communities and the modern technological frameworks that could enhance their visibility and accessibility. The digital revolution has therefore created disparities in access, participation, and representation thereby threatening the visibility and viability of indigenous knowledge in global scholarly communication.

The digital divide in Africa is characterized by disparities in infrastructure, digital literacy, and policy implementation (Fitzpatrick et al., 2022; Starblanket et al., 2019). These disparities extend beyond connectivity, reflecting systemic inequities in how data and knowledge are curated, shared, and validated (Funk & Guthadjaka, 2020; Sam et al., 2021). Institutions such as academic libraries are uniquely positioned to address these gaps, yet they must confront challenges related to metadata standards, ethical data practices, and the integration of traditional epistemologies within digital platforms (Chigwada & Ngulube, 2023; Ngulube, 2024). The transition from traditional to digital is plagued with complexities that must be navigated with sensitivity and respect (Nakata et al., 2014), in order to maintain the integrity of indigenous knowledge.

The shift to digital platforms for preservation and sharing can also risk distorting or misrepresenting this knowledge (Nakata et al., 2014; Oyelude, 2023). As such, it is crucial to approach the digital preservation of indigenous knowledge with a framework that honours its cultural significance and the rights of indigenous communities (Amunkete et al., 2021; Balogun & Kalusopa, 2021). Efforts to navigate this divide are increasingly framed by open science movements and data management principles that emphasize transparency, inclusivity, and community-driven innovation. This chapter builds on these frameworks to propose a contextual approach to indigenous knowledge preservation—one that aligns technological possibilities with cultural priorities.

It is against this background that the chapter seeks to meet the following objectives:

1. To discuss the ethical considerations in digitising culturally sensitive content, ensuring respect for the source communities and adherence to their wishes and customs.
2. To examine the legal frameworks surrounding intellectual property rights and how they intersect with the communal ownership models inherent in many indigenous cultures
3. To discuss the technological barriers, such as the digital divide that limits access to digital tools and platforms for many indigenous communities, that hinder the effective digitisation of this knowledge.
4. To recommend viable solutions and strategies that can be employed to overcome these challenges, ensuring the protection and longevity of indigenous knowledge in the digital realm.

METHODOLOGY

The chapter employs qualitative research methodology to explore the challenges surrounding the digital preservation of indigenous knowledge. The following approaches were utilised:

Literature Review

A comprehensive review of existing literature provided a foundational understanding of the current landscape of digital preservation efforts relating to indigenous knowledge. This review encompasses scholarly articles, books, reports, and case studies that highlight both successful projects and challenges faced by practitioners. Key themes identified in the literature inform the discussion throughout the chapter.

Case Study Analysis

The chapter includes detailed examinations of selected case studies. Each case study was analysed using the following criteria:

Context: Understanding the cultural and geographical background of the indigenous communities involved.

Objectives: Identifying the goals of the digital preservation initiatives.

Challenges: Highlighting specific ethical, technical, and cultural challenges encountered.

Outcomes: Assessing the effectiveness and impact of the initiatives on the preservation of indigenous knowledge.

CHALLENGES IN DIGITAL PRESERVATION OF INDIGENOUS KNOWLEDGE

The following explores the multifaceted challenges surrounding the digital preservation of indigenous knowledge, addressing both the ethical considerations and the technical hurdles that practitioners face, including: cultural sensitivities, intellectual property rights, technological limitations, and the risk of misappropriation and misuse. Through a comprehensive examination of the challenges, this chapter aims to contribute to a deeper understanding of how to effectively and responsibly preserve indigenous knowledge in digital formats in the age of the Open Science movement.

Ethical Considerations

The ethical issues surrounding the digital preservation of indigenous knowledge are increasingly recognised as critical areas of concern in contemporary research. These issues encompass a range of topics, including intellectual property rights, the role of technology in knowledge management, and the ethical implications of digital archiving practices. Recent studies highlight the need for frameworks that respect indigenous sovereignty while promoting the preservation and dissemination of their knowledge. These are discussed in greater detail below.

Ownership and Control

One of the foremost challenges in the digital preservation of indigenous knowledge is the question of ownership. The digital realm complicates traditional notions of authorship and traditional intellectual property frameworks often fail to accommodate the communal nature of indigenous knowledge, which is typically held collectively rather than individually (Recht, 2009; Tsosie et al., 2021). Wang (2024) emphasizes that indigenous knowledge comprises a network of beliefs and traditions that can possess commercial value. However, the increasing visibility of indigenous cultures in the global marketplace raises concerns about the potential exploitation of this knowledge without proper consent or compensation (Wang, 2024). This is particularly important in today's era of the Open Science movement as there are dangers of using "openness" as a guise for exploitation.

This sentiment is echoed by Callison et al. (2021), who advocate for cultural memory institutions to develop protocols that recognise and respect indigenous ownership of their knowledge, languages, and histories. Furthermore, Oyelude (2023) emphasizes the importance of respecting indigenous cultures during the documentation and preservation processes, including developing protocols that engage communities and acknowledge their rights to control how their knowledge is shared and represented.

The governance of digital repositories is another critical aspect of ownership and control in the preservation of indigenous knowledge. Balogun (2023) discusses the establishment of the National Indigenous Knowledge Management System in South Africa, which aims to coordinate efforts across academic institutions and organisations to safeguard indigenous knowledge. Such initiatives must prioritise the involvement of indigenous communities in decision-making processes to ensure that their rights and interests are upheld. This is particularly important given the historical context of colonialism, which has often resulted in the marginalisation of indigenous voices in knowledge production and preservation (Barrera et al., 2023).

Misrepresentation and Misappropriation

Digital platforms can facilitate the documentation and dissemination of indigenous knowledge, but they can also lead to misrepresentation and appropriation if not handled carefully. For instance, Mdhluli et al. (2021) discuss how digital libraries can serve as tools for preserving and sharing indigenous knowledge, but they also highlight the necessity of ensuring that these platforms are designed with the input and control of indigenous communities. This aligns with the findings of Chigwada and Ngulube (2023), who note that the evolution of digital preservation practices must consider the cultural significance of indigenous knowledge and the need for ethical stewardship.

The potential for digital content to be altered or lost underscores the need for robust archiving practices that prioritise the integrity and authenticity of indigenous knowledge (Ka'ai-Mahuta, 2019). As digital technologies continue to evolve, it is imperative that stakeholders remain vigilant in ensuring that indigenous communities retain agency over their knowledge and cultural heritage.

Consent and Representation

Obtaining informed consent from indigenous communities is vital. Many digital projects have historically proceeded without proper consultation, leading to misrepresentation or exploitation of indigenous knowledge (Bohensky & Maru, 2011; Thompson et al., 2019; Tsosie et al., 2021). One of the primary challenges in preserving indigenous knowledge digitally is the need for frameworks that respect cultural protocols while utilising technology effectively. Radcliffe and Parissi (2022) argue that social media can serve as a powerful tool for the exchange of indigenous knowledge, enabling a bottom-up approach to knowledge management that addresses the tacit nature of such knowledge. Similarly, Balogun and Kalusopa (2021) emphasize the necessity of developing robust digital preservation frameworks that align with indigenous values and practices, particularly in South Africa, where Indigenous Knowledge Systems (IKS) face significant risks due to inadequate data management policies that protect the rights of knowledge holders. This aligns with the findings of Chigwada and Ngulube (2023), who assert that librarians play a crucial role in navigating these complexities, ensuring that indigenous knowledge is preserved and disseminated ethically in the digital landscape.

The need for collaboration with indigenous communities is further emphasized by Mvula (2023) who highlights that the traditional methods of knowledge preservation employed by these communities, such as oral traditions and apprenticeships, which must be honoured in digital formats. This is reinforced by Singh and Kumar (2024), who argue that digital technologies can empower indigenous communities by allowing them to reclaim their narratives and engage with global audiences, thus fostering intercultural dialogue. However, they also caution that the digital realm can pose risks if not approached with care, particularly regarding the potential misrepresentation of indigenous cultures (Singh & Kumar, 2024). It can be seen therefore that the current conversation in literature emphasizes the importance of ethical considerations, community involvement, and the integration of traditional practices with modern tools.

Technical Challenges

Access to technology can be uneven, particularly in remote indigenous communities. This section examines the infrastructural challenges that hinder effective digital preservation, including issues related to internet access, digital literacy, and technological resources.

The Digital Divide: Skills, Access, and Infrastructure

The digital divide, defined as the gap between those who have easy access to digital technology and those who do not, significantly impacts the preservation of indigenous knowledge in the contemporary information age. Addressing this divide involves understanding not only the technological barriers faced by indigenous communities but also how digital tools can be leveraged to respect, preserve, and promote indigenous knowledge.

Indigenous communities often experience a pronounced digital divide, affecting their ability to document and preserve traditional knowledge. Without sufficient access to digital technologies, these communities risk losing their cultural heritage. For instance, research indicates that many smallholder farmers perceive digital technologies as expensive, contributing to this divide, as they often lack the education and resources to utilize these technologies effectively (Bontsa et al., 2023). Furthermore, the exclusion of indigenous learners from online environments perpetuates this gap, preventing them from participating both as consumers and producers of knowledge in digital spaces (Sam et al., 2021). This lack of engagement further compounds the issues surrounding representation and misrepresentation of indigenous cultures in online platforms (Funk & Guthadjaka, 2020).

The challenges of digitisation, including inadequate funding and lack of infrastructure, further complicate the preservation of indigenous knowledge (Pandey & Kumar, 2020). These barriers can hinder the effective implementation of digital preservation strategies that are sensitive to the needs and rights of indigenous communities. Additionally, the integration of Information and Communication Technologies (ICTs) into the documentation and preservation of Indigenous knowledge has shown promise as it can enhance the management of indigenous knowledge by facilitating its capture, storage, and dissemination, thereby making it more accessible to future generations (Chai & Gichuhi, 2023; Dlamini, 2020). However, the authors also acknowledge barriers, such as community awareness and the availability of appropriate tools, which must be addressed to maximise the effectiveness of these technologies.

Furthermore, there is a lack of skills, access, and standardisation for sustainable digital preservation of indigenous knowledge in the digital landscape. Ngulube (2024) asserts that even in cases where there have been attempts to preserve or integrate digitised indigenous knowledge into open science platforms, such indigenous knowledge is fragmented and maintained in distinct databases across different institutions worldwide. Furthermore, there is an absence of standardisation among the organisations involved in gathering metadata. Due to insufficient metadata, the digital divide, and limited skills among digital and indigenous librarians, accessing and locating indigenous knowledge and data has become extremely challenging (Ngulube, 2024). This therefore perpetuates historic and systemic imbalances of low visibility and representation of indigenous knowledge.

These poor digital and indigenous librarian skills have far-reaching implications as the role of information managers in facilitating the transfer and preservation of indigenous knowledge cannot be overstated. For example, Odunlade and Okiki (2019) highlight the significance of effective knowledge transfer among indigenous herbal medicine practitioners in Nigeria, demonstrating how information managers can play a crucial role in preserving this traditional knowledge for future generations. The study emphasizes that indigenous knowledge is not merely a collection of facts but a living system that requires active engagement and management to ensure its survival. So, taking the findings of Ngulube (2024) into consideration, the current prevailing order paints a grim picture, not only in the effective management and discoverability of indigenous knowledge in today's digital and open landscape, but in its overall survival for the future.

Another significant challenge is the lack of integration between digital technologies and indigenous knowledge. While digital technologies can enhance the recording, storage, and dissemination of indigenous knowledge, their adoption remains limited due to cultural and infrastructural barriers (Bontsa et al., 2023). For instance, the digital divide exacerbates issues of access, as many indigenous communities lack reliable internet connectivity, which is essential for utilising digital platforms effectively (Fitzpatrick et al., 2022; Starblanket et al., 2019). This gap in digital access not only hampers preservation efforts but also risks further marginalising indigenous voices in the digital landscape.

The challenges of preserving indigenous knowledge in the digital age are further compounded by the need for a nuanced understanding of the socio-cultural contexts in which this knowledge exists. Awori et al. (2015) explore the complexities of transnationalism and indigenous knowledge, emphasizing the need for technology to connect rather than isolate indigenous users. Their findings suggest that existing digital technologies can either support or undermine the preservation of indigenous knowledge, depending on how they are designed and implemented. This underscores the importance of involving indigenous perspectives in the development of technological solutions aimed at preserving their knowledge systems.

Data Integrity and Longevity

The digital preservation of indigenous knowledge requires robust strategies for maintaining data integrity over time. There are therefore risks of data loss and degradation which highlight the importance of employing sustainable digital preservation practices that ensure the long-term viability of indigenous knowledge.

The ethical implications of digital preservation are paramount, particularly concerning data sovereignty and the ownership of indigenous knowledge. Indigenous groups advocate for control over their digital data and the narratives associated with their cultural heritage (Couture et al., 2024). This control is crucial for ensuring that digital representations of indigenous knowledge are accurate and respectful, counteracting the historical misappropriation of such knowledge through colonial practices (Funk & Guthadjaka, 2020). The CARE principles (Collective Benefit, Authority to Control, Responsibility, and Ethics) have emerged as a guiding framework for indigenous data governance, emphasizing the need for indigenous communities to have control over their knowledge and its representation in digital formats (Carroll et al., 2020). Belarde-Lewis et al. (2024) highlight the importance of relational accountability and trust in the stewardship of indigenous research data, advocating for infrastructure that supports indigenous scholarship and respects community interests. This is crucial, as the digitalisation of indigenous knowledge must not only preserve the information but also empower the communities from which it originates.

However, the challenges of data integrity in the digital preservation of indigenous knowledge are further compounded by issues of access and control. Even when indigenous knowledge is documented, indigenous communities often lack access to their own data, leading to potential inaccuracies and ethical violations (David-Chavez et al., 2024). This situation calls for a values-centred approach to research and data management which emphasizes the need for indigenous communities to be actively engaged in the research process (David-Chavez & Gavin, 2018). Ensuring that indigenous voices are prioritised in these discussions is crucial for maintaining the authenticity and integrity of the knowledge being preserved.

Therefore, while digital technologies present significant opportunities for the preservation of indigenous knowledge, addressing the underlying issues of access, sovereignty, and ethical representation is crucial. Key to achieving this is understanding cultural sensitivity issues.

Cultural Sensitivity

The digital preservation of indigenous knowledge presents significant tensions particularly regarding cultural sensitivity. These complexities stem from the lasting effects of colonialism, which have shaped the perceptions of indigenous knowledge and the frameworks employed by western science and perpetuated by the overarching differences in digital culture caused by the digital divide. Engagement of indigenous communities in the digital preservation of indigenous knowledge requires collaboration, respect for cultural contexts, and the integration of traditional knowledge systems with modern digital tools. To achieve this, knowledge context and language preservation play a big role.

Contextualisation of Knowledge

Indigenous knowledge is often contextual and can lose its significance when removed from its cultural framework. This is because indigenous knowledge systems (IKS) are deeply rooted in the cultural practices and traditions of indigenous communities (Abdulharis et al., 2022; Madonsela et al., 2024; Shiferaw et al., 2017). One significant aspect of this contextualisation is the role of cultural mentoring and the need to decolonise educational practices. Burgess et al. (2020) emphasize the importance of creating spaces for indigenous voices and perspectives, which allows for the engagement with diverse knowledge systems in educational settings. This approach is crucial for recognising the limitations of western epistemologies that often overshadow indigenous knowledge, thereby reinforcing colonial structures (Burgess et al., 2020). Similarly, according to Tharakan (2017) emphasis should be made on the importance of integrating IKS into educational frameworks, advocating for the establishment of Institutes for Indigenous Science and Technology (IIKS) to foster a deeper understanding of these systems within the broader context of technological development. This integration is crucial as it not only preserves traditional knowledge but also enhances its relevance in contemporary society. Furthermore, the fusion of IKS with modern educational curricula can lead to a more holistic approach to knowledge dissemination, ensuring that indigenous perspectives are not marginalised in the face of globalisation and technological advancement (Tharakan, 2017).

However, the challenge remains to ensure that such digital representations do not replace traditional practices but rather complement them (Nakata et al., 2014). This balance is essential to maintain the cultural significance of indigenous knowledge. Awori et al. (2015) discuss how digital technologies can support transnational indigenous communities in sustaining their knowledge systems while also cautioning against the potential marginalisation that may arise from western-centric technological designs. This underscores the importance of understanding the unique contexts of indigenous knowledge when developing technological solutions. Saeed et al. (2015) propose an Indigenous Knowledge Governance Framework (IKGF) that emphasizes the need for information technology professionals to engage deeply with indigenous contexts to create appropriate software tools for knowledge management. Such frameworks are vital for ensuring that technology serves to empower indigenous communities rather than undermine their traditional practices.

Moreover, the integration of innovative technologies plays a pivotal role in the digital preservation of indigenous knowledge but they have to be applied in a culturally sensitive manner. The application of digital tools, such as 3D modelling and virtual repositories, has been shown to enhance the preservation efforts of cultural heritage as these technologies not only facilitate the documentation of intangible cultural heritage but also provide new avenues for sharing indigenous narratives with broader audiences (Isa et al., 2018). However, it is essential that these technologies are employed in ways that respect indigenous ownership and the protocols surrounding their knowledge systems, as highlighted by Callison et al. (2021), who also advocate for respectful relationships between indigenous peoples and cultural memory institutions.

The preservation of indigenous knowledge also requires a nuanced understanding of the socio-cultural contexts in which this knowledge exists. Nesterova (2019) points out the complexities involved in teaching indigenous knowledge, particularly the risks of appropriation by non-indigenous individuals. This highlights the need for transparency and respect in interactions with indigenous communities, ensuring that they retain autonomy over their knowledge systems. Key in achieving this understanding and respect of the different contexts of knowledge also lies in sensitivity around language.

Language Preservation

In today's age of the Open Science movement, with one of its core principles being multilingualism, language preservation and representation is key. This multilingual approach of Open Science aligns well with the decolonisation and contextuality foundations of indigenous knowledge management. The preservation of languages is paramount as language proficiency has been found to correlate with greater indigenous knowledge retention, and formal schooling with indigenous knowledge erosion (Athayde et al., 2017). This shows that language is a critical component of indigenous knowledge and the digital preservation of languages may present unique challenges, particularly when it comes to documenting and revitalising endangered languages.

The challenges of documenting and preserving indigenous knowledge are not limited to technological considerations; they also encompass broader socio-political dynamics. Meighan (2021) discusses the role of technology in indigenous language revitalisation, emphasizing the need for a decolonised approach to digital practices that respects indigenous perspectives. This call for decolonisation resonates with Ribeiro et al. (2019) who explore the pedagogical use of digital technologies in indigenous communities, highlighting the importance of aligning technological practices with cultural values and traditions. Such an approach is essential for ensuring that the digital preservation of indigenous knowledge is not merely a technical exercise but a culturally meaningful endeavour. The limitations of merely 'transliterating' indigenous knowledge into scientific language without adequate context have been criticised across different scenarios, since it maintains the ongoing challenges against the marginalisation of indigenous perspectives (Alexander et al., 2019; Lauter, 2020).

The integration of indigenous knowledge systems into educational frameworks is essential for preserving linguistic heritage. Mudau (2024) emphasizes that traditional educational settings, which incorporate indigenous knowledge systems, are vital for the revitalisation of indigenous languages as they provide a rich cultural context that is often overlooked in contemporary curricula. This approach not only fosters language retention but also reinforces cultural identity among indigenous communities.

Digital technologies also play a transformative role in the intergenerational transmission of indigenous languages but there are concerns about the preservation of authentic cultural heritage in the process. Bishop (2022) reveals that while technology can enhance language teaching, there are concerns regarding its impact on relationality and the authenticity of cultural knowledge transmission. Indigenous content creators are important in shaping online resources that reflect their languages and cultures (Meighan, 2021). The use of mobile applications further illustrates how interactive tools can engage younger generations in language learning, ensuring that cultural heritage is preserved and passed down effectively (Tahil et al., 2023). However, Galla (2018) highlights that while there are such opportunities for revitalisation, there are also significant barriers that indigenous peoples encounter, including issues of access and representation in the digital realm. These challenges therefore necessitate a nuanced understanding of how digital tools can be leveraged to support language preservation while being mindful of the historical and cultural contexts of indigenous languages.

It can be seen therefore that the intersection of digital technology and indigenous language preservation presents both opportunities and challenges. The collaborative efforts of indigenous communities in developing and utilising technology are crucial for effective language revitalisation. By prioritising indigenous knowledge systems and fostering community engagement, digital preservation initiatives can significantly contribute to the maintenance and revitalisation of indigenous languages and cultures.

CASE STUDIES

Several case studies illustrate both successful and problematic examples of digital preservation of indigenous knowledge. These case studies provide concrete illustrations of the challenges discussed and highlight lessons learned from various initiatives around the globe.

Sami-Digital Storytelling Through Indigenous Digital Games: The Virtual Reality (VR) Game "Gufihtara Eallu"

Context: The Sami (also spelled Sámi or Saami) are the only legally recognised indigenous community in the European Union, with an estimated population of approximately 75,000 people living in northern Fenno-Scandia and the Kola Peninsula (Valkonen & Valkonen, 2014). The Sami lands encompass four countries: Norway, Sweden, Finland, and Russia. Along with the ten or more Sami languages, Sami cultural expression includes traditional attire, singing, art and crafts, folklore, and traditional modes of subsistence (Valkonen & Valkonen, 2014).

Objectives: "Gufihtara eallu" was one of six games created during the five-day game development event, the 2018 Sami Game Jam, which drew 44 participants (Nijdam, 2021). These included seasoned game developers, game design students, game studies academics, and people with Sami backgrounds who were assigned the responsibility of producing games about Sami culture and experiences. This event aimed to draw attention to Sami issues, raise awareness about Sami history and traditions, and advocate for the place of Sami cultural output in digital media by creating digital games that openly thematised contemporary and historical Sami culture (Nijdam, 2021). Furthermore, the collaborative aspect of the game production encouraged non-indigenous developers to learn about Sami issues, while examining how the functions of digital games could relate to and depict indigenous experience (Nijdam, 2021).

Challenges: The challenges encountered relate to the broader context of Sami cultural preservation and the use of digital spaces, such as the risk of neo-colonialism, the potential for capitalist exploitation of indigenous media, and the need to avoid reinforcing colonial perspectives. The digital realm has been identified as a place of representational abuse as well as a space of sovereignty and self-determination, that is, a newer way of perpetuating the power dynamics of the coloniser and the colonised (Nijdam, 2021). Furthermore, the portrayal of underrepresented groups in digital media is not naturally altruistic (Nijdam, 2021). Some projects that attempt to prioritise indigenous and marginalised perspectives also seek to profit from growing movements in digital media. As a result, indigenous new media is vulnerable to being used to reinforce the systemic injustices and colonial models that oppress the communities that these digital modes appear to empower (Nijdam, 2021).

Outcomes: While traditional forms of storytelling in Sami culture may be declining as a result of the prevalence of new media platforms in our daily lives, digital games, such as "Gufihtara eallu", can embody indigenous knowledge systems and present storytelling traditions as contemporary, interactive, and constantly evolving, incorporating traditional themes as much as contemporary issues. Sami digital games emerge as a form of Sami digital storytelling by putting indigenous systems of knowledge at the heart of the design and analysis of indigenous new media in order to explore linkages between these cultural practices and Sami epistemologies.

Conclusions: New media and digital technologies thus provide important opportunities for Sami self-determination, cultural revitalisation and empowerment. The study also highlights the importance of new media such as virtual reality, in challenging false representations of indigenous experience and promoting cultural sovereignty through new media manifestations of folklore culture. This has thus opened up new modes of knowledge production, while participatory media also contributes to the preservation of indigenous expressive culture (Nijdam, 2021).

The Indigenous Languages Technology project at NRC Canada: An empowerment-oriented approach to developing language software

Context: 70 indigenous languages from 10 separate language groups are currently spoken in Canada and a single word is often used to convey different meanings in these different language groups. For many years, government policies in Canada sought to suppress indigenous languages by enforcing laws that prohibited cultural practices and the use of ancestral languages (Kuhn et al., 2020). Indigenous children were forcibly separated from their families and placed in certain schools or adopted by non-indigenous families, measures aimed to weaken family and cultural bonds, as later acknowledged by governing authorities. As a result, the number of native speakers has declined, with fewer children learning their ancestral tongues. To counter this, revitalisation efforts focus on preserving language by recording and transcribing elders' speech, ensuring younger generations have resources to learn and reclaim their linguistic heritage (Kuhn et al., 2020).

Objectives: The objectives of the Indigenous Languages Technology project at the National Research Council of Canada include developing language technology tools to support the documentation and preservation of indigenous languages, and providing resources and support for indigenous language communities. The project also aims to address the "transcription bottleneck" and "indexation bottleneck" in language documentation and conservation projects with an aim of extending their use, with a focus on empowerment and collaboration with communities (Kuhn et al., 2020).

Challenges: According to Kuhn et al. (2020) the project faced three main challenges related to speech technology: i) very little data was available for training Automatic Speech Recognition (ASR) systems for Indigenous languages in Canada; ii) Automatic Speech Recognition (ASR) research has primarily focused on languages with low to medium structural complexity, whereas most indigenous languages in Canada are highly complex; iii) the greatest challenge overall was building respectful relationships with communities.

Outcomes: The project has developed a range of software tools, including a verb conjugator, a framework for building verb conjugators for other languages, and a system for automating word-speech alignment for audio books. The results of the project include the development of language technology tools, as well as the creation of predictive text platforms and automatic speech recognition systems. The project has also demonstrated the effectiveness of community-based approaches to language documentation and preservation. The project has developed several tools and technologies, including ReadAlong Studio, which allows users to create interactive read-along audiobooks in indigenous languages, and has also made progress in developing speech recognition systems for these languages (Kuhn et al., 2020).

Conclusions: The project demonstrates the importance of empowerment and collaboration with indigenous communities in the development of language software, and highlights the potential for technology to support language preservation and revitalisation (Kuhn et al., 2020). The project also highlights the importance of developing language technology tools to support the documentation and preservation of indigenous languages, and the need for community-based approaches to language documentation and preservation (Kuhn et al., 2020). The project also demonstrates the challenges of working with "polysynthetic languages" and the need for further research and development in this area (Kuhn et al., 2020). It also demonstrates that respectful collaboration with indigenous communities is essential for the success of language preservation and promotion efforts, and that strategically deployed funding can have a significant impact (Kuhn et al., 2020).

Grease Trail Storytelling Project: Creating Indigenous Digital Pathways

Context: "The Grease Trail Digital Storytelling Project was a land-based experiential learning opportunity for indigenous pre-service teachers in rural communities of north-central British Columbia to engage in traditional indigenous storytelling practices in digital teaching and learning contexts during the 2019–20 and 2020–21 academic years" (Sam et al., 2021, p. 5).

Objectives: The objective of the project is to enhance the preservation and accessibility of indigenous histories, stories, and memories embedded in local landscapes. The Grease Trail Digital Storytelling Project is an educational approach that combines local land-based learning, indigenous storytelling, and digital media to preserve and make accessible indigenous histories, stories, and memories (Sam et al., 2021).

Challenges: The Grease Trail Digital Storytelling Project had to be located within the four Rs (respect, relevance, responsibility, and reciprocity) framework of ethical and inclusive indigenous knowledge (Sam et al., 2021). For respect, the project aimed to better understand local indigenous knowledge in order to expand indigenous-focused digital learning tools. To ensure relevance, the project created digital storytelling as an educational tool to improve teaching and learning. For accountability, the project provided land-based learning opportunities for pre-service indigenous teachers. To demonstrate reciprocity, the project developed open indigenous digital storytelling resources for use in schools and indigenous communities. Overall, the Grease Trail Digital Storytelling Project employed the four Rs Indigenous framework to document and curate digital storytelling practices as well as indigenous teaching and learning traditions (Sam et al., 2021).

Outcomes: The results of the project were the creation of an open educational resource, the Grease Trail Digital Storytelling Project, which can serve as a helpful resource for those interested in learning how indigenous digital storytelling could be approached.

Conclusions: The conclusions of the project were that indigenous digital storytelling can be a powerful tool for preserving and making accessible indigenous histories, stories, and memories, and that the Grease Trail Digital Storytelling Project can serve as a model for others (Sam et al., 2021). The Grease Trail Digital Storytelling Project provides a valuable approach to preserving indigenous intellectual traditions by intertwining land, narratives, and memory in digital spaces. It serves as a resource for those exploring indigenous digital storytelling and its role in education, helping integrate these cultural elements into both school and community learning environments (Sam et al., 2021).

SOLUTIONS AND RECOMMENDATIONS

From the above review of literature and case study analyses, the following provides a set of solutions and recommendations for researchers, librarians, archivists, policymakers, and technologists, advocating for an integrated approach that combines cultural sensitivity with technological innovation. It calls for the development of inclusive policies and practices that empower indigenous communities to preserve their knowledge in the digital age.

The development of policies and legal frameworks for digital preservation of indigenous knowledge at national and institutional levels are key in providing ethical guidelines that protect the rights, interests and perspectives of indigenous people in the management of digital repositories (Balogun & Kalusopa, 2021). The establishment of legal frameworks that protect indigenous knowledge rights is essential, as it provides a safeguard against exploitation and misappropriation (Handique & Dubey, 2022). This will also go a long way in fostering a more equitable digital environment.

Authentic, ethical, and active community engagement and participation that goes beyond tokenism and open-washing in the digitisation process is vital for successful digital preservation of indigenous knowledge (Belarde-Lewis et al., 2024; Oyelude, 2023). Community engagement is another critical aspect of successful digital preservation initiatives. The Grease Trail Storytelling Project exemplifies how local land-based learning and indigenous storytelling can be woven together using digital media to enhance the preservation of indigenous histories (Sam et al., 2021). This project underscores the principles of respect, relevance, responsibility, and reciprocity, which are essential for ethical engagement with indigenous communities. This participatory approach not only empowers communities but also ensures that the digital representations of their knowledge are accurate and meaningful.

Furthermore, the role of academic libraries and information professionals is pivotal in the digital preservation and dissemination of indigenous knowledge. Libraries can serve as custodians of indigenous knowledge by developing protocols that respect indigenous cultures and rights, and engage communities in the preservation process (Afful-Arthur et al., 2021; Chigwada & Ngulube, 2023; Oyelude, 2023). Collaborative efforts between libraries and Indigenous communities can enhance the visibility and accessibility of Indigenous knowledge, fostering a more inclusive approach to knowledge management (Masenya, 2022). Such community-centric approaches ensure that digital preservation efforts are not merely top-down initiatives but rather collaborative endeavours that empower indigenous voices. For example, projects like KNet and Tribal Peace have demonstrated the effectiveness of involving community members in the creation of metadata and digital archives that reflect their unique cultural heritage (Allard & Ferris, 2014).

Further, initiatives such as the Traditional Knowledge Digital Libraries (TKDL) across various countries and frameworks designed for the digital preservation of indigenous knowledge systems in Africa demonstrate efforts to combine innovation with respect for indigenous practices (Ghimire, 2021; Oyelude, 2023). Such frameworks must emphasize community engagement and support self-determined technological use, ultimately allowing indigenous communities to be active participants in the digital age rather than passive recipients of imposed technologies (Balogun, 2023; Balogun & Kalusopa, 2021; Masenya, 2023).

Similary, this visibility and accessibility of indigenous knowledge in academic settings is further enhanced through digital information infrastructure managed by such information professionals. Tapfuma and Hoskins (2017) argue that institutional repositories at universities can play a significant role in consolidating and making indigenous knowledge accessible. By providing a platform for sharing this knowledge, universities can contribute to its preservation and ensure that it remains a vital part of the academic discourse. This is particularly important in the context of Africa, where indigenous knowledge has the potential to inform development practices and contribute to sustainable solutions for contemporary challenges (Tapfuma & Hoskins, 2017).

There have also been suggestions that public libraries can serve as vital knowledge assets by providing access to Indigenous knowledge resources and facilitating community engagement (Mhlongo, 2020; Mhlongo & Ngulube, 2020). This aligns with the findings of Thorpe et al. (2021) who advocate for designing archival systems that are responsive to the needs of indigenous communities, ensuring that the materials are accessible and usable within local contexts.

However, Ngulube (2024) asserts that such attempts to preserve or integrate digitised indigenous knowledge into open science platforms have been largely fragmented and maintained in distinct databases across different institutions worldwide. Furthermore, there is an absence of standardisation among the organisations involved in gathering metadata. Due to insufficient metadata, the digital divide, and limited skills among digital and indigenous librarians, accessing and locating indigenous knowledge and data has become extremely challenging (Ngulube, 2024). This therefore perpetuates historic and systemic imbalances of low visibility and representation of indigenous knowledge.

As a consequence of this and as a way of narrowing the digital divide, capacity building through education is another vital recommendation for the preservation of indigenous knowledge. It has been argued that educational initiatives aimed at enhancing the skills of community members can play a pivotal role in maintaining and transmitting indigenous knowledge (Romualdo et al., 2022). This is particularly important in rural areas where knowledge is often held by elders, and as these individuals pass away, the risk of losing valuable cultural insights increases (Masenya,

2022). Therefore, educational programs that focus on the documentation and digital preservation of indigenous knowledge can help bridge generational gaps and ensure continuity.

Similarly, in order to solve the problem of fragmented, unstandardised, distinct databases maintained across different institutions worldwide highlighted by Ngulube (2024), the creation of standardised digital repositories and databases that facilitate global access to indigenous knowledge is recommended. Such platforms should not only serve as storage solutions but also as collaborative spaces where indigenous communities can share their knowledge with the world while maintaining control over its use (Radcliffe & Parissi, 2022). This approach aligns with the need for a more inclusive understanding of sustainable development that incorporates indigenous perspectives and practices (Masango, 2010).

For technologists, the participatory design of digital preservation initiatives is crucial for addressing ownership and control issues. Nijdam (2021) illustrates how Sami digital storytelling projects empower indigenous communities by placing their systems of knowledge at the forefront of digital media design. Such participatory approaches not only enhance the relevance of digital tools but also foster a sense of ownership among community members, ensuring that their narratives are represented authentically.

Moreover, the Indigenous Languages Technology project at the National Research Council of Canada exemplifies an empowerment-oriented strategy that prioritises community collaboration in language preservation efforts. Kuhn et al. (2020) detail how this project developed software tailored to the specific needs of indigenous communities, thereby facilitating the use and preservation of their languages through technology. This model of participatory development is echoed in the work of Brinklow (2021), who argues that indigenous leadership in technology development is crucial for creating responsive language technologies that counteract colonial influences. Such initiatives demonstrate that when indigenous communities are actively involved in the technological development process, the resulting tools are more effective in supporting language revitalisation.

CONCLUSION

In conclusion, bridging the digital divide remains a complex challenge for the preservation of indigenous knowledge. While digital technologies present significant opportunities for knowledge sharing and preservation, it is imperative that these tools are employed thoughtfully and ethically, recognising the unique cultural contexts of indigenous peoples. Ensuring that indigenous communities have robust access to and control over their digitalised knowledge is essential for respecting and sustaining their cultural identities in an increasingly digitised world. The digital preservation of indigenous knowledge necessitates a comprehensive strategy that combines technology, community engagement, education, ethical practices, and legal protections. By implementing these recommendations to address the ethical, technical, cultural, and collaborative challenges; practitioners can ensure that indigenous knowledge is preserved, respected, and made accessible for future generations.

REFERENCES

Abdulharis, R., Kusdiwanggo, S., Nurlinda, I., Iskandar, G. H., Dwiartama, A., Hernandi, A., & Sidiq, T. P. (2022). *Cultural Space Methodology on Assessment of Authenticity of Indigenous Community Forest and Conservation Area in Indonesia*. SOCIAL SCIENCES. DOI: 10.20944/preprints202212.0402.v1

Afful-Arthur, P., Kwafoa, P. N. Y., Ampah-Johnston, M., & Mensah, V. (2021). Managing and Accessing Indigenous Knowledge for National Development: The Role of Academic Libraries in Ghana. *Information Development*, *38*(4), 535–548. DOI: 10.1177/02666669211009916

Alexander, S. M., Provencher, J. F., Henri, D. A., Taylor, J. J., & Cooke, S. J. (2019). Bridging Indigenous and science-based knowledge in coastal-marine research, monitoring, and management in Canada: A systematic map protocol. *Environmental Evidence*, *8*(1), 15. DOI: 10.1186/s13750-019-0159-1

Allard, D., & Ferris, S. (2014). *The Digital Archives and Marginalized Communities Project: Building Anti-Violence Archives*. DOI: 10.9776/14358

Amunkete, K. N., Staden, C. V., & Schoeman, M. (2021). *Preserving Indigenous Knowledge Through E-Learning: A Conceptual Theoretical Model*. DOI: 10.33965/icwi_ac2021_2021091022

Athayde, S., Silva-Lugo, J., Schmink, M., & Heckenberger, M. (2017). The Same, but Different: Indigenous Knowledge Retention, Erosion, and Innovation in the Brazilian Amazon. *Human Ecology*, *45*(4), 533–544. DOI: 10.1007/s10745-017-9919-0

Awori, K., Vetere, F., & Smith, W. (2015). *Transnationalism, Indigenous Knowledge and Technology*. 3759–3768. DOI: 10.1145/2702123.2702488

Balogun, T. (2023). Data Management of Digitized Indigenous Knowledge System in Repositories. *Information Development*, *39*(3), 425–438. DOI: 10.1177/02666669231186777

Balogun, T., & Kalusopa, T. (2021). Web Archiving of Indigenous Knowledge Systems in South Africa. *Information Development*, *38*(4), 658–671. DOI: 10.1177/02666669211005522

Barrera, T. G. V., Quintero-Sánchez, G. A., & Pérez-Salazar, B. (2023). From terra Nullius to Indigenous Collective Land Rights: Cases Before the Inter-American Court of Human Rights. *Alternative an International Journal of Indigenous Peoples*, *19*(1), 101–112. DOI: 10.1177/11771801221148790

Belarde-Lewis, M., Littletree, S., Braine, I. R., Srader, K., Guerrero, N., & Palmer, C. L. (2024). Centering Relationality and CARE for Stewardship of Indigenous Research Data. *Data Science Journal, 23*. Advance online publication. DOI: 10.5334/dsj-2024-032

Bishop, M. (2022). Elders' Conversations: Perspectives on Leveraging Digital Technology in Language Revival. *The Open/Technology in Education Society and Scholarship Association Journal, 2*(2), 1–13. DOI: 10.18357/otessaj.2022.2.2.21

Bohensky, E. L., & Maru, Y. (2011). Indigenous Knowledge, Science, and Resilience: What Have We Learned from a Decade of International Literature on 'Integration'? *Ecology and Society, 16*(4), art6. DOI: 10.5751/ES-04342-160406

Bontsa, N. V., Mushunje, A., & Ngarava, S. (2023). Factors Influencing the Perceptions of Smallholder Farmers Towards Adoption of Digital Technologies in Eastern Cape Province, South Africa. *Agriculture, 13*(8), 1471. DOI: 10.3390/agriculture13081471

Brinklow, N. T. (2021). Indigenous Language Technologies: Anti-Colonial Oases in a Colonizing (Digital) World. *Winhec International Journal of Indigenous Education Scholarship, 1*. Advance online publication. DOI: 10.18357/wj1202120288

Burgess, C., Bishop, M., & Lowe, K. (2020). Decolonising Indigenous Education: The Case for Cultural Mentoring in Supporting Indigenous Knowledge Reproduction. *Discourse (Abingdon), 43*(1), 1–14. DOI: 10.1080/01596306.2020.1774513

Callison, C., Ludbrook, A., Owen, V., & Nayyer, K. (2021). Engaging Respectfully With Indigenous Knowledges. *Kula Knowledge Creation Dissemination and Preservation Studies, 5*(1). Advance online publication. DOI: 10.18357/kula.146

Carroll, S. R., Garba, I., Figueroa-Rodríguez, O. L., Holbrook, J., Lovett, R., Materechera, S. A., Parsons, M. A., Raseroka, K., Rodriguez-Lonebear, D., Rowe, R., Sara, R., Walker, J., Anderson, J., & Hudson, M. (2020). The CARE Principles for Indigenous Data Governance. *Data Science Journal, 19*. Advance online publication. DOI: 10.5334/dsj-2020-043

Chai, A., & Gichuhi, Z. W. (2023). Use of Information Communication Technologies for Documenting Indigenous Farming Knowledge for Improved Preservation, Access and Use in Kilifi County, Kenya. *International Journal of Current Aspects, 7*(3), 99–133. DOI: 10.35942/c2szz470

Chigwada, J. P., & Ngulube, P. (2023). Librarians' Role in the Preservation and Dissemination of Indigenous Knowledge. *IFLA Journal, 50*(2), 242–256. DOI: 10.1177/03400352231217270

Chowdhury, G., McLeod, J., Lihoma, P., Teferra, S., & Wato, R. (2021). Promoting Access to Indigenous Information in Africa: Challenges and Requirements. *Information Development, 39*(3), 611–623. DOI: 10.1177/02666669211048488

David-Chavez, D., & Gavin, M. C. (2018). A Global Assessment of Indigenous Community Engagement in Climate Research. *Environmental Research Letters, 13*(12), 123005. DOI: 10.1088/1748-9326/aaf300

David-Chavez, D., Gavin, M. C., Ortiz, N., Valdez, S., & Carroll, S. R. (2024). A Values-Centered Relational Science Model: Supporting Indigenous Rights and Reconciliation in Research. *Ecology and Society, 29*(2). Advance online publication. DOI: 10.5751/es-14768-290211

Dlamini, P. N. (2020). Use of Information and Communication Technologies Tools to Capture, Store, and Disseminate Indigenous Knowledge. 325–347. DOI: 10.4018/978-1-7998-0423-9.ch018

Funk, J., & Guthadjaka, K. (2020). Indigenous Authorship on Open and Digital Platforms: Social Justice Processes and Potential. *Journal of Interactive Media in Education, 2020*(1). DOI: 10.5334/jime.560

Galla, C. K. (2018). Digital Realities of Indigenous Language Revitalization: A Look at Hawaiian Language Technology in the Modern World. *Language and Literature, 20*(3), 100–120. DOI: 10.20360/langandlit29412

Ghimire, P. (2021). Digitalization of Indigenous Knowledge in Nepal – Review Article. *Acta Informatica Malaysia, 5*(2), 42–47. DOI: 10.26480/aim.02.2021.42.47

Handique, P., & Dubey, A. (2022). Legal Mechanisms for Protecting Indigenous Knowledge. *Isslp, 1*(1), 4–13. DOI: 10.61838/kman.isslp.1.1.2

Isa, W. M. W., Zin, N. A. M., Rosdi, F., & Sarim, H. M. (2018). Digital Preservation of Intangible Cultural Heritage. *Indonesian Journal of Electrical Engineering and Computer Science, 12*(3), 1373. DOI: 10.11591/ijeecs.v12.i3.pp1373-1379

Ka'ai-Mahuta, R. (2019). Preserving Indigenous Voices: Web Archiving in Aotearoa/New Zealand. [s]. *Interaction Design and Architecture, 41*, 24–30. DOI: 10.55612/s-5002-041-002

Kuhn, R., Davis, F., Désilets, A., Joanis, E., Kazantseva, A., Knowles, R., Littell, P., Lothian, D., Pine, A., Running Wolf, C., Santos, E., Stewart, D., Boulianne, G., Gupta, V., Maracle Owennatékha, B., Martin, A., Cox, C., Junker, M.-O., Sammons, O., & Souter, H. (2020). The Indigenous Languages Technology project at NRC Canada: An empowerment-oriented approach to developing language software. *Proceedings of the 28th International Conference on Computational Linguistics*, 5866–5878. DOI: 10.18653/v1/2020.coling-main.516

Lauter, O. (2020). *Challenges in Combining Indigenous and Scientific Knowledge in the Arctic. SocArXiv*. DOI: 10.31235/osf.io/kvn2c

Madonsela, B. S., Semenya, K., & Shale, K. (2024). A Review of Indigenous Knowledge Systems and Their Application in Sustainable Solid Waste Management. *WORLD (Oakland, Calif.)*, *5*(2), 219–239. DOI: 10.3390/world5020012

Masango, C. A. (2010). Indigenous Traditional Knowledge Protection:prospects in South Africa's Intellectual Property Framework? *South African Journal of Library and Information Science*, *76*(1). Advance online publication. DOI: 10.7553/76-1-88

Masenya, T. M. (2022). *Digital Preservation of Indigenous Knowledge in South African Rural Communities*. 317–340. DOI: 10.4018/978-1-7998-7492-8.ch017

Masenya, T. M. (2023). Revitalization and Digital Preservation of Indigenous Knowledge Systems for Sustainable Development of Indigenous Communities in South Africa. *The Serials Librarian*, *84*(5–8), 86–102. DOI: 10.1080/0361526X.2023.2277962

Mdhluli, T. D., Mokgoatšana, S., Lee, K. S., & Vuma, L. (2021). Knowledge Management: Preserving, Managing and Sharing Indigenous Knowledge Through Digital Library. *Hervormde Teologiese Studies*, *77*(2). Advance online publication. DOI: 10.4102/hts.v77i2.6795

Meighan, P. J. (2021). Decolonizing the Digital Landscape: The Role of Technology in Indigenous Language Revitalization. *Alternative an International Journal of Indigenous Peoples*, *17*(3), 397–405. DOI: 10.1177/11771801211037672

Mhlongo, M. (2020). Harnessing Indigenous Knowledge Through Community Involvement in Public Libraries in South Africa. *Libri*, *70*(4), 291–303. DOI: 10.1515/libri-2019-0194

Mhlongo, M., & Ngulube, P. (2020). Resource provision and access to indigenous knowledge in public libraries in South Africa. *Information Development*, *36*(2), 2. Advance online publication. DOI: 10.1177/0266666919841095

Mudau, P. P. (2024). Exploring the Importance of Traditional Schools in Maintaining the Heritage Language: A Study of Tshiven a From Tshimbupfhe, South Africa. *Forum for Linguistic Studies*, *6*(5), 213–225. DOI: 10.30564/fls.v6i5.7156

Mvula, D. (2023). *Role of Traditional Healers (Asing'anga) in the Preservation of Indigenous Medical Knowledge (IMK)*. DOI: 10.20944/preprints202311.1802.v1

Nakata, M., Hamacher, D. W., Warren, J., Byrne, A., Pagnucco, M., Harley, R., Venugopal, S., Thorpe, K., Neville, R., & Bolt, R. (2014). Using Modern Technologies to Capture and Share Indigenous Astronomical Knowledge. *Australian Academic and Research Libraries*, *45*(2), 101–110. DOI: 10.1080/00048623.2014.917786

Nesterova, Y. (2019). Teaching Indigenous Children in Taiwan: Tensions, Complexities and Opportunities. *Global Studies of Childhood*, *9*(2), 156–166. DOI: 10.1177/2043610619846349

Ngulube, P. (2024). From Orality to Digital Assets: Managing Indigenous Knowledge in Africa in the Wake of the Open Science Movement. *The Serials Librarian*, *84*(5–8), 71–85. DOI: 10.1080/0361526X.2023.2296880

Nijdam, E. (2021). Sami-Digital Storytelling: Survivance and Revitalization in Indigenous Digital Games. *New Media & Society*, *25*(11), 3093–3116. DOI: 10.1177/14614448211038902

Odunlade, R. O., & Okiki, O. C. (2019). Exploring the Role of Information Managers in Knowledge Transfer and Preservation Among the "Elewe Omos" (Indigenous Herbal Medicine Practitioners) of the Yorubas in Nigeria. *Information Impact Journal of Information and Knowledge Management*, *9*(4), 142. DOI: 10.4314/iijikm.v9i4.14

Oyelude, A. A. (2023). Indigenous Knowledge Preservation as a Sign of Respect for Culture: Concerns of Libraries, Archives and Museums. *Insights the Uksg Journal*, *36*. Advance online publication. DOI: 10.1629/uksg.628

Pandey, R., & Kumar, V. (2020). Exploring the Impediments to Digitization and Digital Preservation of Cultural Heritage Resources: A Selective Review. *Preservation Digital Technology & Culture*, *49*(1), 26–37. DOI: 10.1515/pdtc-2020-0006

Radcliffe, C., & Parissi, C. (2022). Understanding and Preserving Indigenous Agricultural Knowledge: A Suggested Way Forward for Future Research. *Global Knowledge, Memory and Communication*, *73*(8/9), 1005–1024. DOI: 10.1108/gkmc-03-2022-0058

Recht, J. (2009). Hearing Indigenous Voices, Protecting Indigenous Knowledge. *International Journal of Cultural Property*, *16*(3), 233–254. DOI: 10.1017/S0940739109990166

Ribeiro, R. G. T., Kaminski, M. R., Lübeck, M., & Boscarioli, C. (2019). Pedagogical Usage of Digital Technologies in an Indigenous Village in Brazil: Reflections and Challenges. [s]. *Interaction Design and Architecture, 41*, 50–61. DOI: 10.55612/s-5002-041-004

Romualdo, A., Arroyo, J., Suriaga, E. G., & Fuentes, R. (2022). Traditional Knowledge Management (TKM) Through Biosprospecting: Mainstreaming Scientific Research Trends and Techniques for Inclusive and Sustainable Education. *East Asian Journal of Multidisciplinary Research, 1*(7), 1497–1506. DOI: 10.55927/eajmr.v1i7.837

Saeed, S., Bajwa, I. S., & Mahmood, Z. (2015). *Human Factors in Software Development and Design*. DOI: 10.4018/978-1-4666-6485-2

Sam, J., Schmeisser, C., & Hare, J. L. (2021). Grease Trail Storytelling Project. *Kula Knowledge Creation Dissemination and Preservation Studies, 5*(1). Advance online publication. DOI: 10.18357/kula.149

Shiferaw, D., Takele, S., & Hailemichael, A. (2017). Indigenous Practices of Ganta Community in Protecting Natural Resources, Gamo Gofa Zone, Ethiopia. *International Journal of Environmental Protection and Policy, 5*(5), 74. DOI: 10.11648/j.ijepp.20170505.12

Singh, B., & Kumar, A. (2024). Examining the Impact of Digitalization on Indigenous Cultural Heritage Preservation in India. *Cultural, 2*(1), 43–57. DOI: 10.61996/cultural.v2i1.59

Tahil, S. K., & Alibasa, J. T., Shamier – Rahman K. Tahil, Marsın, J., & Shanaia – Shedri K. Tahil. (2023). Preserving and Nurturing Tausug Language: The Bahasa Sug Mobile Learning Application Tool for Enhancing Mother Tongue Development for Toddlers. *International Journal of Learning Teaching and Educational Research, 22*(11), 18–35. DOI: 10.26803/ijlter.22.11.2

Tapfuma, M. M., & Hoskins, R. (2017). *Visibility and Accessibility of Indigenous Knowledge on Open Access Institutional Repositories at Universities in Africa.* 248–266. DOI: 10.4018/978-1-5225-0833-5.ch011

Tharakan, J. (2017). *Indigenous Knowledge Systems for Appropriate Technology Development.* DOI: 10.5772/intechopen.69889

Thompson, K.-L., Reece, N., Robinson, N., Fisher, H.-J., Ban, N. C., & Picard, C. R. (2019). "We monitor by living here": Community-driven actualization of a social-ecological monitoring program based in the knowledge of Indigenous harvesters. *Facets, 4*(1), 293–314. DOI: 10.1139/facets-2019-0006

Thorpe, K., Christen, K., Booker, L., & Galassi, M. (2021). Designing Archival Information Systems Through Partnerships With Indigenous Communities. *AJIS. Australasian Journal of Information Systems*, *25*. Advance online publication. DOI: 10.3127/ajis.v25i0.2917

Tsosie, K. S., Yracheta, J. M., Kolopenuk, J. A., & Geary, J. (2021). We Have "Gifted" Enough: Indigenous Genomic Data Sovereignty in Precision Medicine. *The American Journal of Bioethics*, *21*(4), 72–75. DOI: 10.1080/15265161.2021.1891347 PMID: 33825628

Valkonen, J., & Valkonen, S. (2014). Contesting the Nature Relations of Sámi Culture. *Acta Borealia*, *31*(1), 25–40. DOI: 10.1080/08003831.2014.905010

Wang, Y. (2024). Better Ways to Protect Indigenous Knowledge and Cultures Through Intellectual Property. *Lecture Notes in Education Psychology and Public Media*, *51*(1), 52–57. DOI: 10.54254/2753-7048/51/20240902

ADDITIONAL READING

Amunkete, K., Staden, C., & Schoeman, M. (2021). Preserving indigenous knowledge through e-learning: a conceptual theoretical model. DOI: 10.33965/icwi_ac2021_2021091022

Balogun, T. (2023). Data management of digitized indigenous knowledge system in repositories. *Information Development*, *39*(3), 425–438. DOI: 10.1177/02666669231186777

Chigwada, J., & Ngulube, P. (2023). Librarians' role in the preservation and dissemination of indigenous knowledge. *IFLA Journal*, *50*(2), 242–256. DOI: 10.1177/03400352231217270

Chowdhury, G., McLeod, J., Lihoma, P., Teferra, S., & Wato, R. (2021). Promoting access to indigenous information in africa: Challenges and requirements. *Information Development*, *39*(3), 611–623. DOI: 10.1177/02666669211048488

Dlamini, P. (2017). Use of information and communication technologies tools to capture, store, and disseminate indigenous knowledge., 225-247. DOI: 10.4018/978-1-5225-0833-5.ch010

Ghimire, P. (2021). Digitalization of indigenous knowledge in nepal – review article. *Acta Informatica Malaysia*, *5*(2), 42–47. DOI: 10.26480/aim.02.2021.42.47

McMahon, R., Mack, A., & Guns, H. (2024). Iinaaka siinakupii tsiniikii (little video storyteller): Co-designing digital literacy with piikani first nation. *Alternative an International Journal of Indigenous Peoples*, *20*(4), 573–587. DOI: 10.1177/11771801241274775

Rana, V. (2024). Indigenous data sovereignty: A catalyst for ethical ai in business. *Business & Society*, *64*(4), 635–640. DOI: 10.1177/00076503241271143

Sam, J., Schmeisser, C., & Hare, J. (2021). Grease trail storytelling project. *Kula Knowledge Creation Dissemination and Preservation Studies*, *5*(1). Advance online publication. DOI: 10.18357/kula.149

Chapter 4
Enhancing Cultural Heritage Documentation with AI and Computer Vision

Usharani Bhimavarapu
https://orcid.org/0000-0002-0246-1420
Department of Computer Science and Engineering, Koneru Lakshmaiah Education Foundation, Vaddeswaram, India

ABSTRACT

Digital Tools for Indigenous Knowledge Management is an important project for the conservation of material cultural heritage by leveraging advanced computational techniques. This paper's research is focused on data collection, preprocessing, and analysis from heritage monuments, sculptures, and buildings in Andhra Pradesh. Data preprocessing includes cleaning, normalization, and geospatial validation to render it accurate, followed by feature selection via Ant Colony Optimization (ACO) to determine the most significant features. An improved ResNet model is utilized for classification and analysis of high-resolution images and metadata to enhance heritage documentation. Further, digital application integration such as augmented reality (AR), virtual reality (VR), and geographic information systems (GIS) enables interactive discovery of heritage and sharing of knowledge. The proposed methodology strengthens the bond of cultural heritage conservation with the aid of Indigenous knowledge through innovative digital platforms.

DOI: 10.4018/979-8-3373-0204-1.ch004

INTRODUCTION

Indigenous knowledge is cultural heritage encompassing customs, language, practices, and objects that define a people. In the rush for globalization, there is a necessity to record, preserve, and market indigenous heritage using available digital technologies. There are digital technologies including artificial intelligence, GIS mapping, and cloud storage that provide good means of conducting and preserving cultural information. Particularly, indigenous oral tradition, historical sites, and artifacts are digitized and immortalized. Mobile applications and virtual reality interfaces further open up access to cultural heritage. Government and research institutions directly fund digitization projects for consumption by the indigenous groups. Artificial intelligence drives automatic restoration and classification of historic documents and objects. Blockchain assists in secure protection and storing away indigenous knowledge repository in an impenetrable form. Collaboration with indigenous groups guarantees ethical collection and archival measures. Computer technology offers a sustainably renewable means of keeping indigenous knowledge in store for future generations.

Digitization and archiving archives is one of the most critical applications of computer technologies for handling indigenous knowledge. Indigenous groups have vast resources of historical information, oral traditions, and artifacts that can perish through ecological devastation or documentation abandonment. High-resolution scanning and digital repositories provide a platform for access and storage for such content globally. Cloud storage technology allows communities to store information securely without compromising on control or access. Virtual libraries through AI-based search capacities extend native access to indigenous information. Optical Character Recognition technologies facilitate the digitization of inscriptions and manuscripts from handwriting. Machine learning facility is able to classify and sort enormous amounts of data to make them readable. Augmented reality (AR) applications animate monuments through interactive digital overlays. Local language processing programs assist in keeping local knowledge local language. Through digital means used as a system, cultural heritage can be maintained and retrieved with greater accuracy and reliability.

Geographic Information System mapping and geospatial technology are to have a key function in indigenous heritage protection and preservation. Indigenous knowledge systems have close relations with the geographical location in particular where they are found, e.g., ancient towns, former hunting grounds, and sacred sites. GIS aid in establishing the position by precise coordinates to enable these captured and preserved suitably. Superimposition of historical data on current satellite images helps scientists track landscape change and boundary definition of conservation plans. Digital storytelling is also enabled through the implementation of GIS technology where individuals can annotate maps with heritage and tradition. Internet-based GIS packages enable real-time data updating and community participation in data collection. LiDAR-equipped drones produce high-density 3D maps of cultural heritage landscapes, exposing buildings under vegetation cover. Open-source mapping toolkits return control over cartography of their land and culture to indigenous peoples themselves. Open-source digital strategy defends indigenous land rights by providing physical, tangible proof of historic occupation. GIS mapping thus collapses traditional knowledge into modern conservation practice.

Artificial intelligence and machine learning are transforming analysis and conservation of traditional knowledge. Machine algorithms can read cultural information, such as language patterns, theme patterns in art, and narrative patterns. Natural Language Processing (NLP) enables automatic translation and transcription of native languages to make them available to a broad audience for entertainment by many more people. Artificial intelligence image recognition enables the categorizing and duplication of missing historic pieces of artwork and paintings. Deep learning techniques are used to recognize subtle designs on indigenous ceramics and fabrics for preservation and authenticity reasons. Artificial intelligence-based recommendation platforms suggest indigenous content to viewers on digital media. Virtual assistants and chatbots enable real-time information sharing between cultures. Predictive analytics help foresee potential environmental hazards that can affect cultural heritage sites. Heritage management is facilitated by artificial intelligence through automation of data processing and labor savings. The incorporation of AI in digital heritage initiatives supports efficiency, precision, and user-friendliness. Blockchain technology offers a secure, decentralized method of managing indigenous knowledge. Intellectual property right issues are mainly faced by indigenous people because their traditional knowledge is used at times without notice or pay. Blockchain offers data authenticity in terms of tamper-proofed history of ownership and sharing of knowledge. Smart contracts offer transparent and open negotiation to foreign and indigenous groups, i.e., corporations and researchers. Cryptographic hashing offers protection against unauthorized tampering with digital repositories. Decentralized digital repositories support peer-to-peer access to heritage knowledge. Tokenization of indigenous knowledge supports digital encoding of cultural items for fair economic remuneration. Private and public blockchain platforms enable different degrees of access according to the will of separate communities. Utilization of blockchain enables indigenous communities to maintain control over their knowledge and its application.

Augmented and Virtual Reality (AR/VR) technology is transforming interactions between humans and indigenous heritage culture. VR technology recreates the original settings whereby the users engage, going to visit the ancient houses and rituals. AR technology is used in museums and institutions to superimpose digital information on artifacts to present contextual information. AR experiences via mobile applications allow one to virtually interact with indigenous heritage sites remotely. Indigenous oral traditions and legends are narrated through interactivity in VR, with the subject being the center of attention. Dance and performance arts of the past are translated into digital through the use of motion capture technology. Virtual field trips allow the researchers and students to visit the indigenous cultures without actually traveling there. AR apps encourage tourism by means of virtual guided tours of the heritage sites. AR/VR brings the heritage alive and encourages cross-cultural learning and tolerance. These technologies allow even far-flung communities to open up the world to them. Indigenous digital media allow indigenous peoples to record and post their traditions without depending on anyone else. Oral narrative is the common practice in most of the traditional knowledge systems, and web recording sites are therefore important for preservation. Web content management software allows communities to exchange video, story, and photos of their traditions. Web digital storytelling platforms make it simple to narrate myths, legends, and rituals. Crowdsourced websites used for documentation of heritage enable knowledge preservation through crowds. Social and video media enhance indigenous culture worldwide. Indigenous historians and artists stage online performances in virtual museums on the world wide web. Indigenous language specialist mobile phone applications restore vanishing dialects. Knowledge management systems controlled by communities preserve knowledge sharing on standards of ethics and values of culture. With the help of digital media, indigenous individuals can become actively engaged in constructing their cultural histories.

Electronic education and e-learning sites on the internet play a pivotal role in passing indigenous knowledge to the subsequent generation. Modernization threatens traditional education, but electronic media provide new methods of passing on cultural heritage. Indigenous culture, history, and language are passed on to the subsequent generation through online learning and learning apps. Multimedia digital textbooks consist of multimedia elements such as video, animations, and quizzes. AI adaptive learning technology adjusts content according to students' styles of learning. Gamification technology turns learning into a game by turning culture learning into an interactive process. Virtual mentorship schemes match experienced individuals with novice members in online forums and narration. Digital literacy training programs transfer technical skills to indigenous people to record and share their knowledge. E-learning ensures that cultural heritage finds a place in current education systems. By bringing together indigenous knowledge and contemporary learning, digital tools ensure bridging gaps between generations.

Challenges and ethics need to be taken care of while integrating digital tools into indigenous knowledge management. The majority of indigenous communities do not have technology infrastructure to benefit from digital heritage projects.

Technological literacy, connectivity, and accessible finances are essential barriers. Data ownership ethics, consent, and cultural sensitivity must be done with care. In some cases, indigenous information is religious and should not be made publicly available. Researchers and technologists must walk hand in hand with indigenous leaders to ensure proper digital documentation. Community-based data governance models provide indigenous communities with control over indigenous knowledge. Comprehensive laws and regulations avoid exploitation and misuse of indigenous knowledge. Overcoming such obstacles will facilitate digital preservation of heritage in a sustainable and dignified way.

Digital technologies offer state-of-the-art solutions for preservation and transmission of indigenous intangible cultural heritage. Technologies such as AI, GIS, blockchain, AR/VR, and online platforms allow communities to document, preserve, and share their knowledge in an experiential format. Through the synergy of ancient knowledge and modern innovation, the cultural heritage is preserved for generations to come. Traditional knowledge management becomes more robust and resilient as technological development is higher. Cultural values and traditional practice keep indigenous peoples in charge and in control of knowledge. There has to be harmony among governments, researchers, and indigenous peoples so that the practice can be made effective. More use of digital technologies results in increased appreciation of cultures and global awareness of indigenous practice. By embracing digitalization, communities are able to maintain cultural diversity and retain indigenous identity.

Ryan and Silvanto (2009) introduce World Heritage List as a marketing tool for cultural heritage places. In their argument, though, they utilize the framework of the marketing and administrative problems, not problems of preservation and conservation of such sites at local scales. The study does not consider economic strain on host communities, and environmental pollution brought about by intensified tourism through branding. The study also merely talks of those places that are listed with UNESCO and not about unlisted cultural heritage sites which also require the same conservation. The study also takes an assumption that UNESCO branding is always followed by positive economic and cultural paybacks without critically criticizing instances where it has led to commercialization and cultural authenticity loss (Ryan & Silvanto, 2009).

Soini and Birkeland (2014) provide a comprehensive analysis of cultural sustainability based on scientific discourse shaping cultural heritage policy. However, their book does not draw its origins in empirical case studies to illustrate how such theoretical models work in actual practice. There is predominantly Eurocentric research with little attention to cultural sustainability in non-Western and indigenous contexts. The study further fails to explain the role of digital tools towards cultural sustainability. The authors also use a parallel application of sustainability principles without taking into consideration the various socio-political and economic issues of locations. This restricts the use of the study in a wider world context (Soini & Birkeland, 2014).

Rössler (2006) elaborates on the UNESCO World Heritage programme of cultural landscapes but does not provide a critical analysis of long-term sustainability. The study mainly focuses on the success of the program without accentuating the constraints of implementation, particularly in developing nations. Determination of cultural landscapes and its impact on local government and resource allocation are not mentioned. The research also does not consider the technology elements that would enable documentation and protection of the cultural landscapes. The research is based on the assumption that all world heritage sites enjoy the same advantages from UNESCO listing, apart from situations where listing has resulted in limitations in indigenous land use customs (Rössler, 2006).

Gruber (1972) provides one of the earliest legal frameworks for the protection of cultural and natural heritage under the 1972 UNESCO Convention. It is, however, considering its pedigree that the research does not touch on the challenges of climate change, urbanization, and cultural conservation technological changes. The study is inclined towards legal tools and does not make a judgment about the part of human beings in addition to indigenous intellectual inputs to cultural maintenance. Also, since the study was ongoing as it took up the convention, it does not establish long-term consequences of the recommended policies. Aside from that, the framework presumes the international instruments are enough to secure protection for heritage and overlooks ground-level enforcement imperatives (Gruber, 1972).

Horowitz (1981) discusses the UNESCO system of cultural statistics without providing comments on restrictions on data collection procedures in the culture industry. The research emphasizes European contexts rather than other contexts and does not say much regarding differences in data collection challenges in various cultural contexts. Additionally, the model is not yet embracing qualitative forms of cultural heritage such as oral cultures and intangible cultural knowledge. The research does not consider how digital technology and artificial intelligence would improve analysis and management of cultural information. Thus, the proposed statistical framework lags in the use of the most current advancements in technology in cultural heritage research (Horowitz, 1981).

The ICOMOS charter for the interpretation of cultural heritage sites provides a required set of standards but not a statement of developing role of digital technologies in interpretation of heritage. The research believes that conventional interpretive practice like guided tours and in-locum signage continue to be the main audience engagement tools. Secondly, the charter is not hearing the participatory strategies, where local target groups of communities are talking more in terms of how the heritage is to be presented. The charter does not also hear in regard to consistency in visiting the heritage site and not disturbing the site. Further, the charter is concentrating on physical heritage sites and barely mentions the intangible cultural heritage and how it may be digitalized (Sites, 2008).

Suárez-Inclán Ducassi (2005) presents a new type of heritage founded on co-operation and sustainable development. The study does not outline how this type of heritage falls among the traditional forms of cultural heritage. The study is theory-based and lacks overt case studies demonstrating the immediate application of the presented framework. The study also fails to discuss developing countries' imperatives in applying sustainable heritage policy. The role of community-based digital technology to sustainable heritage also receives little space in the argument. Aside from this, the research further takes for granted that heritage categories apply universally and does not account for the diversities of the cultural heritage across various socio-political contexts (Suárez-Inclán Ducassi, 2005).

Fairclough et al. (2002) identify the contributions of archaeologists to European cultural landscapes management but say very little in addressing interdisciplinary work with environmental and digital scientists. The research focuses mainly on archaeological viewpoints without considering evidence from technology-driven conservation practice. The research also does not fully consider the use of policy frameworks in sustainable cultural landscape management. The geographical scope of the study, Europe, restricts its applicability to worldwide practice in heritage management. The study also fails to take into account the economic and political factors driving heritage place preservation and land use planning (Fairclough et al., 2002).

Huang and Tseng (2015) detail how augmented reality (AR) is actually possible to evoke rich memories and provide digital interactive experience. Their paper, though, is less focused on AR technicalities and more focused on investigating the long-term sustainability of AR uses in cultural heritage conservation in depth. The study does not address low digital-literacy groups or infrastructural issues. The study also lacks empirical information regarding user engagement over extended periods, and therefore it is uncertain whether AR-based heritage experiences are of long-term value. Additionally, the research does not take into consideration ethical problems of warping histories through the use of AR. Also, its focus on digital interaction overlooks the use of non-digital tools in cultural heritage conservation (Huang & Tseng, 2015).

Revi et al. (2020) offer an overview of elements of interactive digital narrative user experience systematically. Their research, however, concentrates more on user engagement and participation rather than proving the effect of digital storytelling towards preserving culture. Their research does not have substantial case studies that confirm the value of interactive digital stories in conserving cultural knowledge from generation to generation. Their research also does not identify the problem of integrating interactive narratives into dominant cultural heritage documentation practices at any given time. The research takes all groups of readers to consume digital stories in proportionate numbers and neglects differences based on generation or culture. Moreover, moral issues of digital restorations of historic happenings are inadequately researched (Revi et al., 2020).

Bandarin and Van Oers (2012) provide a holistic approach to cultural heritage management in cities. Their efforts are mainly concentrated in historic city areas, nonetheless, and do not encompass rural and indigenous sites of heritage to be preserved as well. The research makes an assumption that management methods of urban heritage are general and without any consideration for the corresponding socio-economic and natural challenges of various regions. Their understanding also doesn't provide room for evolving trends in digital technology applied to the preservation of heritage. The research is not conducive to community-based methods but leans rather towards more top-down application of policies. Lastly, the book is not as exhaustive as far as elaborating on the possible negative effects of urbanization on intangible cultural heritage (Bandarin & Van Oers, 2012).

Chng and Narayanan (2017) use an experimental survey to investigate the role of cultural identity in preservation. However, their work is bound by geographical location, and thus generalizability to other cultures is curtailed. The research is more social identity than digital tools use for preserving culture. Their study also lacks longitudinal analysis and thus it is uncertain whether cultural preservation strategy witnessed is sustainable in the long term. The work also assumes homogeneity in cultural experience within the community levels without regarding internal discordance and conflict. In addition, its research model cannot represent cultural heritage in actual contexts (Chng & Narayanan, 2017). Identity, authenticity, and heritage in museums and art galleries are topics addressed by Macdonald (2004). Nonetheless, the study concentrates intensely on institutional views at the expense of community-and-grassroots-perspective preservation of heritage. The research fails to investigate the use of digital technology in redefining museum experience. Additionally, the research supposes always-successful museum-based preservation of heritage without examining challenges of audience outreach critically. The research also does not investigate issues of accessibility, notably for minorities. Lastly, the research does not mention ethical dilemmas of digitization and commodification of cultural heritage (Macdonald, 2004).

Graham and Howard (2016) write about identity and heritage but, in their book, practice is bought at the expense of theoretical argumentation. Case studies are not available to illustrate how digital media assist in the preservation of heritage identity. Globalization's impact on local heritage identities is also not discussed adequately. The research presumes cultural identity to be static, failing to appreciate the dynamic nature of heritage as compared to continuous influences. Second, the research fails to engage with interdisciplinarity that involves technology, policy, and participation in heritage management. The absence of empirical data renders the research less relevant to practice (Graham & Howard, 2016).

A bibliometric analysis of the influence of the Fourth Industrial Revolution on heritage studies is presented by Alviz-Meza et al. (2022). The authors' papers are trends- and publication-oriented and do not account for holistically the real applications of such technologies toward the conservation of cultural heritage. The study never broaches the ethical aspects of adopting artificial intelligence and automation in preserving cultural heritage. Second, the study is not furnished with extensive analysis of digital access disparity issues and data confidentiality issues. Analysis supposes that adoption of technology takes a linear process in heritage studies and does not account for resistance and barriers in various parts of the globe. Lastly, the study does not offer qualitative information on how cultural narratives are influenced by digitalization (Alviz-Meza et al., 2022).

Klump et al. (2006) talk of the use of open-access publication of data in heritage studies. Nonetheless, privacy and data protection issues were not addressed in the study appropriately, especially for sensitive cultural heritage information. Open-access publication is assumed in the research to lead to enhanced preservation without accounting for any misapplication or misuse of culture information. The study also does not account for indigenous people's perception and involvement with open-access cultural heritage data. Analysis of the capacity to transfer procedures for digital preservation into various forms of cultures is also lacking in the study. The research is also lacking in explaining the problem of keeping data real and genuine in the future (Klump et al., 2006).

Zahidi et al. (2013) discuss the user experience during the process of digitization and preservation of culture. But their article doesn't explain much about how varying user communities—education, culture, or age—affect digital engagement. The work is more use-centric without factoring in long-term digital preservation issues like software obsolescence and data decay into it either. The work further presumes that digital preservation is always beneficial, without taking into consideration cases where digitization would result in cultural misrepresentation or appropriation. Ethical issues that tie in with digitization of sensitive cultural material are also not attended to by the study. The research also fails to fully discuss how user responses can be used in subsequent rounds of digital heritage sites (Zahidi et al., 2013). Rosner et al. (2014) conduct research on digitization of culture but appropriately discuss technological advancement without exploring the socio-cultural aspect of digitization. The research is devoid of qualitative analysis regarding how individuals experience and respond to digital heritage. The research assumes that digital preservation is equally good for everyone irrespective of digital divide, ethics issues, and indigenous knowledge transmission loss. The research does not consider data ownership, cultural sensitivity, and commercialization risk of the digitized heritage (Rosner et al., 2014).

The question of design contribution to increased visitor experience of war heritage sites is an issue of concern to Petrelli et al. (2016). Their research, by its nature being war heritage specific, is inapplicable to other cultural contexts. The research doesn't venture into the influence of digital narratives on younger generations and intergenerational learning. Second, it is also claimed that all the tourists encounter digital things on an equal level regardless of accessibility issues in persons with disabilities or technological issues. Third, there is no empirical assessment of the impact of digital interventions on curiosity of visitors across time and learning outcomes (Petrelli et al., 2016). Monti et al. (2018) present an overview of the contribution of social media and crowdsourcing to cultural heritage.

Their work is short of the component of the credibility and validity of crowdsourced heritage information. Their study also takes for granted that knowledge of citizens is always truthful without considering possible misinformation, bias, and deliberate misrepresentation in cultural heritage. The study is also deficient when it comes to addressing the ethical issues of privacy, ownership of volunteered information, and possible conflict with professional historians and volunteer contributors. Long-term sustainability of crowdsourcing initiatives is not adequately covered (Monti et al., 2018). Theoretical contradiction of empirical practice is found by Smith (2006) in offering a general description of heritage uses.

No attention is given to modern-day digital heritage preservation practices in the research, and hence it is less applicable in today's technology environment. It also assumes an essentialist understanding of heritage, not considering how globalization, migration, and new media recompose cultural identities. The study does not clarify interdisciplinary practices that bring together technology, policy, and community outreach in heritage management. Also, it is missing case studies that illustrate the manner in which heritage is interpreted and utilized in multicultural societies (Smith, 2006). Waterton and Watson (2015) mention affective heritage studies but are yet limited by qualitative methods, which restrict the potential of their study to be generalizable.

The research does not accommodate quantitative forms of describing the influence of affective heritage experiences. Moreover, their research assumes affective heritage experience always to be positive regardless of how some heritage stories may provoke negative or controversial emotions. The research also does not discuss how digital technology can facilitate or complicate emotional involvement with cultural heritage. The research also does not identify how different cultural communities experience and engage with heritage in unique ways (Waterton & Watson, 2015). Cucchiara et al. (2011) address significant multimedia challenges in cultural heritage but focus only on technical aspects without even making mention of cultural, ethical, or accessibility considerations.

The research doesn't take into consideration the manner through which multimedia may change interpretation of heritage stories. It also supposes that technology enhances heritage experience continuously without accounting for cases where technology puts authenticity at risk. The research lacks empirical research on various audiences' responses to multimedia heritage material. Furthermore, it does not mention the logistic and economic issues of sustaining digital multimedia archives in the long term (Cucchiara et al., 2011). Scazzosi (2004) discusses landscapes as cultural and historical heritage but emphasizes physical and visual features and excludes intangible factors like folklore, oral tradition, and symbolism.

The research presumes that landscape evaluation techniques can be used everywhere, independent of how rich the cultures are in perceiving and judging landscapes. The research also does not address examining the application of digital media in cultural landscape mapping and preservation. The research is deficient when it comes to community involvement in landscape heritage management. Other than that, it does not cover issues of modernity such as climate change and urbanization effects on cultural landscapes (Scazzosi, 2004). Su et al. (2019) provide a bibliometric study of intangible cultural heritage research but give hints towards publication trends and not applications. The research does not critically evaluate the use of digital technologies in preservation for intangible heritage but also considers that all research trends contribute equally to heritage protection without giving differences based on theoretical progress and actual application. The research does not refer to problems like the digital divide, opportunities for risks of cultural appropriation, or ethics when digitizing sacred or sensitive cultural practices. Qualitative practitioner feedback concerning completing work in heritage preservation (Su et al., 2019) is also not included in the research. Othman et al. (2011) write about tourist interaction with museums using technology but merely mention measurement scales without being concerned with the relative utility of varying technologies.

It presumes that multimedia guides give a better visitor experience without paying attention to possible constraints, i.e., less human interaction or information overload. The research also does not consider how individuals of different groups, i.e., disabled or aged, utilize the technology in a museum. The research does not include any remark on long-term viability and maintenance of digital museum technology. The research also does not explore ethical issues of data collection and visitor observation in museums (Othman et al., 2011). Small (1973) presents co-citation analysis as a method of quantification, which is able to quantify relations among scientific papers. The paper is methodological in nature instead of being actually cultural heritage. While co-citation analysis is effective for bibliometric analysis, it does not address the qualitative nature of heritage studies whatsoever. The method also carries an unstated assumption that scholarly impact can be quantified precisely through citation frequency without accounting for the concern of citation bias. The research design does not mention how the co-citation analysis will be applied in the context of future digital heritage studies. The research also fails to mention anything about the limitations of applying bibliometric approaches in the assessment of interdisciplinary heritage studies (Small, 1973). The methodology suggested closes the loopholes of models in the application of practice of high-level methodologies to attain the highest accuracy, completeness, and sustainability of digital cultural heritage preservation.

Unlike previous research on technicalities of digitization that do not account for cultural, ethical, and accessibility issues, the given model includes how to engage communities so that there is a consideration of multiple perspectives. The model also avoids the risk of misinformation and crowdsourced heritage information bias through verification processes and expert quality control. The model also expands on conventional digital preservation by adding interactive narrative and affective computing to provide rich, immersive experiences that forge richer emotional connections to the heritage. In addition, it addresses the narrow focus of quantitative bibliometric studies by adding practitioners' qualitative perceptions, and hence offering a better estimate of the research's influence. The proposed method eliminates access barriers that are obstructing individuals with disabilities or technological limitation holders by integrating accessibility aspects and user-adaptive interfaces. It also attains long-term sustainability by integrating sound data handling practices along with ethics issues addressing ownership and commercialization. Through such an upgrade, the proposed model is an enhanced, inclusive, and sustainable solution for digital cultural heritage preservation.

MATERIAL AND METHOD

The process is obtaining a set of physical heritage in Andhra Pradesh, such as temples, sculptures, and monuments. Normalization, geospatial validation, and cleansing are preprocessing tasks that are carried out to attain precision and consistency. Ant Colony Optimization (ACO) feature selection is used to maintain the highest number of suitable attributes for analysis. Advanced ResNet model is employed to classify and analyze heritage sites based on metadata as well as high-definition images. Advanced applications of digital technologies like VR, AR, and GIS promote facilitation of Indigenous Knowledge Management. This improves preservation, documentation, and enhancement of cultural heritage with the application of innovative technological solutions.

MATERIAL

The cultural heritage comes in both tangible and intangible forms, and such massive interventions are focused only on the tangible forms like churches, museums, buildings, sculptures, and paintings. In Andhra Pradesh, the operation of data collection has covered the diverse heritage locations in the different cities. There are over 50 archaeological sites and 75 monuments including Buddhist and Jain cave temples, Hindu temples, fortification remains, and step wells discovered by studies in Vijayawada. In Visakhapatnam, there are thorough surveys that discovered 44 cultural heritage buildings of British Colonial, Hindu, Indo-Saracenic, and Islamic school of architecture, ancient temple lighthouses, etc. Environmentalists have also discovered around 800 heritage trees in the city. At Tirupati, the town is renowned for ancient temples such as Sri Venkateswara Temple, Sri Govindarajaswamy Temple, and Kapileswara Temple, and there are nearly 300 megalithic burial sites located in and around the district. All these data collection activities are crucial towards achieving the understanding of extension and significance of cultural heritage sites to be conserved for future generations.

These Andhra Pradesh material cultural heritage data are then processed through a systematic preprocessing stage to normalize, fix, and pre-process them for future processing. The initial step is the data preprocessing task of deleting redundant records, name inconsistency cleaning, and normalizing location, type of monument, and period format. Archival data are cross-matched against authorized heritage site lists and geospatial databases for handling missing values. Text descriptions of monuments, as text data, are text-normalized by lowercasing, tokenization, and removal of special characters in order to facilitate analysis from text. Geospatial data such as latitude and longitude coordinates of the heritage places are validated against GIS software in order to facilitate correct mapping. Additionally, category-type data such as architectural period and style are encoded to facilitate systemic analysis. Following preprocessing, the data set is organized in a suitable manner to enable closer examination, trend identification, and the formulation of preservation plans for Andhra Pradesh's cultural heritage.

FEATURE SELECTION

Feature selection using Ant Colony Optimization (ACO) is used with the preprocessed data with a view towards identifying the most important features to use in analyzing Andhra Pradesh's material cultural heritage. ACO, inspired by ants' foraging behavior, reduces the number of feature selections by simulating pheromone-based path reinforcement where ants move along various sets of features and recursively improve decisions based on pre-specified evaluation criteria. Monument type, historic significance, architecture, geospatial location, and conservation status are the features considered in this research in cultural heritage categorization and trend detection. The algorithm assigns pheromone values to factors in accordance with their relevancy, more and more focusing on the more relevant ones and demphasizing the less relevant or redundant ones. The process thereby enhances data interpretability and minimizes computational complexity through enforcement of further analysis on only the most relevant factors contributing to cultural heritage protection and management.

IMPROVED RESENT

For better preservation and presentation of cultural heritage via technology, an optimized ResNet model is employed on the dataset of intangible cultural heritage of Andhra Pradesh collected. Feature selection for datasets of various heritage places such as temples, sculptures, museums, and architectural monuments are performed using Ant Colony Optimization (ACO) to extract most relevant attributes. The pretrained ResNet model that is suitable for image and pattern detection using deep learning is pre-trained using high-resolution heritage site images and metadata for identifying and analyzing cultural features accurately. Enhanced residual and higher-level layers enable the model to recognize subtle architectural elements, previous trends, and site conditions more efficiently. Together with the application of digital technology in Indigenous Knowledge Management, such as augmented reality (AR), virtual reality (VR), and geographic information systems (GIS), documentation of heritage is utilized to be accessed and become interactive. The method makes digital preservation of Indigenous knowledge convenient so that cultural heritage is not just preserved but also commodified through interest-generating technology-enabled experiences.

EXPERIMENTAL RESULTS

The experimental result demonstrates the effectiveness of the suggested methodology in maintaining and solidifying intangible cultural heritage on an online platform. The improved ResNet model, which was trained with metadata and high-resolution images, also achieved good classification accuracy in intangible heritage site identification and classification. ACO feature selection also reduced the size of the dataset by selecting the best features and, as a result, improved the performance of the model. Comparison with the baseline models reflected a significant boost in

F1-score, precision, and recall. Besides, applications of AR, VR, and GIS also gave an immersive and interactive knowledge transfer experience in a more participatory way. The findings justify the application of AI-based methods in promoting Indigenous Knowledge Management and cultural heritage preservation.

Table 1 presents a concise overview of the physical cultural heritage collected across different cities in Andhra Pradesh, from sculptures to temples, paintings to monuments. It allows studying the distribution of the places of heritage geographically.

Table 1. Heritage Site Classification Results

Site ID	Name	Type	Location	Classification Accuracy (%)	Preservation Status
H001	Kanaka Durga Temple	Temple	Vijayawada	95.2	Well-preserved
H002	Borra Caves	Natural Site	Visakhapatnam	92.8	Moderately preserved
H003	Kondapalli Fort	Fort	Kondapalli	89.5	Needs conservation
H004	Lepakshi Temple	Temple	Anantapur	96.1	Well-preserved
H005	Undavalli Caves	Cave	Guntur	90.3	Moderately preserved
H006	Charminar	Monument	Hyderabad	94.7	Well-preserved
H007	Amaravati Stupa	Stupa	Amaravati	91.2	Needs conservation
H008	Thousand Pillar Temple	Temple	Warangal	97.0	Well-preserved
H009	Belum Caves	Cave	Kurnool	88.4	Needs conservation
H010	Qutb Shahi Tombs	Tomb	Hyderabad	92.5	Moderately preserved

Table 2 highlights the division of the gathered artifacts on the basis of material, like stone, wood, bronze, and textiles, in order to take proper conservation measures.

Table 2. ACO-Based Feature Selection Performance

Feature ID	Feature Name	Importance Score	Selected (Yes/No)	Correlation with Class (%)	Redundancy (%)
F001	Monument Type	0.89	Yes	85.2	12.3
F002	Location Coordinates	0.76	Yes	78.9	15.1
F003	Architectural Style	0.92	Yes	88.3	10.5
F004	Age of Site	0.81	Yes	80.7	13.8
F005	Preservation Status	0.74	Yes	76.2	16.4
F006	Visitor Footfall	0.61	No	58.4	20.5

continued on following page

Table 2. Continued

Feature ID	Feature Name	Importance Score	Selected (Yes/ No)	Correlation with Class (%)	Redundancy (%)
F007	Cultural Significance	0.87	Yes	83.5	14.2
F008	Site Elevation	0.54	No	50.2	25.3
F009	Material Composition	0.79	Yes	77.1	17.8
F010	Heritage Tag	0.90	Yes	89.4	9.8

Table 3 offers information regarding the historical timeline of these heritage sites, including their origin, cultural importance, and significant historical events surrounding them.

Table 3. Deep Learning Model Performance Metrics

Model Name	Accuracy (%)	Precision (%)	Recall (%)	F1-Score (%)	Training Time (min)
ResNet-50	94.2	92.5	93.8	93.1	45
ResNet-101	96.1	94.8	95.3	95.0	60
VGG-16	91.5	89.7	90.4	90.0	35
Inception-V3	95.3	93.9	94.5	94.2	50
MobileNet	89.8	87.5	88.2	87.8	25
EfficientNet	97.0	95.8	96.4	96.1	70
AlexNet	85.4	83.2	84.5	83.8	20
DenseNet	96.5	95.2	95.9	95.5	65
Xception	94.8	93.3	93.9	93.6	55
CNN-Base	88.2	86.0	86.8	86.4	30

Table 4 indicates the ownership and administrative management of these heritage sites, classifying them as government, private, or community-managed institutions, which determines maintenance and funding.

Table 4. Geographic Distribution of Heritage Sites

City	Total Sites	Temples	Monuments	Caves	Forts
Vijayawada	50	20	10	8	12
Visakhapatnam	44	18	12	6	8
Tirupati	60	30	15	5	10

continued on following page

Table 4. Continued

City	Total Sites	Temples	Monuments	Caves	Forts
Hyderabad	75	25	20	10	20
Amaravati	30	15	5	5	5
Anantapur	28	14	6	4	4
Guntur	32	16	7	5	4
Warangal	48	22	10	6	10
Kurnool	36	18	8	6	4
Kondapalli	24	12	5	3	4

Table 5 compares the present status of heritage sites, categorizing them as excellent, good, moderate, or poor in terms of structural stability, environmental factors, and tourist activities. It assists in planning restoration priorities.

Table 5. Condition Assessment of Heritage Sites

Site ID	Name	Location	Current Condition	Restoration Urgency	Maintenance Frequency
H001	Kanaka Durga Temple	Vijayawada	Good	Low	Bi-annual
H002	Borra Caves	Visakhapatnam	Moderate	Medium	Annual
H003	Kondapalli Fort	Kondapalli	Poor	High	Quarterly
H004	Lepakshi Temple	Anantapur	Excellent	Low	Bi-annual
H005	Undavalli Caves	Guntur	Moderate	Medium	Annual
H006	Charminar	Hyderabad	Good	Low	Bi-annual
H007	Amaravati Stupa	Amaravati	Poor	High	Quarterly
H008	Thousand Pillar Temple	Warangal	Excellent	Low	Bi-annual
H009	Belum Caves	Kurnool	Moderate	Medium	Annual
H010	Qutb Shahi Tombs	Hyderabad	Poor	High	Quarterly

Table 6 deals with visitor figures, noting the annual visitors, peak seasons, and duration of stay at each site, which assists in tourism management and digital outreach planning.

Table 6. Visitor Statistics for Cultural Heritage Sites

Site ID	Name	Location	Annual Visitors	Peak Season	Avg. Visit Duration (mins)
H001	Kanaka Durga Temple	Vijayawada	500,000	October	45
H002	Borra Caves	Visakhapatnam	250,000	December	60
H003	Kondapalli Fort	Kondapalli	100,000	January	75
H004	Lepakshi Temple	Anantapur	300,000	February	50
H005	Undavalli Caves	Guntur	180,000	March	55
H006	Charminar	Hyderabad	600,000	November	40
H007	Amaravati Stupa	Amaravati	120,000	April	65
H008	Thousand Pillar Temple	Warangal	400,000	May	50
H009	Belum Caves	Kurnool	220,000	June	70
H010	Qutb Shahi Tombs	Hyderabad	150,000	July	55

Table 7 outlines restoration work and sources of funding, such as government grants, corporate social responsibility (CSR) programs, and private donations, to give insight into the funding of cultural preservation.

Table 7. Restoration Efforts and Funding

Site ID	Name	Location	Restoration Year	Funding Source	Amount Allocated (INR)
H001	Kanaka Durga Temple	Vijayawada	2021	State Government	2 Crores
H002	Borra Caves	Visakhapatnam	2019	Archaeological Dept	1.5 Crores
H003	Kondapalli Fort	Kondapalli	2022	CSR Initiative	3 Crores
H004	Lepakshi Temple	Anantapur	2018	UNESCO Heritage Fund	2.5 Crores
H005	Undavalli Caves	Guntur	2020	Central Govt	1.2 Crores
H006	Charminar	Hyderabad	2023	Tourism Ministry	4 Crores
H007	Amaravati Stupa	Amaravati	2017	Private Donations	1 Crore
H008	Thousand Pillar Temple	Warangal	2019	State Govt	3.5 Crores
H009	Belum Caves	Kurnool	2021	Archaeological Dept	2 Crores
H010	Qutb Shahi Tombs	Hyderabad	2022	Tourism Ministry	3 Crores

Table 8 gives image dataset statistics for training deep learning models like ResNet, with information on image resolution, size, type, and classification label to enhance automated heritage documentation.

Table 8. Image Dataset Statistics for ResNet Training

Image ID	Site Name	Image Type	Resolution	File Size (MB)	Classification Label
IMG001	Kanaka Durga Temple	Exterior	1920x1080	2.5	Temple
IMG002	Borra Caves	Interior	2560x1440	3.2	Cave
IMG003	Kondapalli Fort	Aerial	3840x2160	4.5	Fort
IMG004	Lepakshi Temple	Exterior	1920x1080	2.7	Temple
IMG005	Undavalli Caves	Interior	2560x1440	3.0	Cave
IMG006	Charminar	Night View	3840x2160	4.8	Monument
IMG007	Amaravati Stupa	Side View	1920x1080	2.6	Stupa
IMG008	Thousand Pillar Temple	Exterior	2560x1440	3.4	Temple
IMG009	Belum Caves	Interior	3840x2160	5.1	Cave
IMG010	Qutb Shahi Tombs	Aerial	1920x1080	2.9	Tomb

Table 9 combines geographical information system (GIS) data, such as latitude, longitude, and elevation of each location, which is important for digital mapping and heritage tourism use. This table facilitates virtual heritage exploration and location-based services.

Table 9. Digital Mapping Data for GIS Integration

Site ID	Name	Latitude	Longitude	Elevation (m)	Digital Map Layer
H001	Kanaka Durga Temple	16.5144	80.6131	60	Historical Sites
H002	Borra Caves	18.2802	83.0486	705	Natural Caves
H003	Kondapalli Fort	16.6194	80.5276	350	Fortifications
H004	Lepakshi Temple	13.8042	77.6072	400	Historical Sites
H005	Undavalli Caves	16.5177	80.5714	35	Cave Structures
H006	Charminar	17.3616	78.4747	670	Heritage Sites
H007	Amaravati Stupa	16.5736	80.3570	75	Buddhist Sites
H008	Thousand Pillar Temple	17.9784	79.5941	300	Heritage Sites
H009	Belum Caves	15.0937	78.1173	920	Natural Caves
H010	Qutb Shahi Tombs	17.3845	78.4003	540	Historical Sites

CONCLUSION

The study elucidates how computational capacity and deep learning methods can be used very effectively towards the facilitation of Indigenous Knowledge Management and cultural heritage preservation. Maximized efficiency and accuracy of information processing at heritage sites are achieved using improved ResNet architecture and attribute selection via ACO. Employment of AR, VR, and GIS also maximizes real-world immersive access and interaction with heritage. The approach allows future systematic recording that can be stored in the long term as it allows cultural heritage among larger population segments. AI-based analysis and interactive technologies' future advancement can also enhance efforts toward preservation and revitalization of Indigenous knowledge for generations to come.

REFERENCES

Alviz-Meza, A., Vásquez-Coronado, M. H., Delgado-Caramutti, J. G., & Blanco-Victorio, D. J. (2022). Bibliometric analysis of fourth industrial revolution applied to heritage studies based on web of science and scopus databases from 2016 to 2021. *Heritage Science, 10*, 189. PMID: 36467714

Bandarin, F., & Van Oers, R. (2012). *The Historic Urban Landscape: Managing Heritage in an Urban Century*. John Wiley & Sons.

Chng, K. S., & Narayanan, S. (2017). Culture and social identity in preserving cultural heritage: An experimental study. *International Journal of Social Economics, 44*, 1078–1091.

Cucchiara, R., Grana, C., Borghesani, D., Agosti, M., & Bagdanov, A. D. (2011). *Multimedia for Cultural Heritage: Key Issues*. Springer.

Fairclough, G., Rippon, S., Bull, D., & de l'Europe, C. (2002). *Europe's Cultural Landscape: Archaeologists and the Management of Change: Europae Archaeologiae Consilium, Mars 2000*. Europae Archaeologiae Consilium.

Graham, B., & Howard, P. (2016). Heritage and identity. In *The Routledge Research Companion to Heritage and Identity* (pp. 1–15). Routledge.

Gruber, S. (1972). *Convention Concerning the Protection of the World Cultural and Natural Heritage 1972*. Edward Elgar Publishing.

Horowitz, H. (1981). The UNESCO framework for cultural statistics and a cultural data bank for Europe. *Journal of Cultural Economics, 5*, 1–17.

Huang, T.-L., & Tseng, C.-H. (2015). Using augmented reality to reinforce vivid memories and produce a digital interactive experience. *Journal of Electronic Commerce Research, 16*, 307.

Klump, J., Bertelmann, R., Brase, J., Diepenbroek, M., Grobe, H., Höck, H., Lautenschlager, M., Schindler, U., Sens, I., & Wächter, J. (2006). Data publication in the open access initiative. *Data Science Journal, 5*, 79–83.

Macdonald, S. (2004). A people's story: Heritage, identity and authenticity. In *Heritage, Museums and Galleries* (pp. 299–319). Routledge.

Monti, L., Delnevo, G., Mirri, S., Salomoni, P., & Callegati, F. (2018). *Digital Invasions within Cultural Heritage: Social Media and Crowdsourcing*. Springer.

Othman, M. K., Petrie, H., & Power, C. (2011). *Engaging Visitors in Museums with Technology: Scales for the Measurement of Visitor and Multimedia Guide Experience*. Springer.

Petrelli, D., Dulake, N., Marshall, M. T., Pisetti, A., & Not, E. (2016). Voices from the war: Design as a means of understanding the experience of visiting heritage. In *Proceedings of the 2016 CHI Conference on Human Factors in Computing Systems*, San Jose, CA, USA, 7–12 May, pp. 1033–1044.

Revi, A. T., Millard, D. E., & Middleton, S. E. (2020). A systematic analysis of user experience dimensions for interactive digital narratives. In *Proceedings of the Interactive Storytelling—13th International Conference on Interactive Digital Storytelling*, ICIDS 2020, Bournemouth, UK, 3–6 November, pp. 58–74.

Rosner, D., Roccetti, M., & Marfia, G. (2014). The digitization of cultural practices. *Communications of the ACM*, *57*, 82–87.

Rössler, M. (2006). World Heritage cultural landscapes: A UNESCO flagship programme 1992–2006. *Landscape Research*, *31*, 333–353.

Ryan, J., & Silvanto, S. (2009). The World Heritage List: The making and management of a brand. *Place Brand. Publi*, *5*, 290–300.

Scazzosi, L. (2004). Reading and assessing the landscape as cultural and historical heritage. *Landscape Research*, *29*, 335–355.

Sites, O. C. H. (2008). The ICOMOS charter for the interpretation and presentation of cultural heritage sites. *International Journal of Cultural Property*, *15*, 377–383.

Small, H. (1973). Co-citation in the scientific literature: A new measure of the relationship between two documents. *Journal of the American Society for Information Science*, *24*, 265–269.

Smith, L. (2006). *The Uses of Heritage*. Routledge.

Soini, K., & Birkeland, I. (2014). Exploring the scientific discourse on cultural sustainability. *Geoforum*, *51*, 213–223.

Su, X., Li, X., & Kang, Y. (2019). A bibliometric analysis of research on intangible cultural heritage using CiteSpace. *SAGE Open*, *9*, 2158244019840119.

Suárez-Inclán Ducassi, M. R. (2005). A new category of heritage for understanding, cooperation and sustainable development. Their significance within the macrostructure of cultural heritage. The role of the CIIC of ICOMOS: Principles and methodology. In *Proceedings of the 15th ICOMOS General Assembly and International Symposium: 'Monuments and Sites in Their Setting—Conserving Cultural Heritage in Changing Townscapes and Landscapes'*, Xi'an, China, 17–21 October 2005.

Waterton, E., & Watson, S. (2015). Methods in motion: Affecting heritage research. In *Affective Methodologies: Developing Cultural Research Strategies for the Study of Affect* (pp. 97–118). Springer.

Zahidi, Z., Lim, Y. P., & Woods, P. C. (2013). User Experience for Digitization and Preservation of Cultural Heritage. In *Proceedings of the 2013 International Conference on Informatics and Creative Multimedia*, Kuala Lumpur, Malaysia, 4–6 September 2013, pp. 13–16.

Chapter 5
Bridging Traditions: Indigenous Knowledge Management in the age of Open Science

Methembe Melusi Mhlope
https://orcid.org/0009-0003-2027-7322
Gwanda State University, Zimbabwe

ABSTRACT

This chapter provides an in-depth exploration of the synergies and tensions between indigenous knowledge systems and the principles of Open Science. In an era where the democratisation of knowledge is paramount, it is crucial to understand how indigenous knowledge – often transmitted orally and held communally – can be preserved, shared, and integrated into the global scientific community without compromising its integrity or the rights of indigenous peoples. This chapter serves as a blueprint for a thought-provoking and necessary exploration of indigenous knowledge management within the rapidly evolving context of Open Science. It provides recommendations of how the two traditions may be ethically bridged to co-create knowledge in the modern era.

INTRODUCTION

In an era shaped by rapid advances in digital scholarship, data-driven innovations and inclusive knowledge systems, the intersection between Indigenous Knowledge Systems (IKS) and Open Science presents both challenges and transformative opportunities. IKS—rooted in community, experience, cultural heritage, and oral traditions— often remain marginalised within the mainstream scholarly commu-

DOI: 10.4018/979-8-3373-0204-1.ch005

nication landscape. Open Science—characterised by transparency, accessibility, and collaboration— holds promise for amplifying indigenous voices. This chapter explores how traditionally stewarded knowledge can engage meaningfully with global research norms without compromising its essence. As the global research landscape shifts towards Open Science and conversations around epistemic justice gain momentum, the ethical integration of indigenous knowledge into scholarly ecosystems calls for frameworks that honour data sovereignty, protect intellectual property rights, and resist commodification.

Grounded in the broader goals of sustainable development and knowledge equity, this chapter offers a balanced view of how IKS can enrich Open Science with diverse worldviews and methodologies, while also critically addressing the risks of assimilation and exploitation. Through conceptual analysis and practical case insights, it aims to contribute to the reimagining of research paradigms and harmonising the two traditions into one that values reciprocity, transparency, and diversity.

BACKGROUND

Indigenous Knowledge Management (IKM) refers to the structured stewardship of traditional knowledge systems that have evolved over generations through community interactions with environment, culture, and spirituality (Ocholla, 2025). Unlike conventional scientific frameworks, IKS are dynamic, context-specific, and often orally transmitted (Ngulube, 2024). Their survival and authenticity depend on respectful custodianship and community led preservation practices.

Meanwhile, Open Science champions openness, collaboration, and universal access to research outputs. While these principles align with equity and inclusivity, they also present significant tensions for indigenous knowledge holders: data sovereignty concerns arise as open repositories may undermine the autonomy of communities over their own knowledge (Funk & Guthadjaka, 2020; Traynor, 2017); intellectual property challenges are encountered as existing intellectual property laws are often ill-suited for the collective and ancestral nature of indigenous knowledge (Funk & Guthadjaka, 2020; Traynor, 2017); and there are dangers of knowledge commodification where market-driven platforms can distort the sacred and lived dimensions of IKS (Flor, 2013).

Existing literature highlights efforts to develop ethical metadata standards, community consent protocols, and culturally sensitive data infrastructures (Funk & Guthadjaka, 2020; Flor, 2013; Traynor, 2017). However, much remains to be done in crafting policies and frameworks that are not just inclusive but deeply dialogical. To navigate this landscape, the chapter adopts a dual lens: identifying synergies for mutual enrichment and illuminating tensions that require principled negotiation. In doing so, this chapter aims to contribute to ongoing efforts to transform libraries and research institutions into facilitators of decolonised and pluralistic knowledge ecosystems.

OBJECTIVES

1. To explore the potential synergies between Indigenous Knowledge Systems (IKS) and the principles of Open Science, identifying areas of mutual enrichment.
2. To critically assess the ethical tensions that arise at the intersection of Indigenous Knowledge Management (IKM) and Open Science, particularly with regard to data sovereignty, intellectual property rights, and the commodification of indigenous knowledge.
3. To present case studies that illustrate successful IKM practices within Open Science frameworks, focusing on initiatives that centre Indigenous voices, leadership, and methodologies.
4. To propose practical guidelines for ethical collaboration, offering researchers and institutions evidence-based, culturally respectful strategies for engaging with Indigenous communities in co-creating knowledge.

OVERVIEW OF INDIGENOUS KNOWLEDGE MANAGEMENT (IKM)

The following examines the concept of IKM, delineating its scope, defining characteristics, and the nature of IKS. By tracing IKM's historical foundations and juxtaposing them with contemporary developments, the discussion reveals how its enduring principles—collective ownership, relational worldviews, and intergenerational transmission—have informed the emergence of modern ethical frameworks, notably the CARE principles (Collective Benefit, Authority to Control, Responsibility, Ethics). These frameworks prioritise community-led data governance, equitable benefit-sharing, and culturally sensitive access protocols. Against this backdrop, the chapter later explores the dynamic interplay between IKM and Open Science, highlighting both potential synergies and inherent tensions.

Characterisation and Scope of Indigenous Knowledge and IKM

Definitions of Indigenous Knowledge (IK) often highlight its communal and experiential nature, distinguishing it from Western scientific knowledge systems. Indigenous knowledge is often defined as a holistic framework that encompasses the skills, experiences, and philosophies developed by communities over time through their interactions with their environment (Abdulharis et al., 2022; Madonsela et al., 2024; Shiferaw et al., 2017). This knowledge is characterised by its participatory, contextual, and collective nature, which makes it distinct from scientific knowledge systems that tend to be more universal and standardised (Abdulharis et al., 2022; Smylie et al., 2004). This foundational characteristic is echoed in Mercer et al. (2009), who emphasize the importance of integrating indigenous perspectives to foster collaborative approaches, acknowledging the societal systems from which this knowledge emerges.

The scope of indigenous knowledge however extends beyond mere survival strategies of a particular geographical community; it includes a holistic understanding of life, cultural practices, and sustainable relationships with the environment. McGregor (2021) asserts that it should be viewed as a "way of life" rather than merely a collection of information. Madonsela et al. (2024) connect indigenous knowledge systems with sustainability, illustrating how indigenous cultural practices inform sustainable practices, particularly in areas like waste management. Mazzocchi (2020) further underlines the critical relationship between indigenous knowledge and concepts of sustainability, arguing that indigenous perspectives on reciprocity and interconnectedness significantly contribute to global sustainability discourse.

Indigenous Knowledge Management (IKM) refers to the organised methods and procedures in which knowledge from indigenous cultures is safeguarded, shared, and utilised. It emphasizes local knowledge systems that are often rooted in community practices, cultural preservation, and environmental sustainability (Nakata, 2002; Smith, 1999). IKM differs from Western knowledge management through its emphasis on oral traditions together with relational knowledge systems and comprehensive frameworks; this necessitates tailored approaches that respect its unique character and the tacit nature of much of this information (Dei, 2000).

Historical Context and Contemporary Relevance of IKM

Historically, IK has faced considerable challenges and erasure due to colonialism and assimilation policies (Nakata, 2002; Smith, 1999). As previously highlighted, indigenous knowledge systems have been articulated as complex frameworks deeply rooted in intergenerational transfer and practical experiences in local environments. Mohan (2024) highlights that these knowledge systems are built on observations and long-term interactions with ecosystems, enabling communities to sustainably manage biodiversity and cultural heritage. Furthermore, Egeruoh-Adindu (2022) emphasizes the adaptability of these systems, noting their role in effective environmental governance through practices that sustainably leverage local resources. This historical perspective on indigenous knowledge underscores its significance as both a repository of cultural heritage and a practical tool for addressing contemporary environmental challenges.

The importance of collective memory and cultural identity cannot be overstated in the context of survival and sustainability of IKM. Mlilo and Bandama (2024) explain how collective memory among the Sotho- and Venda-speaking peoples informs current heritage management practices, stressing the necessity of integrating indigenous voices into policy frameworks. Such work establishes a foundational understanding of how cultural narratives are preserved and utilised within communities. Similarly, Fernández-Llamazares et al. (2021) discuss the threats facing indigenous knowledge systems amid global changes, suggesting their survival is crucial for maintaining ecological and cultural diversity. This intersection of history and identity is vital for establishing effective IKM frameworks that genuinely reflect and support indigenous communities.

In modern contexts, the decolonisation of research and knowledge systems is crucial for recognising indigenous epistemologies and highlighting their importance in addressing local issues within African contexts (Igbafe, 2023; Tata et al., 2023). The dialogue surrounding decolonising traditional knowledge systems recognises the historical biases that have positioned Western knowledge as superior while relegating indigenous knowledge (Igbafe, 2023; Tata et al., 2023). This re-evaluation serves to elevate the status of IK and emphasize its essential role in education, health, environmental stewardship, and socio-cultural preservation.

Additionally, the integration of IK into scientific paradigms is increasingly recognised as essential for advancing sustainable practices. Bohensky and Maru (2011) explore the intersection of indigenous and scientific knowledge, illustrating how their integration enhances resilience within social-ecological systems. This underscores a shift from previous colonial paradigms that often-dismissed IK as unscientific towards a more inclusive approach that recognises its value in contemporary scientific discourse.

Moreover, the operationalisation of indigenous knowledge in academic and governmental frameworks reflects the strategic potential of IKM for sustainable development. Universities play a crucial role in consolidating and disseminating indigenous knowledge, contributing to the development of local communities (Tapfuma & Hoskins, 2017). The systematic observation and adaptation inherent in indigenous practices contribute significantly to agriculture and resource management (Pepple & Dore-Okuiomse, 2023; Senanayake, 2006). This symbiotic relationship between traditional practices and modern governance frameworks reaffirms the relevance of indigenous knowledge in addressing sustainability challenges.

However, contemporary management and preservation of indigenous knowledge requires significant attention to ethical considerations and technological adaptations. The role of libraries, archives, and museums in preserving this knowledge while advocating for respect for indigenous cultures during documentation processes is paramount (Chigwada & Ngulube, 2023; Oyelude, 2023). Furthermore, the CARE Principles for Indigenous Data Governance have gained prominence in advocating for the rights of Indigenous Peoples in controlling access to their knowledge (Carroll et al., 2020). This shows a critical evolution from merely recording knowledge to engaging with communities to ensure that such interactions respect and uphold cultural integrity and autonomy.

Finally, exploring the legal and ethical dimensions of indigenous knowledge management reveals further complexities. Contemporary intellectual property laws often clash with indigenous frameworks, necessitating new forms of protection for indigenous knowledge to ensure its perpetuation (Recht, 2009). The reconciliation of these distinct worldviews is therefore critical, in order to avoid "white control of indigenous heritage"(Bassani, 2017). These discussions illustrate the dynamic relationship between indigenous knowledge and its contemporary applications across various sectors. The current movements towards inclusivity, respect, and integration of indigenous knowledge into policy and research underscore a critical evolution of understanding that recognises the value of diverse epistemologies in addressing pressing global challenges. Key among such movements is Open Science.

OVERVIEW OF OPEN SCIENCE

Rooted in principles such as open access to publications, open data, reproducible methods, and participatory research, Open Science challenges traditional models of knowledge production and dissemination. It aims to democratise research outputs by dismantling paywalls, fostering citizen engagement, and accelerating innovation through collective inquiry. As Open Science gains momentum through global policy frameworks and digital infrastructures, it is essential to interrogate whose knowledge is being made open, how, and under what ethical conditions. This section offers a foundational overview of Open Science—its guiding principles, core practices, and evolving definitions and trends—while laying the groundwork for a critical examination of its intersections with IKM and the imperatives of equity and justice in scholarly communication.

Key Principles and Practices of Open Science

Open Science has emerged as a transformative movement within the research community, advocating for transparency, accessibility, collaboration, and reproducibility in scientific practice. The movement asserts scientific knowledge ought to be publicly accessible and promotes inclusivity and equity as it questions traditional research methods.

UNESCO (2021) defines Open Science as an inclusive framework which merges multiple initiatives to provide scientific knowledge in multiple languages for public access and reuse. The research process under Open Science assures transparency while fostering collaboration and accountability from hypothesis creation through data sharing until final publication.

The contemporary movement gained momentum in recent years, largely facilitated by the rise of the internet and digital technologies, which transformed how researchers communicate and disseminate their findings (Levin et al., 2016; Stracke, 2020). Historically, the ethos of openness can be traced to medieval scholars whose collaborative efforts laid the groundwork for communal knowledge sharing—a contrast to the proprietary attitudes that dominated later scientific practices (Stracke, 2020). It was through this gradual evolution that Open Science began to address significant issues in research credibility and accessibility, spurring a demand for policies that facilitate open access to research assets (Cook et al., 2021). It is interesting to note that communal knowledge, as is inherent in IKM and Open Science, was originally the norm rather than the exception in scientific practice.

The Amsterdam Call for Action on Open Science further clarified the objectives of the movement, promoting principles that enable cooperation across disciplines and advocate for the democratisation of science (Mielkov, 2023). Policy frameworks across various countries, especially in Europe, have been instrumental in driving the movement forward, showcasing a commitment from governments to endorse and implement Open Science initiatives (Schöpfel et al., 2016; [illegible]urcan et al., 2023). These frameworks are often designed to enhance collaboration among stakeholders, including researchers, funding agencies, and institutional management, thereby fostering a culture of engagement with Open Science practices (Zečević et al., 2020). This is important in the drive to advocate for the democratisation of science and the rise of citizen science.

Despite its growth, however, the adoption of Open Science is uneven across disciplines and regions, generating disparities in how this movement is embraced. The Open Science movement has also highlighted issues of power and inequality within scientific discourse, emphasizing the need for inclusive practices that ensure diverse participation and consider the perspectives of underrepresented regions (Albornoz & Chan, 2018). In light of international initiatives promoting Open Science, it is critical to address these dynamics to pave the way for equitable access to research opportunities and resources (Zellner et al., 2024). Ultimately, the future trajectory of the Open Science movement hinges on continued commitment from all stakeholders to embrace transparency and foster an inclusive scientific community that can effectively respond to societal challenges through collaborative efforts (Guzzo et al., 2022; Karega et al., 2023). This move for diversity, equity, and inclusion is key in integrating IK in the global knowledge landscape.

Challenges and Ethical Considerations in Open Science

The advancement of Open Science has brought about a transformative shift in how research is conducted and shared, yet it poses significant challenges and ethical considerations that must be navigated carefully. Advocacy for Open Science stems primarily from the perception that it enhances transparency, fosters collaboration, and boosts public trust in scientific endeavours (Khalil et al., 2022; Song et al., 2021). However, various barriers currently hinder its widespread adoption.

One fundamental challenge to Open Science is the existing academic culture that often does not incentivise open practices. Early career researchers are particularly vulnerable, facing systemic pressures that discourage engagement with Open Science practices due to fears of negative impacts on publication prospects, concerns about sharing sensitive data, and a lack of institutional support (Cobey et al., 2024; Gownaris et al., 2022). Studies have shown that many researchers worry about how Open Science practices, like preprints or data sharing, could compromise their intellectual property rights or professional credibility, hindering their willingness to engage fully with Open Science (Zečević et al., 2020). Moreover, the existing reward systems within academia, which often prioritise traditional metrics of success over collaborative or transparent practices, reinforce these challenges (Zečević et al., 2020; Pathak et al., 2024). It is interesting to note that there seems to be a sharing of common challenges between IKM and Open Science in terms of intellectual property rights and going against conventional research practices.

In exploring the ethical landscape of Open Science, important considerations arise around confidentiality, privacy, and the potential harm to marginalized communities (Khalil et al., 2022). While Open Science promotes inclusivity and accessibility, it raises concerns regarding the exposure of vulnerable groups to risks through unchecked data sharing and the lack of comprehensive guidelines to protect participants in research. Ethical discourse surrounding Open Science necessitates that researchers weigh the benefits of transparency against the potential risks involved (Guzzo et al., 2022).

Furthermore, while there is an increase in the perceived value of Open Science among researchers—suggesting a cultural shift—actual implementation often lags behind. Surveys indicate that while many agree on the merits of Open Science, the transition from advocacy to practice is uneven, with significant barriers such as lack of training, funding, and institutional frameworks hindering the full realization of Open Science principles (Ng et al., 2024; Mwangi et al., 2021). Scholars advocate for developing robust training programs and incentivizing Open Science practices, urging institutions and funding bodies to prioritize Open Science compliance to facilitate participation across varying levels of research maturity (Christensen et al., 2019).

Additionally, Open Science brings forth a juxtaposition between promoting transparency and navigating the complexities of scientific rigor and methodological integrity. Some critiques argue that the drive towards openness may inadvertently lead to oversimplified interpretations of scientific practices, thereby compromising the nuanced and rigorous approaches that form the backbone of robust research methodologies (Morgan et al., 2022). These ethical and practical challenges prompt ongoing discourse regarding the future of Open Science, emphasizing the necessity for a balanced approach that safeguards scientific integrity while advancing collaborative practices.

The promise of Open Science can also be compromised by practices such as "open washing," where institutions superficially claim adherence to Open Science principles without actual engagement (Spitschan et al., 2021). Similarly, Khalil et al. (2022) highlight the ethical dilemmas that arise from the push for openness, which, while beneficial in many respects, could lead to privacy concerns and potential misuse of shared data. It is essential that researchers navigate these dilemmas carefully, as they can either substantiate or undermine the overarching goals of the Open Science movement (Nawroth & Krause, 2021).

Moreover, the political context significantly influences the effectiveness of Open Science initiatives. Besançon et al. (2021) highlight that governmental support for Open Science practices is essential for fostering an environment conducive to scientific collaboration and integrity, particularly in regions with constraints on research freedom. This interplay between political structure and scientific practice underscores the multifaceted nature of Open Science and the need for commitment from various stakeholders.

The complexities surrounding the challenges and ethical dimensions of Open Science necessitate a multifaceted approach that values transparency, supports researchers at all stages of their careers, and prioritises ethical considerations to protect vulnerable populations.

Current Trends and Future Directions of Open Science

Recent research on Open Science illustrates a notable shift towards increased transparency, accessibility, and collaboration in the scientific community, driven by various stakeholders, including researchers, institutions, and funding bodies. One of the most significant trends is the acknowledgement of Open Science as a transformative movement that enhances the quality and impact of research outputs across multiple disciplines. For instance, studies have highlighted that Open Science practices lead to reductions in barriers for knowledge exchange, thereby improving research engagement and collaboration (Ragnier & Dacos, 2023; Tzanova, 2020; Zečević et al., 2020).

Open Science encompasses several interconnected principles such as open data, reproducible research, and open educational resources which necessitate new skills that may be a barrier to wider uptake of Open Science (Lindemann & Häberlein, 2023; Ragnier & Dacos, 2023). Recent initiatives, particularly in resource-constrained regions like Kenya, have focused on building capacity in Open Science skills, demonstrating that targeted training and awareness programs can facilitate the adoption of open practices among researchers (Karega et al., 2023). Moreover, frameworks and training programs, such as the Framework for Open and Reproducible Research Training (FORRT), are being implemented to support the effective integration of these practices (Zečević et al., 2020). Academic librarians have also been a target for such capacity building initiatives, for example, the African Library and Information Associations and Institutions (AFLIA) Openness Course for Academic Librarians. This is especially relevant as librarians and information professionals have a role as strong advocators and pioneers of openness, who should take up the role of training and advocacy adopting it into their existing information and digital literacy training programs in academic libraries.

Looking ahead, the literature indicates a clear trajectory for Open Science, with ongoing discussions centred on enhancing its systemic integration across various contexts and research domains. Thaldar et al. (2023) recommend frameworks for integrating Open Science within national policies, especially in countries like South Africa, to align with global standards. Moreover, ongoing efforts to foster a mindset geared toward openness amongst researchers are crucial for sustaining the momentum of the Open Science movement (Hagger, 2021; Mendez et al., 2020). The establishment of policies at conferences and the promotion of open access by organizations like UNESCO indicate institutional recognition of Open Science's importance in enhancing research practices (Oliveira et al., 2021). Furthermore, measuring the impact of Open Science through citation rates and research dissemination metrics has become a key area of focus, with evidence indicating that open practices are likely to affect the visibility and uptake of research findings (Markowitz et al., 2021; Reichmann & Wieser, 2022).

While there is optimism surrounding the potential of Open Science to democratise scientific knowledge and foster citizen science engagement, challenges such as biases and entrenched power dynamics remain prevalent (Cunningham & Kim, 2024). The call for equitable participation in Open Science is echoed in discussions about the demands of modern research dissemination, particularly in leveraging social media for wider outreach (Vignoli & Rörden, 2019). Library and information professionals as knowledge stewards should therefore be active agents of change in this regard.

The future direction of Open Science appears to hinge on establishing strong collaborative networks, further embedding these practices within institutional frameworks, and developing robust metrics to evaluate their effectiveness. Efforts to increase the availability and usability of research data are critical as they underpin the very essence of Open Science, promoting discoverability and reuse (Ragnier & Dacos, 2023; Yu et al., 2024). Continued inter-sectoral partnerships and stakeholder engagement are essential to ensure that Open Science realises its promised transformative effects on research practices globally. With this direction of Open Science and the transformation of research practices globally anchored on diverse, equitable, and inclusive creation, use, and re-use of various forms knowledge, it becomes relevant to analyse the role IKM plays in this endeavour.

SYNERGIES BETWEEN IKM AND OPEN SCIENCE

The integration of Indigenous Knowledge Systems (IKS) and Open Science presents transformative potential in how knowledge is created, shared, and utilised especially in developing inclusive and effective approaches for addressing contemporary challenges.

Recent studies highlight the philosophical underpinning that seeks to legitimise indigenous knowledge not merely as complementary to scientific knowledge, but as an essential component that enhances collaborative practices. Bingham et al. (2021) argue that productive interaction between indigenous knowledges and scientific approaches does not require a strict separation but instead thrives on pluralistic strategies that allow for collaborative management of natural resources, as exemplified through First Nations' salmon management practices. This perspective aligns with findings from Mercer et al. (2009), who emphasize the necessity of establishing common ground for communication to narrow the gulf between Indigenous and Western scientific worldviews. The co-production of knowledge, driven by mutual respect and collaboration among various stakeholders, is therefore integral to this integration.

Moreover, literature also emphasizes the implementation of indigenous methodologies in research practices, calling for respectful engagement with indigenous communities. Authentic indigenous methodologies effectively position indigenous voices at the centre of research agendas, fostering a more equitable and respectful dialogue between indigenous and western scientific paradigms (Louis, 2007; Stewart, 2020). This methodological shift not only enriches academic discourse but also reclaims indigenous identities and knowledge systems in the context of education and health research.

The concept of indigenous knowledge management is also closely linked to the dialogical processes essential for effective knowledge mobilisation which is key in the Open Science movement. Hutchinson et al. (2023) describe indigenous knowledge mobilisation as a dynamic process characterised by continuous and reciprocal knowledge flows, promoting accessibility and community engagement in research. This participatory focus is critical because it addresses historical inequities in knowledge creation and application thereby achieving inclusivity.

Similarly, the multilingual approach of Open Science aligns well with decolonisation and contextuality foundations of IKM. Language proficiency has been found to correlate with greater indigenous knowledge retention, and formal schooling with indigenous knowledge attrition (Athayde et al., 2017). As already highlighted in the sections above, language is also important for collective memory and cultural identity as it informs current heritage management practices, highlighting the necessity of integrating indigenous voices into policy frameworks (Mlilo & Bandama, 2024). However, Onyancha (2024) emphasizes the necessity for clearer terminologies to unify understandings of indigenous and traditional knowledge.

Another point of synergy is the communal knowledge sharing component of both IKM and Open Science. Institutions and researchers are encouraged to create supportive environments—structured frameworks and communal resources—to foster a communal culture around Open Science. This includes initiatives aimed at standardising data collection, sharing educational materials, and facilitating collaborative networks (Manco, 2023; Murphy et al., 2020). The development of guidelines, such as the Transparency and Openness Promotion (TOP) guidelines, further underscores the movement's commitment to improving research methodologies and expanding access to scientific work (Garrett-Ruffin et al., 2021).

IKM in this era of Open Science also becomes easier because of the principle of transparency which gives appreciation to the various forms of knowledge or research output across the different stages of scientific work. This means that indigenous knowledge in all its forms is better regarded and accepted. For example, the rising prominence of pre-registration of study designs and outcomes—in accordance with the principles of open reproducible research—further exemplifies the push for methodological transparency in research processes (Hagger, 2021).

Technological advancements also facilitate the development of IKM through open datasets and the establishment of collaborative frameworks that support transparency. Demchenko et al. (2022) argue for a new infrastructure model termed Platform Research Infrastructure as a Service (PRIaaS), which can adapt to the evolving nature of scientific data management. This model promises to address the increasing data heterogeneity and distribution that modern research demands. Similarly, open licensing practices, as discussed by Guzzo et al. (2022), provide a mechanism for enhancing scientific accessibility and participatory research, striking a balance between protecting intellectual rights and fostering open dissemination.

In general, the intersection of Indigenous Knowledge Management and Open Science offers promising pathways for mutual learning, equitable collaboration, and the development of integrative frameworks that respect cultural contexts. To realise this potential, it is crucial to maintain ongoing dialogue, uphold ethical research practices, and ensure that indigenous peoples contribute their narratives and knowledge in collaborative scientific endeavours.

TENSIONS AND ETHICAL CONSIDERATIONS OF IKM IN THE OPEN SCIENCE ERA

The intersection of Indigenous Knowledge Management (IKM) and Open Science presents significant tensions and ethical considerations, particularly regarding knowledge integration, participation, and power dynamics. These complexities stem from the lasting effects of colonialism, which have shaped the perceptions of indigenous knowledge and the frameworks employed by western science.

A crucial aspect of this tension arises from the differing epistemologies underpinning indigenous knowledge and western scientific approaches. Indigenous Knowledge is inherently contextual and fluid, rooted in local cultures and practices, which may not align with the structured methodologies of Western science (Abdulharis et al., 2022; Smylie et al., 2004). Athayde et al. (2017) emphasize the need for a nuanced understanding of indigenous knowledge that recognises its subjective and context-dependent nature, cautioning against attempts to objectify this form of understanding, which could lead to oversimplification or misrepresentation. Muir et al. (2010) further highlight that indigenous knowledge and western science emerge from fundamentally different worldviews, indicating that harmonising these approaches requires careful navigation of these differences.

Moreover, the integration of these knowledge systems often occurs under power dynamics that can compromise the ethical integrity of collaborative research efforts. Bohensky and Maru (2011) discuss the risks associated with poorly managed integration of indigenous and scientific knowledge, highlighting that such efforts can perpetuate colonial legacies rather than rectify them. Similarly, equitable partnerships involving indigenous communities must go beyond tokenism and genuinely acknowledge indigenous peoples' rights and knowledge as fundamental to the research process (Thompson et al., 2019).

Ethical considerations also underline the imperative to honour knowledge ownership and stewardship as a way to mitigate tensions of data sovereignty and intellectual property. Tsosie et al. (2021) argue that the disregard for the continuous struggle of indigenous people to safeguard and assert their sovereignty, self-determination, and governance results in a troubling characterisation of indigenous involvement in research as simply a matter of "gifting" and "reciprocity." Contemporary intellectual property laws often clash with indigenous frameworks, necessitating new forms of protection for indigenous knowledge to ensure its perpetuation (Recht, 2009). Fortunately, the 'CARE Principles for Indigenous Data Governance'- Collective benefit, Authority to control, Responsibility, and Ethics; and the 'FAIR Guiding Principles for scientific data management and stewardship'- Findable, Accessible, Interoperable, Reusable- have gained prominence in advocating for the rights of indigenous people in controlling access to their knowledge (Carroll et al., 2020).

Furthermore, in an "open" environment there are risks of misappropriation and misrepresentation. Ethical dilemmas can arise concerning the methodologies and motivations behind research that integrates indigenous knowledge. Critiques have surfaced regarding efforts that are primarily driven by the objectives of external researchers rather than initiated by indigenous communities (Lauter, 2020). This can result in outcomes that do not reflect the priorities or interests of indigenous communities, further entrenching disparities in power dynamics that govern research practices (Lauter, 2020; Thompson et al., 2019). Relationships between indigenous knowledge holders and western scientists may replicate colonial patterns unless actively disrupted (Bassani, 2017; Lauter, 2020; Tsosie et al., 2021). This therefore underlines the importance of community consent and participation

Despite the positive implications of integrating indigenous knowledge into scientific frameworks, challenges remain. Traditional knowledge systems may face limitations due to climate change, necessitating a more profound interaction with scientific methodologies to bolster resilience among indigenous communities (Jiri et al., 2016; Khalafzai, 2023). This suggests that both systems can be leveraged to enhance adaptive capacities and promote sustainable practices through shared learning and collaborative action.

It follows therefore that while the convergence of Indigenous Knowledge Management and Open Science holds great potential for collaborative and respectful exchanges of knowledge, it is imperative that such exchanges are approached with critical awareness of historic and systemic power imbalances, ethical responsibilities, and the nature of indigenous traditions. The goal should be to empower and transform the practices and understandings of all involved, fostering equity and inclusivity across communities.

CO-PRODUCTION OF KNOWLEDGE: CASE STUDIES AND EXAMPLES

One prominent aspect of knowledge co-production is the integration of indigenous perspectives in contemporary practices. For instance, McKemey et al. (2022) illustrate a framework termed "Right-way science," which recognises the distinct yet complementary nature of indigenous knowledge and western scientific approaches. This approach has been shown to empower indigenous stakeholders by centring their priorities in discussions about cultural fire management, thereby enhancing ecological management outcomes. The study's findings indicate that adopting indigenous seasonal calendars and kinship systems significantly improves adaptive management strategies for savanna burning in Northern Australia (McKemey et al., 2020; McKemey et al., 2022). This demonstrates the practical benefits of including indigenous knowledge in environmental policy.

Moreover, the methodology surrounding the integration of indigenous and scientific knowledge is critical. Strategies such as the concept of "two-eyed seeing," which recognises the value of both indigenous and western ways of knowing, have emerged as effective frameworks allowing for a parallel approach to research rather than a superficial integration of methods (Chatwood et al., 2015; Reid et al., 2020; Wright et al., 2019). This co-production of knowledge fosters an environment where both cultural and scientific insights inform research processes, thereby ensuring comprehensive outcomes that respect indigenous worldviews and values (Cullen-Unsworth et al., 2011). This philosophy is evident in various case studies related to fisheries management and water governance in Canada, where community-based participatory research frameworks facilitate the respectful sharing of knowledge among indigenous elders and youth (Arsenault et al., 2018; Reid et al., 2020). This reciprocal learning model fosters innovative governance frameworks deeply rooted in indigenous epistemologies, ultimately leading to more effective and culturally relevant solutions for environmental challenges (Arsenault et al., 2018; Zanotti, 2024). This approach emphasises the necessity of understanding knowledge not just as fixed but as dynamic and evolving, a concept supported by Romualdo et al. (2022) in their advocacy for integrating indigenous knowledge systems into education to support sustainable practices. Such frameworks not only enrich research outcomes through broadened perspectives but also empower indigenous people and situate their knowledge within relevant scientific inquiry.

Participatory methodologies have also emerged as essential in fostering collaboration between indigenous communities and scientists. Neale et al. (2019) postulate a decolonising approach utilised in collaborative bushfire management between Dja Dja Wurrung peoples and settler bushfire management interventions on Dja Dja Wurrung country in southeast Australia, reflecting a shift towards participatory frameworks that recognise the relevance of indigenous epistemologies. Such frameworks facilitate dialogue and respect for multiple systems of knowledge, thus addressing historical imbalances, much like those discussed by Chapman and Schott (2020), who argue for the ethical imperative of western scientists to collaborate with indigenous knowledge holders to create mutually beneficial outcomes.

However, challenges remain in ensuring that such partnerships are effective and respectful. For example, Thompson et al. (2019) stress the importance of maintaining indigenous leadership in research projects to ensure that programs are genuinely reflective of community needs and perspectives. They found that successful projects often involve documenting indigenous knowledge through methods that resonate with community practices, such as land visits and oral histories. The drawbacks of simply 'transliterating' indigenous knowledge into scientific discourse without proper context have been criticised in various contexts as it perpetuates the ongoing struggles against the devaluation of indigenous perspectives (Alexander et al., 2019; Lauter, 2020).

The role of a co-developed open science framework is pivotal in recognising indigenous rights to data governance. Hunter (2021) highlights the need for science to be conducted in a manner that respects indigenous knowledge sovereignty while increasing access to scientific outputs. This is based on a national trend in Canada that recognises indigenous peoples' specific rights regarding knowledge and data, aligning with efforts to ensure meaningful engagement and benefit-sharing (Hunter, 2021). Furthermore, the ethical imperatives of inclusion and the benefits derived from indigenous participation in scientific research processes are likely to be more impactful when indigenous perspectives are incorporated and end-users are engaged from the outset (Wright et al., 2022).

It can be seen that the co-production of knowledge is an evolving concept that signals a paradigm shift towards inclusivity and respect for diverse epistemologies. The indications from various case studies underline the tangible benefits derived from such collaborations while also highlighting the challenges that necessitate careful consideration and ethical engagement practices.

LESSONS LEARNT AND GUIDELINES FOR FUTURE COLLABORATIONS

Ethical considerations are fundamental when integrating IKM into scientific endeavours of Open Science. A crucial element supporting these collaborations is the establishment of trust and respect among partners, which can foster a sense of shared purpose and mutual benefit (Pyett, 2002). Establishing enduring relationships, grounded in cultural protocols and ethical standards, is vital for creating frameworks that honour the contributions of both Indigenous and non-Indigenous researchers (Adams et al., 2014; Maldonado & Lazrus, 2019). Such partnerships not only recognise the importance of indigenous voices but also facilitate accountability in research practices that can have lasting impacts on indigenous communities (Maldonado & Lazrus, 2019; Pyett, 2002). By navigating these complex relationships with care and intention, researchers can work toward reconciliation and the co-creation of knowledge that is not only scientifically valid but also culturally relevant.

Additionally, frameworks that emphasize community involvement throughout the research process serve to empower indigenous participants and ensure that their knowledge systems are respected and preserved (Priest et al., 2009). A notable aspect of ethical research is the recognition of indigenous methodologies, which prioritise community involvement at every stage to align research with indigenous values and expected outcomes (Durie, 2004).

Finally, as shown by the literature emphasizes the transformative potential of incorporating indigenous knowledge into academic and scientific communities. This integration challenges traditional paradigms and promotes the development of research infrastructures that are inclusive of diverse knowledge systems (Kennedy et al., 2024). By valuing indigenous knowledge equally with scientific knowledge, academic institutions may enhance research capacities and outcomes, paving the way for more equitable partnerships that address pressing social and environmental issues (Martin, 2016). These considerations play a vital role in ensuring that this integration respects indigenous voices and prioritises community engagement. Continued exploration of these integrative practices can significantly enhance the efficacy of scientific inquiry while fostering reconciliation and respect for indigenous knowledge systems.

RECOMMENDATIONS

Challenges remain, particularly in balancing the power dynamics between indigenous communities and western scientific institutions. Researchers, both indigenous and non-indigenous, often grapple with historical injustices and ongoing colonial frameworks that influence research practices. The imposition of western methodologies can marginalise indigenous voices and erode the intention behind collaborative research efforts (Allard-Tremblay & Coburn, 2024; Williams & Shipley, 2023). This is particularly evident in educational contexts where indigenous students face conflicts between cultural expectations and the demands of western academic structures (Williams & Shipley, 2023). An effective solution proposed in the literature involves developing context-sensitive educational frameworks that embrace indigenous perspectives, thereby enriching science education and promoting better community engagement (Tsosie et al., 2021).

Moreover, there is need for national and institutional policies that guide the integration of indigenous knowledge systems into the current open science paradigm, especially in Africa as the above case studies show little to no practical success stories. The shift towards Open Science presents both opportunities and dilemmas for indigenous communities. On one hand, it advocates for increased access to knowledge and resources; on the other hand, issues surrounding data ownership, control, and the potential misuse of indigenous knowledge arise. Sámi research in Sweden highlights these dilemmas, illustrating the need for policies that respect the sensitivity of indigenous cultural data while promoting cooperative science (Cocq, 2023). Researchers are called to adopt frameworks that ensure indigenous sovereignty over data and representation within collaborative research initiatives (Tsosie et al., 2021). There is therefore need for legal protections to safeguard indigenous intellectual property and prevent unauthorised commercial exploitation.

Heavy investment in digital technologies is also recommended as digital technologies serve as a backbone for facilitating new research methodologies and improving communication among scholars. Vos et al. (2020) highlight how these technologies stimulate the development and utilisation of open science practices such as open-access publications and open-source software. This is consistent with the findings of Hollaway et al. (2020), who assert that technological support is paramount for achieving the goals of Open Science, including enhanced transparency and reduced research costs. This shows that the integration of advanced digital platforms allows for a more accessible scientific lifecycle, thus enabling better understanding and reproducibility of research outputs in any context.

The complexity of digital platforms creates interconnected ecosystems that further promote open innovation. Hilbolling et al. (2020) discuss how digital product platforms can act as ecosystems of open innovation, where independent contributors enhance the collaborative potential of platforms. This interconnectedness is vital for the dissemination and accessibility of research findings in a co-production of knowledge environment. Pizzi (2020) echoes this sentiment, demonstrating how platforms like AiiDA and the Materials Cloud are employed to cultivate a robust Open Science environment, facilitating collaboration and resource sharing among researchers.

Moreover, participatory approaches, made possible through digital platforms, significantly enhance public engagement and the democratisation of science. Borda and Bowen (2020) illustrate this through citizen science initiatives that leverage online tools to involve the public in research. Such efforts resonate with the findings of Morgan et al. (2022), which point to the need for greater inclusivity in research practices, fostering a collaborative environment that both professionals and amateurs can contribute to. Furthermore, while open platforms are positions for transparency, they may also impose challenges concerning intellectual property and ownership that researchers must navigate (Morgan et al., 2022).

Librarians and information professionals also have a crucial role to play as they should enhance accessibility by digitising indigenous knowledge where appropriate while respecting restrictions on sacred or sensitive information. Conducting cultural sensitivity training, providing ongoing education on indigenous knowledge protocols to avoid misrepresentation or misuse is also key. Moreover, participating in community collaboration for community led documentation and curation, working directly with indigenous communities to curate knowledge repositories in ways that align with their cultural norms.

CONCLUSION

The integration of Indigenous Knowledge into the Open Science paradigm is not merely an academic exercise; it has profound implications for the decolonisation of the global knowledge landscape, social justice, environmental sustainability, and cultural resilience. As such, managing indigenous knowledge in an open and respectful manner requires a collaborative approach with key stakeholders involved. A holistic and respectful approach ensures that indigenous knowledge is not merely integrated into dominant systems but recognised and preserved on indigenous peoples' own terms. Respectful Indigenous Knowledge Management requires equity and inclusion, accountability, and systemic change. By centring indigenous sovereignty and adopting anti-colonial practices, stakeholders can foster equity in the global knowledge commons while honouring the past, present, and future of Indigenous peoples.

REFERENCES

Abdulharis, R., Kusdiwanggo, S., Nurlinda, I., Iskandar, G. H., Dwiartama, A., Hernandi, A., & Sidiq, T. P. (2022). Cultural Space Methodology on Assessment of Authenticity of Indigenous Community Forest and Conservation Area in Indonesia. *SOCIAL SCIENCES*. DOI: 10.20944/preprints202212.0402.v1

Adams, M. S., Carpenter, J., Housty, J. A., Neasloss, D., Paquet, P. C., Service, C., Walkus, J., & Darimont, C. T. (2014). Toward increased engagement between academic and indigenous community partners in ecological research. *Ecology and Society*, *19*(3), art5. DOI: 10.5751/ES-06569-190305

Albornoz, D., & Chan, L. (2018). Power and Inequality in Open Science Discourses. *IRIS - Revista de Informação. Memória e Tecnologia*, *4*(1), 70. DOI: 10.51359/2318-4183.2018.238912

Alexander, S. M., Provencher, J. F., Henri, D. A., Taylor, J. J., & Cooke, S. J. (2019). Bridging Indigenous and science-based knowledge in coastal-marine research, monitoring, and management in Canada: A systematic map protocol. *Environmental Evidence*, *8*(1), 15. DOI: 10.1186/s13750-019-0159-1

Allard-Tremblay, Y., & Coburn, E. (2024). "Reviewer 2 must be stopped"; Indigenous Contributions and Epistemic Domination in Political Science. *Appartenances & Altérités*, *5*. Advance online publication. DOI: 10.4000/12kpw

Arsenault, R., Diver, S., McGregor, D., Witham, A., & Bourassa, C. (2018). Shifting the Framework of Canadian Water Governance through Indigenous Research Methods: Acknowledging the Past with an Eye on the Future. *Water (Basel)*, *10*(1), 49. DOI: 10.3390/w10010049

Athayde, S., Silva-Lugo, J., Schmink, M., & Heckenberger, M. (2017). The Same, but Different: Indigenous Knowledge Retention, Erosion, and Innovation in the Brazilian Amazon. *Human Ecology*, *45*(4), 533–544. DOI: 10.1007/s10745-017-9919-0

Bassani, M. (2017). International Cultural Heritage Law and World Heritage Listing: A Vehicle for "White Control of Indigenous Heritage"? *Santander Art and Culture Law Review*, *2*. Advance online publication. DOI: 10.4467/2450050XSNR.17.032.8433

Besançon, L., Peiffer-Smadja, N., Segalas, C., Jiang, H., Masuzzo, P., Smout, C., Billy, E., Deforet, M., & Leyrat, C. (2021). Open science saves lives: Lessons from the COVID-19 pandemic. *BMC Medical Research Methodology*, *21*(1), 117. DOI: 10.1186/s12874-021-01304-y PMID: 34090351

Bingham, J. A., Milne, S., Murray, G., & Dorward, T. (2021). Knowledge Pluralism in First Nations' Salmon Management. *Frontiers in Marine Science, 8.* Advance online publication. DOI: 10.3389/fmars.2021.671112

Bohensky, E. L., & Maru, Y. (2011). Indigenous Knowledge, Science, and Resilience: What Have We Learned from a Decade of International Literature on 'Integration'? *Ecology and Society*, *16*(4), art6. DOI: 10.5751/ES-04342-160406

Borda, A., & Bowen, J. P. (2020, July). *Turing's Sunflowers: Public research and the role of museums. Proceedings of EVA London 2020.* DOI: 10.14236/ewic/EVA2020.5

Carroll, S. R., Garba, I., Figueroa-Rodríguez, O. L., Holbrook, J., Lovett, R., Materechera, S. A., Parsons, M. A., Raseroka, K., Rodriguez-Lonebear, D., Rowe, R., Sara, R., Walker, J., Anderson, J., & Hudson, M. (2020). The CARE Principles for Indigenous Data Governance. *Data Science Journal, 19.* Advance online publication. DOI: 10.5334/dsj-2020-043

Chapman, J. M., & Schott, S. (2020). Knowledge coevolution: Generating new understanding through bridging and strengthening distinct knowledge systems and empowering local knowledge holders. *Sustainability Science*, *15*(3), 931–943. DOI: 10.1007/s11625-020-00781-2

Chatwood, S., Paulette, F., Baker, R., Eriksen, A., Hansen, K. L., Eriksen, H., Hiratsuka, V., Lavoie, J., Lou, W., Mauro, I., Orbinski, J., Pabrum, N., Retallack, H., & Brown, A. (2015). Approaching *Etuaptmumk* – introducing a consensus-based mixed method for health services research. *International Journal of Circumpolar Health*, *74*(1), 27438. DOI: 10.3402/ijch.v74.27438 PMID: 26004427

Chigwada, J. P., & Ngulube, P. (2023). Librarians' Role in the Preservation and Dissemination of Indigenous Knowledge. *IFLA Journal*, *50*(2), 242–256. DOI: 10.1177/03400352231217270

Christensen, G., Wang, Z., Paluck, E., Swanson, N., Birke, D., Miguel, E., & Littman, R. (2019). Open Science Practices Are On The Rise: The State Of Social Science (3s) Survey. https://doi.org/DOI: 10.31222/osf.io/5rksu

Cobey, K., Alayche, M., Saba, S., Barnes, N., Ebrahimzadeh, S., Alarcon, E., & Moher, D. (2024). Cardiology Researchers' Practices and Perceived Barriers to Open Science: An International Survey. *Open Heart*, *11*(1), e002433. DOI: 10.1136/openhrt-2023-002433 PMID: 38233041

Cocq, C. (2023). Open science in Sámi research: Researchers' dilemmas. *Frontiers in Research Metrics and Analytics*, *8*, 1095169. DOI: 10.3389/frma.2023.1095169 PMID: 37124717

Cook, B. G., Johnson, A. H., Maggin, D. M., Therrien, W. J., Barton, E. E., Lloyd, J. W., Reichow, B., Talbott, E., & Travers, J. C. (2021). Open Science and Single-Case Design Research. *Remedial and Special Education, 43*(5), 359–369. DOI: 10.1177/0741932521996452

Cullen-Unsworth, L. C., Hill, R., Butler, J. R. A., & Wallace, M. (2011). A research process for integrating Indigenous and scientific knowledge in cultural landscapes: Principles and determinants of success in the Wet Tropics World Heritage Area, Australia. *The Geographical Journal, 178*(4), 351–365. DOI: 10.1111/j.1475-4959.2011.00451.x

Cunningham, H., & Kim, C. (2024, March 5). Weaving Open Dialogue Using Canada's Open Science Roadmap Framework. *Toronto Metropolitan University.* DOI: 10.32920/25343236.v1

Dei, G. J. S. (2000). Rethinking the role of Indigenous knowledges in the academy. *International Journal of Inclusive Education, 4*(2), 111–132. DOI: 10.1080/136031100284849

Demchenko, Y., De Laat, C., Los, W., & Gommans, L. (2022). Defining Platform Research Infrastructure as a Service (PRIaaS) for Future Scientific Data Infrastructure. In Otto, B., Ten Hompel, M., & Wrobel, S. (Eds.), *Designing Data Spaces* (pp. 241–260). Springer International Publishing., DOI: 10.1007/978-3-030-93975-5_15

Durie, M. (2004). Understanding health and illness: Research at the interface between science and indigenous knowledge. *International Journal of Epidemiology, 33*(5), 1138–1143. DOI: 10.1093/ije/dyh250 PMID: 15218015

Egeruoh-Adindu, I. (2022). Leveraging Indigenous Knowledge for Effective Environmental Governance in West Africa. *Beijing Law Review, 13*(04), 931–947. DOI: 10.4236/blr.2022.134060

Fernández-Llamazares, Á., Lepofsky, D., Lertzman, K., Armstrong, C. G., Brondizio, E. S., Gavin, M. C., Lyver, P. O., Nicholas, G. P., Pascua, P., Reo, N. J., Reyes-García, V., Turner, N. J., Yletyinen, J., Anderson, E. N., Balée, W., Cariño, J., David-Chavez, D. M., Dunn, C. P., Garnett, S. C., & Vaughan, M. B. (2021). Scientists' Warning to Humanity on Threats to Indigenous and Local Knowledge Systems. *Journal of Ethnobiology, 41*(2), 144–169. DOI: 10.2993/0278-0771-41.2.144

Flor, A. (2013). Exploring the Downside of Open Knowledge Resources: The Case of Indigenous Knowledge Systems and Practices in the Philippines. *Open Praxis, 5*(1), 75. DOI: 10.5944/openpraxis.5.1.15

Funk, J., & Guthadjaka, K. (2020). Indigenous Authorship on Open and Digital Platforms: Social Justice Processes and Potential. *Journal of Interactive Media in Education, 2020*(1). DOI: 10.5334/jime.560

Garrett-Ruffin, S., Hindash, A. C., Kaczkurkin, A. N., Mears, R. P., Morales, S., Paul, K., Pavlov, Y. G., & Keil, A. (2021). Open science in psychophysiology: An overview of challenges and emerging solutions. *International Journal of Psychophysiology, 162*, 69–78. DOI: 10.1016/j.ijpsycho.2021.02.005 PMID: 33556468

Gownaris, N., Vermeir, K., Bittner, M., Gunawardena, L., Kaur-Ghumaan, S., Lepenies, R., & Zakari, I. (2022). Barriers to full participation in the open science life cycle among early career researchers. *Data Science Journal, 21*. Advance online publication. DOI: 10.5334/dsj-2022-002

Guzzo, R. A., Schneider, B., & Nalbantian, H. R. (2022). Open science, closed doors: The perils and potential of open science for research in practice. *Industrial and Organizational Psychology: Perspectives on Science and Practice, 15*(4), 495–515. DOI: 10.1017/iop.2022.61

Hagger, M. S. (2021). Developing an open science 'mindset'. *Health Psychology and Behavioral Medicine, 10*(1), 1–21. DOI: 10.1080/21642850.2021.2012474 PMID: 34993003

Hilbolling, S., Berends, H., Deken, F., & Tuertscher, P. (2020). Complementors as connectors: Managing open innovation around digital product platforms. *R & D Management, 50*(1), 18–30. DOI: 10.1111/radm.12371

Hollaway, M. J., Dean, G., Blair, G. S., Brown, M., Henrys, P. A., & Watkins, J. (2020). Tackling the Challenges of 21st-Century Open Science and Beyond: A Data Science Lab Approach. *Patterns (New York, N.Y.), 1*(7), 100103. DOI: 10.1016/j.patter.2020.100103 PMID: 33205137

Hunter, A. (2021). *Co-developing Openness: Indigenous Knowledge and Data Governance and Open Science in Canada* [Master of Arts, Carleton University]. DOI: 10.22215/etd/2021-14620

Hutchinson, P. J., McIlduff, C. D., Legare, M., Keewatin, M., Hagel, M., Chapados, M., & Acharibasam, J. B. (2023). Indigenous knowledge mobilization: Reflection on context, content, and relationship. *AlterNative: An International Journal of Indigenous Peoples, 19*(4), 902–913. DOI: 10.1177/11771801231198082

Igbafe, E. C. (2023). Researching African Indigenous Population in Contemporary Times: What Scholars and Practitioners Need to Know. In Gbendazhi Barnabas, S. (Ed.), *Indigenous and Minority Populations—Perspectives From Scholars and Writers across the World*. IntechOpen., DOI: 10.5772/intechopen.107591

Jiri, O., Mafongoya, P. L., Mubaya, C., & Mafongoya, O. (2016). Seasonal Climate Prediction and Adaptation Using Indigenous Knowledge Systems in Agriculture Systems in Southern Africa: A Review. *Journal of Agricultural Science*, *8*(5), 156. DOI: 10.5539/jas.v8n5p156

Karega, P., Mwaura, D. K., Mwangi, K. W., Wanjiku, M., Landi, M., & Kibet, C. K. (2023). Building awareness and capacity of bioinformatics and open science skills in Kenya: A sensitize, train, hack, and collaborate model. *Frontiers in Research Metrics and Analytics*, *8*, 1070390. DOI: 10.3389/frma.2023.1070390 PMID: 37324282

Kennedy, M., Ninomiya, M. M., & Ninomiya, M. M., Brascoupé (Anishinabeg/ Haudenausanee), S., Smylie (M tis), J., Calma (Kungarakan, Iwaidja), T., Mohamed (Narrunga Kaurna), J., Stewart (Taungurung), P. J., & Maddox (Bagumani, Modewa), R. (2024). Knowledge translation in Indigenous health research: Voices from the field. *The Medical Journal of Australia*, *221*(1), 61–67. DOI: 10.5694/mja2.52357 PMID: 38946651

Khalafzai, M. A. K. (2023). Climate Change-Related Natural Hazards and Risk Communication: Incorporating Traditional Indigenous Knowledge. In Mokhtari, M. (Ed.), *Natural Hazards—New Insights*. IntechOpen., DOI: 10.5772/intechopen.108302

Khalil, A. T., Shinwari, Z. K., & Islam, A. (2022). Fostering openness in open science: An ethical discussion of risks and benefits. *Frontiers in Political Science*, *4*, 930574. DOI: 10.3389/fpos.2022.930574

Lauter, O. (2020). *Challenges in Combining Indigenous and Scientific Knowledge in the Arctic*. *SocArXiv*. DOI: 10.31235/osf.io/kvn2c

Levin, N., Leonelli, S., Weckowska, D., Castle, D., & Dupré, J. (2016). How Do Scientists Define Openness? Exploring the Relationship Between Open Science Policies and Research Practice. *Bulletin of Science, Technology & Society*, *36*(2), 128–141. DOI: 10.1177/0270467616668760 PMID: 27807390

Lindemann, T., & Häberlein, L. (2023). Contours of a research ethics and integrity perspective on open science. *Frontiers in Research Metrics and Analytics*, *8*, 1052353. DOI: 10.3389/frma.2023.1052353 PMID: 37235074

Louis, R. P. (2007). Can You Hear us Now? Voices from the Margin: Using Indigenous Methodologies in Geographic Research. *Geographical Research*, *45*(2), 130–139. DOI: 10.1111/j.1745-5871.2007.00443.x

Madonsela, B. S., Semenya, K., & Shale, K. (2024). A Review of Indigenous Knowledge Systems and Their Application in Sustainable Solid Waste Management. *WORLD (Oakland, Calif.)*, *5*(2), 219–239. DOI: 10.3390/world5020012

Maldonado, J., & Lazrus, H. (2019). A Story of "Rising Voices" and Intercultural Collaboration. *Practical Anthropology*, *41*(3), 34–37. DOI: 10.17730/0888-4552.41.3.34

Manco, A. (2023). Open science policies as regarded by the communities of researchers from the basic sciences in the scientific periphery. *Online Information Review*, *48*(6), 1065–1087. DOI: 10.1108/OIR-03-2023-0135

Markowitz, D. M., Song, H., & Taylor, S. H. (2021). *Tracing the Adoption and Effects of Open Science in Communication Research*. *PsyArXiv*. DOI: 10.31234/osf.io/dsf67

Martin, K. L. (2016). Vision led = vision fed: NIRAKN's role in progressing scholarship in Aboriginal early childhood education, raising the bar in Aboriginal knowledges in higher education and holding ground in Aboriginal research capacity building. *International Journal of Critical Indigenous Studies*, *9*(2), 1–14. DOI: 10.5204/ijcis.v9i2.137

Mazzocchi, F. (2020). A deeper meaning of sustainability: Insights from indigenous knowledge. *The Anthropocene Review*, *7*(1), 77–93. DOI: 10.1177/2053019619898888

McGregor, D. (2021). Indigenous Knowledge Systems in Environmental Governance in Canada. *KULA: Knowledge Creation, Dissemination, and Preservation Studies*, *5*(1). Advance online publication. DOI: 10.18357/kula.148

McKemey, M., Ens, E., Rangers, Y. M., Costello, O., & Reid, N. (2020). Indigenous Knowledge and Seasonal Calendar Inform Adaptive Savanna Burning in Northern Australia. *Sustainability*, *12*(3), 995. DOI: 10.3390/su12030995

McKemey, M. B., Rangers, B., Rangers, Y. M., Costello, O., Hunter, J. T., & Ens, E. J. (2022). 'Right-way' science: Reflections on co-developing Indigenous and Western cross-cultural knowledge to support Indigenous cultural fire management. *Ecological Management & Restoration*, *23*(S1), 75–82. DOI: 10.1111/emr.12532

Mendez, D., Graziotin, D., Wagner, S., & Seibold, H. (2020). Open Science in Software Engineering. In Felderer, M., & Travassos, G. H. (Eds.), *Contemporary Empirical Methods in Software Engineering* (pp. 477–501). Springer International Publishing., DOI: 10.1007/978-3-030-32489-6_17

Mercer, J., Kelman, I., Taranis, L., & Suchet-Pearson, S. (2009). Framework for integrating indigenous and scientific knowledge for disaster risk reduction. *Disasters, 34*(1), 214–239. DOI: 10.1111/j.1467-7717.2009.01126.x PMID: 19793324

Mielkov, Y. (2023). The paradigm of open science and democratization of the research activity of universities. *Paradigm of knowledge, 3*(57). Advance online publication. DOI: 10.26886/2520-7474.3(57)2023.4

Mlilo, T., & Bandama, F. (2024). Collective Memory and Memory-Making within the Mapungubwe and Makgabeng Heritage Landscapes: Indigenous Voices Amongst the Sotho- and Venda- Speaking Groups of Southern Africa. *Arts and Humanities*. DOI: 10.20944/preprints202403.0850.v1

Mohan, D. (2024). Wisdom of the ancestors: Harnessing indigenous knowledge for contemporary challenges. *International Journal of Applied Research, 10*(3), 112–119. DOI: 10.22271/allresearch.2024.v10.i3b.11591

Morgan, J. A., Lindsay, B. L., & Moran, C. (2022). Opening a "closed door": A call for nuance in discussions of open science. *Industrial and Organizational Psychology: Perspectives on Science and Practice, 15*(4), 537–541. DOI: 10.1017/iop.2022.72

Muir, C., Rose, D., & Sullivan, P. (2010). From the other side of the knowledge frontier: Indigenous knowledge, social–ecological relationships and new perspectives. *The Rangeland Journal, 32*(3), 259. DOI: 10.1071/RJ10014

Murphy, M. C., Mejia, A. F., Mejia, J., Yan, X., Cheryan, S., Dasgupta, N., Destin, M., Fryberg, S. A., Garcia, J. A., Haines, E. L., Harackiewicz, J. M., Ledgerwood, A., Moss-Racusin, C. A., Park, L. E., Perry, S. P., Ratliff, K. A., Rattan, A., Sanchez, D. T., Savani, K., & Pestilli, F. (2020). Open science, communal culture, and women's participation in the movement to improve science. *Proceedings of the National Academy of Sciences of the United States of America, 117*(39), 24154–24164. DOI: 10.1073/pnas.1921320117 PMID: 32929006

Mwangi, K., Mainye, N., Ouso, D., Esoh, K., Muraya, A., Mwangi, C., & Kibet, C. (2021). Open Science In Kenya: Where Are We? *Frontiers in Research Metrics and Analytics, 6*. Advance online publication. DOI: 10.3389/frma.2021.669675 PMID: 34056516

Nakata, M. (2002). Indigenous Knowledge and the Cultural Interface: Underlying issues at the intersection of knowledge and information systems. *IFLA Journal, 28*(5–6), 281–291. DOI: 10.1177/034003520202800513

Nawroth, C., & Krause, E. T. (2021). The academic, societal and animal welfare benefits of Open Science for animal science. *Open Science Framework*. DOI: 10.31219/osf.io/dnfvs

Neale, T., Carter, R., Nelson, T., & Bourke, M. (2019). Walking together: A decolonising experiment in bushfire management on Dja Dja Wurrung country. *Cultural Geographies*, *26*(3), 341–359. DOI: 10.1177/1474474018821419

Ng, J., Lin, B., Kreuder, L., Cramer, H., & Moher, D. (2024). Open science practices among authors published in complementary, alternative, and integrative medicine journals: an international, cross-sectional survey. DOI: 10.1101/2024.02.26.24303402

Ngulube, P. (2024). From Orality to Digital Assets: Managing Indigenous Knowledge in Africa in the Wake of the Open Science Movement. *The Serials Librarian*, *84*(5–8), 71–85. DOI: 10.1080/0361526X.2023.2296880

Ocholla, D. (2025). The complex marriage: Interfacing open science and open indigenous knowledge. *Library Management*, *46*(3/4), 252–267. DOI: 10.1108/lm-11-2024-0125

Oliveira, E. J., Chavez, C. V. F. G., Cordeiro, A. F. R., & Feitosa, D. (2021). How do Brazilian Software Engineering Researchers Perceive and Practice Open Science? *Anais Do I Workshop de Práticas de Ciência Aberta Para Engenharia de Software (OpenScienSE 2021)*, 28–33. DOI: 10.5753/opensciense.2021.17142

Onyancha, O. B. (2024). Indigenous knowledge, traditional knowledge and local knowledge: What is the difference? An informetrics perspective. *Global Knowledge, Memory and Communication*, *73*(3), 237–257. DOI: 10.1108/GKMC-01-2022-0011

Oyelude, A. A. (2023). Indigenous Knowledge Preservation as a Sign of Respect for Culture: Concerns of Libraries, Archives and Museums. *Insights the Uksg Journal*, *36*. Advance online publication. DOI: 10.1629/uksg.628

Pathak, K., Marwaha, J., Chen, H., Krumholz, H., & Matthews, J. (2024). Use of open science practices in surgical journals. *JAMA Surgery*, *159*(2), 228. DOI: 10.1001/jamasurg.2023.5389 PMID: 38117492

Pepple, D. G., & Dore-Okuiomse, M. M. (2023). An enabler framework for developing knowledge management practices: Perspectives from Nigeria. *South African Journal of Information Management*, *25*(1). Advance online publication. DOI: 10.4102/sajim.v25i1.1647

Pizzi, G. (2020). Open-Science Platform for Computational Materials Science: AiiDA and the Materials Cloud. In Andreoni, W., & Yip, S. (Eds.), *Handbook of Materials Modeling* (pp. 1813–1835). Springer International Publishing., DOI: 10.1007/978-3-319-44677-6_64

Priest, N., Mackean, T., Waters, E., Davis, E., & Riggs, E. (2009). Indigenous child health research: A critical analysis of Australian studies. *Australian and New Zealand Journal of Public Health*, *33*(1), 55–63. DOI: 10.1111/j.1753-6405.2009.00339.x PMID: 19236360

Pyett, P. (2002). Towards reconciliation in Indigenous health research: The responsibilities of the non-Indigenous researcher. *Contemporary Nurse*, *14*(1), 56–65. DOI: 10.5172/conu.14.1.56 PMID: 16114194

Ragnier, M., & Dacos, M. (2023). *Report of the G7 Open Science—Research on Research Sub-Working Group*. Ministère de l'enseignement supérieur et de la recherche. DOI: 10.52949/32

Recht, J. (2009). Hearing Indigenous Voices, Protecting Indigenous Knowledge. *International Journal of Cultural Property*, *16*(3), 233–254. DOI: 10.1017/S0940739109990166

Reichmann, S., & Wieser, B. (2022). Open science at the science–policy interface: Bringing in the evidence? *Health Research Policy and Systems*, *20*(1), 70. DOI: 10.1186/s12961-022-00867-6 PMID: 35725491

Reid, A. J., Eckert, L. E., Lane, J., Young, N., Hinch, S. G., Darimont, C. T., Cooke, S. J., Ban, N. C., & Marshall, A. (2020). "Two-Eyed Seeing": An Indigenous framework to transform fisheries research and management. *Fish and Fisheries*, *22*(2), 243–261. DOI: 10.1111/faf.12516

Romualdo, A., Arroyo, J., Suriaga, E. G., & Fuentes, R. (2022). Traditional Knowledge Management (TKM) Through Biosprospecting: Mainstreaming Scientific Research Trends and Techniques for Inclusive and Sustainable Education. *East Asian Journal of Multidisciplinary Research*, *1*(7), 1497–1506. DOI: 10.55927/eajmr.v1i7.837

Schöpfel, J., Ferrant, C., André, F., & Fabre, R. (2016). Ready for the future? A survey on open access with scientists from the French National Research Center (CNRS). *Interlending & Document Supply*, *44*(4), 141–149. DOI: 10.1108/ILDS-06-2016-0023

Senanayake, S. G. J. N. (2006). Indigenous knowledge as a key to sustainable development. *Journal of Agricultural Sciences – Sri Lanka*, *2*(1), 87–94. DOI: 10.4038/jas.v2i1.8117

Shiferaw, D., Takele, S., & Hailemichael, A. (2017). Indigenous Practices of Ganta Community in Protecting Natural Resources, Gamo Gofa Zone, Ethiopia. *International Journal of Environmental Protection and Policy*, *5*(5), 74. DOI: 10.11648/j.ijepp.20170505.12

Smith, L. T. (1999). *Decolonizing Methodologies: Research and Indigenous Peoples*. Bloomsbury Academic. https://books.google.co.zw/books?id=Nad7afStdr8C

Smylie, J., Martin, C. M., Kaplan-Myrth, N., Steele, L., Tait, C., & Hogg, W. (2004). Knowledge translation and indigenous knowledge. *International Journal of Circumpolar Health, 63*(sup2), 139–143. DOI: 10.3402/ijch.v63i0.17877

Song, H., Markowitz, D., & Taylor, S. (2021). Trusting on the shoulders of open giants? open science increases trust in science for the public and academics. DOI: 10.31219/osf.io/g328c

Spitschan, M., Schmidt, M. H., & Blume, C. (2021). Principles of open, transparent and reproducible science in author guidelines of sleep research and chronobiology journals. *Wellcome Open Research*, *5*, 172. DOI: 10.12688/wellcomeopenres.16111.2 PMID: 33748432

Stewart, S. L. (2020). One Indigenous Academic's Evolution: A Personal Narrative of Native Health Research and Competing Ways of Knowing. *First Peoples Child & Family Review*, *4*(1), 57–65. DOI: 10.7202/1069350ar

Stracke, C. M. (2020). Open Science and Radical Solutions for Diversity, Equity and Quality in Research: A Literature Review of Different Research Schools, Philosophies and Frameworks and Their Potential Impact on Science and Education. In D. Burgos (Ed.), *Radical Solutions and Open Science: An Open Approach to Boost Higher Education* (pp. 17–37). Springer Singapore. DOI: 10.1007/978-981-15-4276-3_2

Tata, E. F., Yongabi, K. A., & Dzelamonyuy, E., Forbang, & Geh, E. A. (2023). Decolonizing African science: Efforts, challenges, and future directions. *International Journal of Science and Research Archive*, *10*(1), 823–833. DOI: 10.30574/ijsra.2023.10.1.0840

Thaldar, D., Gooden, A., & Steytler, M. (2023). Open science and human genetic data: Recommendations on South Africa's Draft National Open Science Policy. *Frontiers in Genetics, 14*, 1248747. DOI: 10.3389/fgene.2023.1248747 PMID: 37849503

Thompson, K.-L., Reece, N., Robinson, N., Fisher, H.-J., Ban, N. C., & Picard, C. R. (2019). "We monitor by living here": Community-driven actualization of a social-ecological monitoring program based in the knowledge of Indigenous harvesters. *Facets*, *4*(1), 293–314. DOI: 10.1139/facets-2019-0006

Traynor, C. (2017). Data Management Plan: Empowering Indigenous Peoples and Knowledge Systems Related to Climate Change and Intellectual Property Rights. *Research Ideas and Outcomes*, *3*, e15111. DOI: 10.3897/rio.3.e15111

Tsosie, K. S., Yracheta, J. M., Kolopenuk, J. A., & Geary, J. (2021). We Have "Gifted" Enough: Indigenous Genomic Data Sovereignty in Precision Medicine. *The American Journal of Bioethics*, *21*(4), 72–75. DOI: 10.1080/15265161.2021.1891347 PMID: 33825628

urcan, N., Cuciureanu, G., Cujba, R., Cojocanu, I., & Cojocaru, I. (2023). Readiness Toward the Implementation of Open Science Initiatives In the Republic of Moldova. *Central European Journal of Public Policy*, *17*(2), 97–118. DOI: 10.2478/cejpp-2023-0009

Tzanova, S. (2020). Changes in academic libraries in the era of Open Science. *Education for Information*, *36*(3), 281–299. DOI: 10.3233/EFI-190259

Vignoli, M., & Rörden, J. (2019). Why We Need Open Science Communication Experts. *Mitteilungen Der Vereinigung Österreichischer Bibliothekarinnen Und Bibliothekare*, *72*(2), 284–296. DOI: 10.31263/voebm.v72i2.3049

Vos, M. G. D., Hazeleger, W., Bari, D., Behrens, J., Bendoukha, S., Garcia-Marti, I., Van Haren, R., Haupt, S. E., Hut, R., Jansson, F., Mueller, A., Neilley, P., Van Den Oord, G., Pelupessy, I., Ruti, P., Schultz, M. G., & Walton, J. (2020). Open weather and climate science in the digital era. *Geoscience Communication*, *3*(2), 191–201. DOI: 10.5194/gc-3-191-2020

Williams, D. H., & Shipley, G. P. (2023). Indigenous Research Methodologies: Challenges and Opportunities for Broader Recognition and Acceptance. *Open Journal of Social Sciences*, *11*(05), 467–500. DOI: 10.4236/jss.2023.115030

Wright, A. L., Gabel, C., Ballantyne, M., Jack, S. M., & Wahoush, O. (2019). Using Two-Eyed Seeing in Research with Indigenous People: An Integrative Review. *International Journal of Qualitative Methods*, *18*, 1609406919869695. DOI: 10.1177/1609406919869695

Wright, K., Tapera, R. M., Stott, N. S., Sorhage, A., Mackey, A., & Williams, S. A. (2022). Indigenous health equity in health register ascertainment and data quality: A narrative review. *International Journal for Equity in Health*, *21*(1), 34. DOI: 10.1186/s12939-022-01635-2 PMID: 35279132

Yu, W., Chen, J., & Deng, S. (2024). Open Science Under Debate: Disentangling the Interest on Twitter and Scholarly Research. *SAGE Open*, *14*(3). DOI: 10.1177/21582440241271300

Zanotti, L. (2024). Transdisciplinary Research Practice by, with, and for Indigenous Knowledge Holders. *Bulletin of the Ecological Society of America, 106*(2), e2212. DOI: 10.1002/bes2.2212

Zečević, K., Houghton, C., Noone, C., Lee, H., Matvienko-Sikar, K., & Toomey, E. (2020). Exploring factors that influence the practice of Open Science by early career health researchers: A mixed methods study. *HRB Open Research, 3*, 56. DOI: 10.12688/hrbopenres.13119.1 PMID: 33537552

Zellner, P. J., Claus, M., Dolezalova, T., Balogun, R. O., Eberle, J., Hodam, H., Eckardt, R., Meißl, S., Jacob, A., & Anghelea, A. (2024). MOOC Cubes and Clouds—Cloud Native Open Data Sciences for Earth Observation. *The International Archives of the Photogrammetry, Remote Sensing and Spatial Information Sciences, XLVIII-4/W12-2024*, 157–162. DOI: 10.5194/isprs-archives-XLVIII-4-W12-2024-157-2024

ADDITIONAL READING

Adeyeye, B., & Mason, J. (2019). Opening futures for Nigerian education – Integrating Educational Technologies with Indigenous Knowledge and Practices. *Open Praxis, 12*(1), 27. DOI: 10.5944/openpraxis.12.1.1055

Flor, A. (2013). Exploring the Downside of Open Knowledge Resources: The Case of Indigenous Knowledge Systems and Practices in the Philippines. *Open Praxis, 5*(1), 75. DOI: 10.5944/openpraxis.5.1.15

Funk, J. and Guthadjaka, K. (2020). Indigenous Authorship on Open and Digital Platforms: Social Justice Processes and Potential. *Journal of Interactive Media in Education, 2020(1)*. DOI: 10.5334/jime.560

Ocholla, D. (2025). The Complex Marriage: Interfacing Open Science and Open Indigenous Knowledge. *Library Management, 46*(3/4), 252–267. DOI: 10.1108/lm-11-2024-0125

Simpson, A., McCann, J., & Miroff, L. (2023). Learners' Perspectives on ARCH + STEM: Integration of Archaeology and Indigenous Knowledges with Western Knowledges of STEM. *Education Sciences, 13*(5), 450. DOI: 10.3390/educsci13050450

Traynor, C. (2017). Data Management Plan: Empowering Indigenous Peoples and Knowledge Systems Related to Climate Change and Intellectual Property Rights. *Research Ideas and Outcomes, 3*, e15111. DOI: 10.3897/rio.3.e15111

KEY TERMS AND DEFINITIONS

Data Sovereignty: The right of communities to control how data about and from them is collected, stored, and used.

Indigenous Knowledge Management (IKM): The methods and systems used by indigenous communities to preserve, organise, and share their indigenous knowledge in culturally appropriate ways.

Indigenous Knowledge Systems (IKS): Traditional knowledge developed by and among indigenous communities over generations, encompassing all aspects of everyday life and sustainable living.

Open Science: An emerging transformative movement that encourages free access to scientific research, data, and collaboration to improve transparency and innovation.

Synergies: areas where Indigenous Knowledge Systems (IKS) and Open Science complement each other, creating opportunities for mutual enrichment.

Tensions: points of conflict or ethical concern that emerge when integrating indigenous knowledge into Open Science frameworks.

Chapter 6

Preserving Medicinal Indigenous Knowledge Among Zimbabwean Communities Through Digitization

Stephen Tsekea
https://orcid.org/0000-0003-2885-2515
Zimbabwe Open University, Harare, Zimbabwe

Alfred Mapolisa
https://orcid.org/0009-0001-2753-1528
Zimbabwe Open University, Harare, Zimbabwe

ABSTRACT

Traditional knowledge has always played a critical role in the health fraternity of most African countries. However, indigenous communities in Zimbabwe continue to lose their valuable knowledge through such factors as globalization, climate change and death or memory loss of the elders. The threats of extinction of this knowledge particularly medicinal IKS such as herbal medicine and indigenous healing have been facilitated by lack of recordings since most of it relies on oral and memory transmission, too much preference for western health systems and lack of intellectual property protection from pirates. The loss of such critical knowledge has led to continued community diseases and ultimately loss of life in some circumstances. Culture is also threatened whenever there is a loss of its essential system of knowl-

DOI: 10.4018/979-8-3373-0204-1.ch006

edge. This chapter details how traditional medicine such as the use of herbs is a key player in the health of indigenous communities. It also unpacks what efforts can be made to preserve such knowledge systems.

INTRODUCTION

Traditional knowledge has always played a significant part in the development of most African countries. However, the indigenous people of Zimbabwe continue to lose their valuable knowledge through such factors as globalization, climate change and death or memory loss of the elders. The threats of extinction of this knowledge particularly medicinal indigenous knowledge system (IKS) that include herbal indigenous medicine have been facilitated by lack of recordings since most of it relies on oral and memory transmission, too much preference for western health systems and lack of intellectual property protection from pirates. The loss of such critical knowledge has led to continued community diseases and ultimately loss of life in some circumstances. Culture is also threatened whenever there is a loss of its essential system of knowledge. This chapter details how traditional medicine such as the use of herbs is a key player in the health of indigenous communities. It also unpacks what efforts can be made to preserve such knowledge systems. The focus will be more on harnessing digital technologies such as Artificial intelligence tools and applications, social media applications, and other online technologies emphasizing their potentiality in preserving, sharing and utilising such traditional medicinal knowledge systems.

Nations need to capitalize on these technologies through digitization to enable long-term preservation. Technology also has the capability of making indigenous knowledge systems available on the open access platform thereby improving access. Preservation of IKS is critical in that some of the knowledge has the potential to contribute to the sustainable development of many communities. The chapter therefore explores the strategies and efforts used in Zimbabwe to digitalise IKS. It also shows the barriers being faced such as lack of experts, lack of support from the health professional councils, copyright and intellectual challenges, and technological obsolescence. However, the chapter documents how digitization can be the panacea to the long-term preservation of traditional medicinal knowledge systems. The ethical issues surrounding the use of digital technologies are also discussed. The chapter is critical in that it contributes to preserving medical knowledge systems through harnessing emerging digital technologies. Scholars need therefore to find possible ways of promoting indigenous knowledge systems because they have the potential to contribute immensely to a nation's development.

Although most African nations have appeared to shun traditional medicine, statistics have shown that more than 85% of Africans rely on traditional medicine or traditional healers to provide healthcare (Dlamini and Khanyile, 2021). There has been a tendency to focus more on Western practices. Such Western medicinal medicine has its challenges like cost and suitability due to local culture. The situation is even more pronounced in rural communities that have challenges in accessing modern medicine due to monetary constraints and inadequately resourced health infrastructure in developing countries. Pandemics such as COVID-19 showed how exposed developing countries are to the healthcare fraternity as there was a serious challenge to proper health facilities. The population however turned to traditional medicine to try and cure COVID-19. This showed that traditional medicine can be a panacea towards empowering the community's medical provisions. While there seems to be some recognition of the importance of IKS, there is however not much effort being made to take charge of its severe deterioration and demise. The other challenge is that traditional medicinal knowledge systems are mostly present in the minds and brains of the older generation. Failure to harvest such knowledge may lead to the extinction of very important medicinal knowledge. It is therefore necessary to explore such knowledge systems and see how nations can take advantage of emerging technologies. Indigenous knowledge has the potential to enhance sustainable development, hence the need for societies to worry about preservation and conservation. Developing nations need to come up with platforms and digital media applications that can dovetail with the changing traditional knowledge systems.

The chapter makes an important contribution to the knowledge base of indigenous systems. It provides key information necessary for policy and practice. The chapter shows how digital media technologies and applications can be the solution to the challenge of deteriorating indigenous knowledge systems through proper harnessing of such. The chapter also attempts to proffer solutions to challenges that are facing developing nations in their attempts to digitize and preserve their indigenous knowledge systems. In this chapter, IKS will also often be referred to as traditional knowledge (TK). While there have been some significant strides that other nations have made to document and preserve their indigenous knowledge systems, no major attempts have been done in Zimbabwe. This therefore presents the need for igniting the debate on what possible attempts could be made to document medicinal indigenous knowledge systems.

WHAT ARE INDIGENOUS KNOWLEDGE SYSTEMS?

Indigenous medicine is defined by the WHO (2022) as "the total of the knowledge, skills and practices based on the theories, beliefs and experiences indigenous to different cultures, whether explicable or not, used in the maintenance of health, as well as in the prevention, diagnosis, improvement or treatment of physical and mental illnesses". Similarly, the World Intellectual Property Organisation (WIPO) (2013) defined Indigenous knowledge as know-how, skills, innovations, practices, teaching, and learning of indigenous people and local communities. Indigenous knowledge (IK) refers to the knowledge that belongs to a specific group of people which will be unique to that group, society or culture (Ghimire, 2021). According to Gorjestani (2000), IKS belongs to traditional people and is usually oral and passed from one generation to another. Such knowledge is developed over a long period and always plays an important role in any society. These indigenous communities possess a rich tapestry of cultural knowledge, traditions, and languages (Ajani et al., 2024). A particular culture is identified by its indigenous knowledge. Many communities take pride in their IKS and want to make sure that it is preserved for future posterity. A community that values its knowledge system will try to shoulder the responsibility of making sure it goes out and defends its knowledge from extinction so that it passes it on to generations to come. Some common myths that need to be demystified include calling IKS as old-fashioned or outdated. IKS as observed by Twarog (2004) is "tradition-based and multifaceted concept that includes cultural identity, practical knowledge, oral tradition, spiritual and religious beliefs and aesthetics".

EXAMPLES OF COMMUNITY KNOWLEDGE SYSTEMS

- Folklore
- Rituals
- Songs
- Dance
- Customary Practices
- Teachings

CHARACTERISTICS OF IKS

- It is local
- It is easily overlooked
- It is transmitted orally
- Is widely shared among the community members
- More pronounced in the practical experiences
- It is not properly managed

- Is predominantly tacit
- It is an endangered species
- Unwillingness to share by some knowledge holders

DIGITALISING IKS IN AFRICA

There have been some reported successful case studies of attempts at digitizing indigenous knowledge systems in Africa. While, Globally, India has been one notable country that has done tremendously well in digitizing IKS, South Africa has also made strides to document and preserve their traditional knowledge systems. These involve converting oral knowledge systems that traditional healers and elders have and digitizing it into audio and video recordings. Such processes help in preservation and wide sharing of medicinal knowledge systems. The Ulwazi Digital Programme in South African has tried to focus on the documentation and digitazition of oral history. The project was the "brainchild of eThekwini Municipal Library's mission of preserving and disseminating indigenous knowledge of local munities in Durban and its surrounding environments" (Oppenneer, 2025). The programme created a localized Wiki that facilitate online archiving of local knowledge and history.

THE CONCEPT OF TRADITIONAL MEDICINE

Traditional medicine or indigenous medicine includes "herbal medicines composed of herbs, herbal, materials herbal preparations, and finished herbal products that contain active ingredients parts of plants or other plant materials, or combinations thereof" (Tora and Heliso, 2017). "Globally, 85% of the traditional medicines used for primary health care are derived from plants" (Crepeau, 1997). Indigenous medicinal knowledge involves its uses in clinical practices like appropriate dosage and forms of administration as well as the procedures and rituals applied by the healers as part of their traditional healing methods (Torri, 2010). According to the World Health Organisation (WHO), traditional medicine is "the total of the knowledge, skills and practices based on the theories, beliefs and experiences indigenous to different cultures, whether explicable or not, used in the maintenance of health as well as in the prevention, diagnosis, improvement or treatment of physical and mental illnesses" (WHO, 2008). "It is therefore urgent to preserve such knowledge given that external expropriation largely results in the deprivation of both ownership rights and the right of benefit-sharing that arises from the utilization of this knowledge" (Chakrabarty and Hossain, 2024).

Gurib-Fakim (2006) noted that medicinal plants are a most preferred option for patients. He further notes that for most "people, traditional healers offer information, counselling, and treatment to patients and their families in a personal manner" (Gurib-Fakim, 2006). Despite the massive growth in modern Western medicine, African communities continue to rely on traditional medicine for their healthcare needs (Masocha, 2002). Chigora, Masocha and Mutenheri (2007) observed that "traditional medicine continues to provide solutions to the health needs of poor people who cannot afford expensive drugs prescribed in clinics, hospitals and private health facilities". The main players in indigenous medicine are the traditional healers who play an important role in the primary health care systems of local people. Yirga (2010) observed that there is a need for these traditional healers to transfer their knowledge to interested persons in the communities to allow the preservation of such knowledge systems.

Traditional indigenous medicine is so popular that in some Asian and African countries, up to 80% of the population relies on traditional indigenous medicine for their primary health care, and patients will only be referred to hospitals for modern medicines when there are complications. This popularity of traditional indigenous medicines is based on the fact that there is unlimited application of the medicine. All diseases can be cured using traditional indigenous medicines. In some cases, patients using modern medicines to treat chronic diseases such as Diabetes, Hypertension and Cancer end up opting for traditional indigenous medicines which are cheaper and often effective if properly administered. Almost half the population in many industrialized countries now regularly uses some form of Traditional and Contemporary Medicine (TC & M) (United States, 42%; Australia, 48%; France, 49%; Canada, 70%); considerable use of some form of T&CM exists in many other countries, such as Chile (71%), Colombia (40%) and up to 80% in some African countries (WHO, 2023).

Traditional indigenous medicine is not only practised in Africa, but throughout the World, for example in Europe it is called traditional European medicine, in China it is known as traditional Chinese medicine, in Korea, it is known as traditional Korean medicine and in Africa it is called traditional African medicine (Mahomoodally, 2013). African traditional medicine is very important and is commemorated annually on 31 August as African Traditional Medicine day.

For time immemorial, indigenous, traditional and ancestral knowledge has been an integral resource for health in households and communities, and it continues to form a significant part of healthcare in many regions (WHO, 2023). As such, "173 of WHO's 194 member states have all reported on the use of herbal medicines, acupuncture, yoga, indigenous therapies and other forms of indigenous traditional medicines. Many countries recognize traditional medicine as a valuable source of healthcare and have taken steps to integrate practices, products and practitioners into their national systems" (WHO, 2023). Today, traditional indigenous medicine has become a global phenomenon with very high demand. The 2018 Declaration of Astana on primary health care acknowledges the need to include traditional medicine knowledge and technologies in the delivery of primary health care in pursuit of health for all (WHO, 2018).

Traditional Medicine in Zimbabwe

In Zimbabwe, reliance on traditional medicine has been there since time immemorial. The coming in of the colonizers in 1890 who introduced modern medicine did not entirely stop the use of traditional medicines. The local communities continued to rely on the traditional healers and elders who possessed significant knowledge on treatment of ailments. The main sources of herbal medicine in Zimbabwe are plants and wildlife. The Zimbabwe Traditional Medical Practitioners Act governs the operations of the traditional healers in the country. The act is the major legislation that oversees the herbal medicine preparation and dissemination. Traditional medicine is also governed by the National Medicine Policy of 2011. This policy governs the way medicines including traditional one is managed.

Forms of Traditional Indigenous Medicines in Africa

Africa is a treasure trove of medicinal plants and the following seven are the most popular as given by Aremu and Makunga (2022):

- African wormwood *(Artemisia afra)*- This plant species is the only one regarded as indigenous to the continent and has the potential to become Africa's "flagship plant because of its popularity and diverse uses in African traditional indigenous medicines" (Sundaymail, 2022) . The African wormwood has been used to treat coughs, colds, influenza and malaria. However, with insufficient clinical data, it's not yet known whether African wormwood is a treasure chest of new drugs.
- Cape periwinkle, graveyard plant, bright eyes (*Catharanthus roseus*)- It is also known as Madagascar periwinkle and is native to Madagascar. The plant is used to treat many health conditions such as diabetes, rheumatism, and skin-related venereal diseases.
- Griffonia, Atooto, gbogbotri, kajya, kanya, kwakuo-aboto (*Griffonia simplicifolia)*

This is a tree-climbing shrub, native to West African countries like Cameroon, Liberia, Togo, Benin, Cote d'Ivoire, Ghana and Nigeria (Sundaymail, 2022). The seeds of the plant have several medicinal effects such as a remedy for diarrhea, stomach ache and dysentery. It is also used as an aphrodisiac.

- Kanna, Kougoed, (*Sceletium tortuosum*)- This a succulent plant which is endemic in South Africa and is sparsely distributed in semi-arid areas. It is mainly used to treat conditions such as stress, depression, pain and anxiety. It is also used as a dietary supplement with the potential to elevate mood, relieve stress and improve cognition.
- Climbing oleander (*Strophanthus gratus*)- This evergreen climbing shrub is found in Tropical regions of Africa such as Senegal and DRC. It is used as medicine to treat snake bites, sores, gonorrhoea, constipation and fever. Its roots are also an aphrodisiac.
- African geranium, (*Pelargonium sidoides*)- this plant is indigenous to South Africa and is also found in the Lesotho highlands. The root tubers of the plant are harvested and used to treat diarrhoea, colic, gastritis, tuberculosis, cough, liver disorders, menstrual complaints, and gonorrhoea.
- African ginger, wild ginger, (*Siphonochilus aethiopicus)*- This is a plant native to southern and western tropical Africa. It is used to treat respiratory problems such as cough, influenza as well as other ailments such as pain, inflammation and malaria. This plant has a wide range and this explains why it is widely used. In South Africa, because of over-harvesting for traditional medicine, it is now classified as critically endangered.

Indigenous Medicine in Zimbabwe

Across the whole country, there is a high use of traditional indigenous medicines because of their affordability and easy accessibility. These traditional medicines can be administered by anyone, however, in most cases, there are some cultural and spiritual considerations hence traditional healers are consulted and administer these medicines. Besides traditional healers, there are also some herbalists specializing in different traditional medicines and different ailments who administer these medicines effectively.

Table 1 is an extract from Maroyi (2023) with a sample of plants that are used to treat various diseases and ailments in Zimbabwe.

Table 1. Medicinal Plants in Zimbabwe

Indigenous or local name	Botanical name	Diseases cured	Part of the tree used	How it is administered
Mugan'acha	*lannea discolor*	Fever & constipation in children, reduce menstrual flow	Bark and roots	An infusion of the bark and roots for fever and constipation. Fibre extract for menstrual flow
Mutsambatsi	*lannea edulis*	Bilharzia, diarrhoea, gonorrhoea	Roots	Root is prepared and drunk
Mumango	*mangfera indica*	All kinds of respiratory ailments, Desentry & diarrhoea	Leaves, Bark extract	Tender leaves are boiled for respiratory ailments. Bark extracts for dysentery and diarrhea
Mubedu	*ozoroa insignis*	STIs, diarrhoea	Roots	Root extract is prepared and drunk
Mufokosiyana	*rhus longipes*	Infertility in woman & to dilate the birth canal	Roots	Root extract is prepared and drunk
Mupfura	*sclerocarya caffra*	Eyes	Roots	Roots are crushed and boiled and steam directed to sore eyes
Mubikasadza	*searsia dentata*	Ulcers, diarrhoea and stomach problems	Leaves	Leaves are chewed while fresh and sap is swallowed
Muroro	*annona stenophylla*	Boils,Chest pains, STIs, Snake repellent	Roots	Root paste is applied to boils, Root extract is drunk as remedy for chest pains and STIs, Roots can be mixed with *mafufu* roots and sprinkled around the homestead as a snake repellent
Muruguru	*carissa bispinosa, carisa edulis*	Cough, diarrhoea	Roots	Root extract is drunk
Chirindamatongo	*catharanthus roseus*	Dysentery, diarrhoea, ingestion	Roots	Root extract is drunk
Rukato	*asparagus africanus*	Diarrhoea, pneumonia, dilate birth canal	Roots	Root extract is drunk
Konje	*sansevieria aethiopica*	Painful ear	Leaves	Leaves are put into a fire so that the sap exuded is squeezed into a painful ear

continued on following page

Table 1. Continued

Indigenous or local name	Botanical name	Diseases cured	Part of the tree used	How it is administered
Mushayamhanda	*sansevieria hyacinthoides*	Aching tooth, dehydrated child, dilate birth canal	Leaves, rhizome	Warm rhizome is put in the mouth of a person with aching tooth to keep up to 2 min before spitting out the mixture, Leaf or rhizome sap is given to a dehydrated chiid. Rhizome is added to none alcoholic beverages taken by pregnant woman to dilate the birth canal.
Gavakava	*aloe greathedii*	Constipation, gonorrhoea	Leaf	Leaf extract is drunk
Mupasa	*branchylaena discolor*	Ulcers & abdominal disorders	Leaves	Leaves are chewed and juice swallowed
Chifumuro	*dicoma anomala*	Wide range of illnesses	Bulb	the bulb extract is drunk or commonly tied on a string and worn around the neck, waist or wrist of a child to protect againstspiritually induced illness (expose spells)
Ruhwahwa	*schkuhria pinnata*	Stomach pains	Whole plant	An infusion of the whole plant is drunk

Adopted from Maroyi (2013)

How traditional indigenous medicines can be preserved.

In Zimbabwe, preserving traditional indigenous medicines is of paramount importance because traditional knowledge and medicines are interwoven with cultural identity and community heritage. Zimbabwe has a rich heritage of traditional medicinal knowledge often passed down orally from generation to generation. Therefore, preserving such traditional indigenous medicines can provide valuable health solutions as evidenced by how Zumbani (*lippia javanica*), a medicinal plant became a remedy of choice among Africans during the COVID-19 pandemic.

Documentation of traditional medicinal practices is an essential step in preserving and safeguarding this knowledge for future generations. In Zimbabwe, where much of traditional knowledge remains undocumented, there is a very high risk of losing critical knowledge and inadvertently compromising public safety. Through documentation, traditional knowledge can be systematically recorded, categorized and reviewed for safe and informed public use. More so, documented knowledge can be protected as intellectual property, ensuring that communities retain control over their heritage and benefit from its usage.

Traditional Indigenous Medicines and Intellectual Property Rights

The World Intellectual Property Organization (WIPO) was set up in 1967 to protect traditional knowledge systems. WIPO, which become part of the UN systems, is concerned with copyrights, patents and trademarks. Its mandate was the promotion of creative intellectual activity in order to accelerate economic, social and cultural development in the world (Thairu, 2007). In Africa, intellectual property rights are protected by the African Regional Intellectual Property Organization (ARIPO). ARIPO aims to harmonise IP laws in Africa. Such laws that falls under ARIPO include patents and copyrights. Intellectual Property Rights (IPRs) are a means of obtaining ownership for specific assets that are elusive and for the most part, include the security of some type of innovation made by the human mind (Mdhluli et al., 2021). Similarly, Tsekea (2016) defined intellectual property rights as legislative and policy frame towards the protection of knowledge, ideas, innovations, inventions and others. In Zimbabwe, the Zimbabwe Intellectual Property Rights Office is mandated to protect intellectual property law for rights such as copyrights, patents and trade secrets. Patenting in Zimbabwe is protected by the Patents Act [Chapter 26: 03]. The Act defines patent as an exclusive right that is granted for a novel invention which can be a product or process. Copyright is protected under the Copyright and Neighboring Rights Acts [Chapter 26:05]. This act defines copyright as a real right which subsists in a work and which entitles its owners exclusively to do in Zimbabwe

Countries like India and China offer valuable global insights into traditional medicine management. For example, India has established the Traditional Knowledge Digital Library (TKDL), which is a comprehensive database of documented traditional medicinal practices. The TKDL is accessible to national and international patent offices which prevents unauthorized patenting of traditional remedies and ensure that this knowledge remains within the public domain (Makonya, 2024). Likewise, in China, traditional medicines have been integrated into its national healthcare system which allows systematic regulation, quality control and standardized usage of traditional remedies. Through the Traditional Chinese Medicines (TCM), framework, China has formalized traditional medicine to ensure both the preservation of its cultural heritage and the safety of its citizens (Makonya, 2024).

Zimbabwe can draw lessons from these examples by establishing a centralized documentation system for traditional indigenous medicines and knowledge which could serve as a resource for both safeguarding heritage and promoting safe usage. Zimbabwe's legislative framework, specifically the Traditional Healers Act, Chapter 27:14 (1981), seeks to regulate traditional medicine practitioners to some extent since the act mandates all traditional healers to be registered and work within ethical guidelines set by the Zimbabwe National Traditional Healers Association (ZINA-

THA). However, the Act does not extensively cover the documentation, verification or regulation of specific traditional medicines and practices (Makonya, 2024).

By standardizing traditional medicine usage, the framework can mitigate health risks, while promoting informed use through public education and avoid disasters such as the one which occurred in 2024 when four men died after drinking an over-dosed consortium (*guchu*) in Chiadzwa, Marange. The system could be governed by relevant intellectual property laws that allow traditional communities to maintain ownership and control over their knowledge. This framework will serve several purposes as this will involve creating a centralized database to record traditional medicinal practices, ensuring the knowledge is preserved for future generations (Makonya, 2024). Zimbabwe must establish a comprehensive framework that not only protects traditional medicinal knowledge, but also regulates its application. Thus, by following the examples set by China and India, Zimbabwe can promote the safe and sustainable use of traditional indigenous medicines, ensuring that tragic incidents are minimized while preserving the invaluable cultural asset (Makonya, 2024).

There are however concerns about the protection of indigenous knowledge. For example, Wilder (2000) noted that patent protection of indigenous knowledge is challenging because of concerns of "novelty" or "inventive steps". Some certain parts of traditional medicine may be known and do not present any novel or new things. Traditional knowledge is also held collectively and no single person may claim ownership. Trade secrets protection applies slightly to traditional medicines such as plants and animal matter used by traditional healers but is of greater applicability to non-medicinal practices by traditional healers such as ceremonies, methods or processes (Wilder, 2000).

DIGITIZING INDIGENOUS KNOWLEDGE

Akinwale (2012, p.2) define digitization as "the process of codifying information or knowledge so that it can be accessed globally and on a long-term basis". According to Bandi et al. (2015) digitization involves the making of an electronic version of a real-world object or event enabling the object to be stored, displayed and manipulated on a computer and disseminated over networks or the World Wide Web Digitization is the process of converting an analogue material into a digital format. Once material is digitized it will now be capable of being used by a computer. Computers, as digital machines, can only distinguish two values, 0 and 1 or off and on. When material is digitized, a digital surrogate is created. That digital surrogate can now be processed and stored by computers. Indigenous knowledge which is usually in oral format can be digitized for storage and use by a computer. The medicinal indigenous knowledge should be converted first into digital objects that computers will be able to encode digitally, that are as a series of zeros and ones. The oral knowledge can be transferred to print material readying it for digitization. According to IFLA (2012), "digital technologies offer the opportunity of preserving the original by providing access to the digital surrogate; of separating the informational content from the degradation of the physical medium".

Why Digitize Indigenous Knowledge Systems

Traditional medicinal knowledge plays a significant role in world health and continues to be used to treat a vast array of conditions and complaints (Tora and Heliso, 2017). In Zimbabwe, since time immemorial, plants, animals and other natural endowments have been used as a source of medicine to treat different ailments. The rapid loss of IK within communities is cause for concern and failure to find solutions to preserve it would contribute to the further loss (Sraku-Lartey et al., 2017). Documentation and digitization of traditional knowledge are also necessary to protect bio-piracy as well and expropriation without compensation by multi-national entities (Christian, 2009). Digitization and using ICT technologies to store, preserve and access IKS can be a panacea to this since most of the knowledge is perpetuated orally within families and small communities of Zimbabwe.

According to IFLA (2002: 6) the following are some of the reasons for digitization:

- To increase access; this seems the primary reason in instances where there is a high demand from users. In most developing countries where there is more and more need for reliance on medicinal knowledge, digitization would be the way to go.
- To reduce the handling and use of fragile or heavily used original material and create a 'backup' copy for endangered material. For example, medicinal knowledge that would have been written on paper can be digitized to make additional copies while the original copies are safely kept.
- To seek partnerships with other institutions and other projects

In addition, digitization creates the following:

- Constant availability of material anytime. Material that is digitized and available in electronic format will now be accessible 24/7 allowing users to download it anytime.
- Material in electronic format can be accessed simultaneously by multiple numbers of users.

- Digitized material can be easily searched. A user can conduct a full-text search on the treatment of an ailment once IKS is digitized. Hyperlinks can also be created on a particular document to link it to related items or articles.
- Access to digital materials can also be protected. Those not authorized to access the resources will not be able to do so. For example, authentication can be a requirement to determine the bonafide users of the facility. If a community wants to make its resources available on open access, then no restrictions to access will be made.
- Material digitized can also be transformed back into analogue through scanning, printing and photocopying.

Indigenous knowledge is so critical for the development of a nation and nations must ensure that such knowledge systems are properly archived for access and future prosperity. In Africa, indigenous communities possess a wealth of traditional knowledge passed down through generations encompassing such vast domains as medicine, agriculture, ecology and storytelling (Ajani et al., 2024). If the potential of indigenous knowledge is to be realized, it must be available on the Internet. This helps in the preservation of tacit medicinal knowledge. According to Sabbagh et al., (2012), countries at the most advanced stage of digitalization derive 20% more in economic benefits than those at the initial stage. Materials that can be digitized include recorded information, photographs, manuscripts and other artistic work. However, Hippolyte and Namaseb (2011) noted that during the digitization process, it is necessary to ensure that processes applied to adapt to internationally acceptable guidelines, policies and practices.

TOOLS, EQUIPMENT AND APPLICATIONS FOR DIGITIZING IKS

For communities to fully digitize their medicinal IKS, we recommend the following as some of the required ICT tools and applications:

Computers

A successful digitalization and preservation of indigenous knowledge must consider the proper computers needed to capture and manage digital content. For the whole project to succeed there must be proper computers that are reliable in speed and storage. The computer's Random Access Memory (RAM) should be as large as possible to allow the computer to quickly process the stored materials which can include large scanned materials, audio and video materials. The hard disk space storage should also be as large as possible because most of the digitized indigenous medicinal knowledge material will be in formats that take a lot of space. External hard drives with a capacity of up to a minimum of 1 terabyte for data storage and backup services are also required. It is not advisable to have storage on removable optical media such as CDs or DVDs as they are not suitable for large-scale digitization projects since they can be cumbersome because of their small capacity size as compared to what will be stored. Backups can be done on-site or offsite or some combination of both. The computers should also have internal sound systems of external speakers. This will allow the users of the system to play audio or video of the resources. The computers should have monitors with high resolutions capable of showing text or graphics. The contemporary forms of monitors are the Liquid Crystal Display (LCD) which tend to be smaller, thinner and more energy efficient than the Cathode Ray Tube (CRT) monitors.

Scanners

Scanning is a process of capturing digital images using scanners (Parekh, 2000). Scanners and digital cameras are some of the most important equipment that one would require in converting documents from analogue to digital. These are important for converting documents that are not born digital. According to Parekh (2000) scanners and digital cameras consist of a charge-coupled device or CCD electronic array which converts lights into electrical signals. There are different types of scanners and these include drum scanners, portable, flatbed and some 3D scanners. The flatbed is the commonly used scanner in many organisations. It is ideal for digitizing documents as it can scan a wider variety of records such as photographs, any paper records and books. The flatbed scanners provide quality scans and are easy to operate. With scanners, you can be able to generate other forms of document formats such as Portable Document Format (PDF) or Joint Photographic Expert Group (JPEG). The resulting file can be formatted into the required standard. When scanning, one can use a predefined resolution for quality output. The flatbed scanner is easy to operate and has fast scanning speeds and produces high-quality documents. With a flatbed scanner, you can easily see the document and progress of work through the transparent screen. Instead of scanning, digital photography can be taken by a high-end digital camera. Scanning documents results in the creation of images of the pages of documents. However, no editing, cutting or pasting will be possible from the scanned document. It is possible to further convert the document. The process through which a page is transformed into a text file is called optical Character Recognition (OCR).

Audio and Video Players

Medicinal knowledge can be transcribed into audio and various audio players that can be used to play and store the information. Examples of media players where we can be store IKS include CD, flash drives and cassettes. A database can store the material in audio formats such as MP3, MP4 and others. Users accessing the IKS databases will need different software audio players such as Windows media players, VLC Media Player and AIMP3. When deciding on which video players to install on the computer, it is advisable to consider the compatibility of the device to ensure that the player can read the format of your media and the image quality, that is, how the video will be coming out.

Interactive software

A wide variety of software that could be used to document and archive IKS is available in the market. In digitizing material, software that could be used includes document image processing and document management systems. Such software usually results in the generation of a Tag Image File Format (TIFF) which can be converted to PDF (Portable Document Format). Scanned documents cannot be edited or manipulated. As highlighted above, if there need for conversion to other formats, there will be a need for the use of Optical Character Recognition (OCR) software that can allow conversion to formats such as Word. The advantage of using OCR software is that it will allow documents to be easily searchable and accessible.

Various indigenous knowledge systems are low-cost and simple to enable communities to manage their own digital collections (Ghimire, 2021). Such software will allow captured or uploaded files to be indexed using MyXQL or XML database. Examples of software that could be used include DSpace, Fedora Commons and Greenstone. Some of the software can be customized to allow the capturing and storage of indigenous knowledge systems. In the Republic of South Africa, they have used the National Knowledge Management System (NIKMAS) to capture, store and make available indigenous knowledge systems. An Indigenous knowledge management system should have provisions for allowing the capturing of metadata, storage of the documented information and search interface. Metadata is data about data. It will provide important information that will allow the description of the resource. A resource will be easy to search once important metadata is captured. According to Parekh (2000), the function of metadata is to provide users with a standardized means for intellectual access to digitized materials. The Dublin Core (DC) and the Machine Readable Cataloguing (MARC) are some of the metadata standards used in the archiving of records. The Dublin Core (DC) is one of the most widely accepted metadata standards in the world as they are simple, international and cross-sectorial (IFLA, 2002). The management system should enable data to be described using customized metadata. In the system, there should also be some access rights management where each content manager can upload materials. Each content manager will have their login details such as username and password. The system must allow different file formats of data to be served. For example, medical knowledge can be transcribed and recorded in a Word document that can easily be converted into a PDF document. Some medicinal knowledge can be captured in pictures so the software must allow the capturing of documents in formats such as Graphic Interchange Format (GIF) and Joint Photographic Expert Group (JPEG). Thus the database can allow storage of materials such as images, video clips, video clips and text. The search interface must use a standard web browser technology such as Internet Explorer or Google Chrome. For example, software such as DSpace and Greenstone can allow capturing and searching of medical knowledge systems and they provide some search interfaces that users can use to access the stored material. According to the Open Archival Information Systems (OAIS) framework, when creating a digital repository like an indigenous knowledge system, there are some minimum attributes required. The OAIS Reference Model is composed of several activities like metadata management, data storage and preservation, access management, and document of preservation strategies (Balogun, 2023). Donaldson (2020) noted these attributes as "compliance with the Reference Model, administrative responsibility, organisational viability, financial viability, technological and procedural suitability".

Geographic Information Systems (GIS) technology can assist in mapping of traditional medicinal knowledge such as forest, animal, land and other environmental resources. Communities can actively engage in this modern technology and have their knowledge documented. GIS can assist in gathering information on plant identification in certain communities. A record of the plants and even their images can be captured and also showing what ailments they treat.

Social Media Applications

Ajani et al. (2024) noted that digital media technologies play a pivotal role in the preservation of indigenous knowledge systems such as languages by offering a diverse array of tools and platforms that broaden storage, dissemination and accessibility. The online digital platforms allow the preservation and dissemination of cultural medicinal knowledge in appropriate ways. They also provide an opportunity to revitalize and reclaim IKS regarding land, medicines and foods as more communities may provide ongoing feedback to build and validate the data (Jennings, Johnson-Jennings and Little, 2020). Digital technologies such as wikis, podcasts and blogs can allow communities to capture their indigenous medical practices ensuring the long-term preservation and accessibility of such knowledge systems by a wider audience. The digital media technologies as observed by Austin (2019) can offer innovative ways to preserve these traditional knowledge systems. Communities can create digital online platforms, can also use mobile applications, social media technologies such as blogs to store and broadcast their medicinal knowledge systems. Social media can offer the communities opportunities to contribute personally on the IKS database. Discussion forums can be created where people can share their medicinal knowledge. This will benefit a wider audience. Even in poor communities, social media applications such as WhatsApp can be used. In these communities, the mobile app has gained widespread popularity and can be easily used as some platforms for the storage and dissemination of medicinal information. According to Adam, Othman and Halim (2022), digital platforms can offer indigenous communities opportunities to document and share their traditional ecological knowledge, cultural practices and medicinal plant information. Similarly, Ajani et al. (2024) noted that online platforms enable the dissemination of traditional materials, recordings and learning resources significantly enhancing their visibility and impact. One of the most important platforms that have gained popularity in developing countries is Facebook and YouTube. Facebook allows individuals to post videos, share interests, make connections and join groups with similar interests. According to Essoungou (2011), more than 15% of people online in Africa will be using Facebook. This could be a great platform where pages could be created and communities contribute the knowledge on treatment of diseases and other health education topics. In Zimbabwe, for example, a group called Herbal Medicine and Natural Remedies on Facebook has over 160, 000 members who share medicinal information on their platform.

YouTube, for example, allows users to upload, share, record, and view their community medicinal knowledge. This is an important platform for social networking and education allowing users to post their user-generated content. There are millions of YouTube users in Africa. Users can post content on how to use certain herbs and also some general health educational videos, tutorials and other resources. For example, communities can view educational videos on how traditional medicine can be used to cure ailments and also observe medicine coming from other tribes. On YouTube, users can create their channels and invite users to like and follow them. Users can also search on YouTube for their favorite videos of medicinal knowledge systems.

Google Docs and Twitter can also be handy in the storage and dissemination of indigenous knowledge. Google Docs allows users to create documents, spreadsheets, forms and presentations with the application itself or to import them through a web interface (Owiny, Mehta and Maretzki, 2014). Multiple users using indigenous knowledge documents are capable of editing content and sharing the documents. Such platforms can be handy in sharing medicinal knowledge systems documented in a Word document.

Other digital technological applications such as Blogs and Wikis can be useful for storing and sharing knowledge. Blogs allow the sharing of knowledge online to a wider audience. Blogs can be easily accessed by members of a particular community, allowing the sharing and dissemination of medicinal knowledge. Some virtual hubs and indigenous knowledge experts can have dedicated pages to broadcast information the traditional medicine that can be used to cure particular ailments and health-related shrubs. In South Africa, Ajani et al. (2024) observed that strides to document and preserve indigenous knowledge have been witnessed through the introduction of an innovative application called Indigenous Digital Archive (IDA). With this application, the local communities collaborate to digitize their rich indigenous knowledge including oral histories, songs, and cultural artefacts.

THE ROLE OF ARTIFICIAL INTELLIGENCE

The modern world is embracing artificial intelligence at a faster rate. These emerging technologies can offer great potential in the preservation of indigenous knowledge systems. According to Ajani (2024), these cutting-edge technologies can introduce novel approaches to enhance knowledge documentation and foster cultural immersion, thereby opening up fresh opportunities for indigenous communities. Artificial intelligence is defined by Tredinnick (2017) as a "cluster of technologies and approaches to computing focused on the ability of computers to make flexible rational decisions in response to unpredictable environmental conditions". Similarly, Harisanty et al. (2023) define AI as "a group of technologies that allow computers to solve problems in a more dynamic way than previous methods". So AI is the use of computers and related technologies to do things that require human intelligence. AI has the potential therefore to make communities harvest their medicinal knowledge and preserve it for generations to come.

THE ROLE OF BLOCKCHAIN TECHNOLOGIES

Blockchain technology is another powerful tool that can be used for the preservation of indigenous knowledge ensuring proper secure and transparent platform. Blockchains are tamper evident and tamper resistant digital ledgers implemented without a central repository and usually without a central authority (Yaga et al., 2019). Blockchain allows the indigenous communities to have their knowledge digitized and decentralized with them having a collective ownership of their knowledge systems and having a transparent system and retaining control over its use. However, challenges of scalability, that is, when the database becomes too big to absorb other materials and installing of tight security mechanism are often encountered when preserving digital materials.

METADATA APPLICATIONS

The preservation of indigenous medicinal knowledge systems can be enhanced through having proper metadata. Proper metadata can assist content managers in properly describing digital objects and other related indigenous digital materials like digital footage and audio recordings so that it is easily discovered and transparent. The description of such indigenous knowledge systems can be done through respecting local traditions and culture. For example, properly designed metadata can allow use of local languages and terms to be added in the databases thus allowing the system to be discoverable. Properly designed metadata can also assist in bridging the digital divide as some fields can be simplified to make even the average technology literate individuals to search, discover and retrieve materials within the medicinal indigenous repository

CHALLENGES IN DIGITIZING AND PRESERVING IKS

While there are a lot of advantages of digital preservation of medical knowledge as presented above, there are also some challenges with digitization. Perry (2014) notes the issue of obsolescence of older technologies and formats. Technology is moving at a fast rate, faster than what communities may be able to cope with. In such circumstances, a digitalized IKS system can become a white elephant if technology changes and there are no proper conversion mechanisms to the latest technologies. However, Corrado (2019) noted that there is a need to ensure that there is sustainable funding and policies in place to sustain digital preservation.

There are also challenges in the lack of policies and strategies for indigenous medicinal knowledge in developing countries like Zimbabwe. The issues of policy and digital preservation frameworks are major problems in Africa (Mutula, 2014). It is also difficult to identify the rightful owners of traditional medicine. No significant policies have been put in place on how the digital IKS libraries or repositories should be created and managed. There are also challenges with customary laws and practices. Due diligence should be done in determining whether customary laws and practices in a particular society allow the sharing of indigenous knowledge systems. Many African governments have also not properly provided supervision over the exploitation of traditional medicinal herbs.

Community attitude towards digitization is another significant challenge that is likely to be encountered. Many indigenous communities are generally reluctant to share their knowledge systems with outsiders. Furthermore, documenting indigenous knowledge can result in misinformation. Mawere (2011) observed that "lies, when documented, resemble truth". If misinformed knowledge is documented and digitized, there is a possibility of further sharing false information. Thus Ndhluli (2021) notes that, "the digitization of IKS should not manipulate, misinform or influence the prospective customer deliberately or otherwise.

There are also challenges in addressing copyright and intellectual property rights issues. Greyling and McNuty (2011) observed that using intellectual property laws to protect indigenous knowledge is challenging since most of the rights issues like patents and copyrights are Eurocentric, where they place too much emphasis on individuality and material, contrary to traditional and indigenous cultural norms. There are also some security concerns with databases of digital repositories that would have been created as they are susceptible to unauthorized access, corruption or deletion. The content managers entering the data may also encounter some metadata issues involving inaccurate metadata which can hinder discoverability and access.

Digitization is a costly exercise. A community intending to preserve their indigenous medicinal knowledge would need to conduct a cost-benefit analysis of the project. There are a lot of expensive tools and equipment required such that a community will need to evaluate whether the costs and time involved with be able to cover the costs. Managing digital materials requires frequent updates of software and hardware which are usually a bit costly to acquire and maintain (Balogun, 2023). Balogun (2023) recommended that in developing countries where there are funding challenges, such projects like digitization of indigenous management systems require that funding is sourced through multinationals, international organization or private individuals who are interested in supporting the cause.

The digital divide is another big challenge, especially in developing countries. Access to technology and skills required to operate the ICT gadgets is a significant barrier to many African communities. Digitization can require experts with knowledge in various fields such as website designing and management, database design and management, programming, photography and content or metadata management. However, Balogun (2023) noted that it is not only ICT skills that are necessary in digital preservation. He cautioned that "ICT skills are not enough to prove that an individual has knowledge about issues such as digital preservation" (Balogun, 2023). On the other hand some social media technologies, for example, require some specialized skills or training on how to use them and they will therefore be accessible by a few literate individuals. However, other platforms like YouTube can be used even by illiterate community members as they will need to download and listen to audio and video materials. Technical skills required in the process are also a significant challenge.

CONCLUSION

The coming of the Fourth Industrial Revolution has brought about the urgent need for African communities to digitize and preserve their knowledge systems. The need is also necessitated by the serious misappropriation, loss and lack of transmission or transfer of African indigenous knowledge (Thairu, 2007). Traditional knowledge is disappearing due to communities adopting modern ways of living, lack of preservation strategies and also lack of government policies on the preservation and protection of indigenous medicine. Despite the fact that indigenous knowledge is usually oral, it is imperative that there be some national and community efforts to preserve and make the knowledge accessible. This chapter has explored the role of indigenous medicinal knowledge in African communities. Indigenous knowledge systems have been seen as powerful knowledge among communities that they can use to advance their communities. Medicinal traditional knowledge, which is getting support from international bodies such as WHO is also handy in supporting health programmes of communities. It has been seen in this chapter that there are different medicinal plants and animals that can be used as medicine. However, the major challenge of traditional knowledge is that it is tacit knowledge embedded in the practices and experiences of its holders and usually transferred from generation to generation. The need therefore of digitizing such knowledge has also been discussed. There are various digital tools and equipment that is required. Such tools include computers, scanners, cameras, and software. However, digitization is a very costly exercise (Ezema, 2013) and requires consultation with the local communities to get their consent. It has also been noted that there are challenges with legal (Biyela et al., 2016, Ngulube, 2012) the obsolescence of technology (Masenya and Ngulube, 2019, Balogun, 2023) and ethical and moral issues (Adu and Ngulube, 2017). Nevertheless, digital preservation of indigenous knowledge systems offers solutions for the storing of knowledge ensuring its availability for future generations.

REFERENCES

Adam, A. A., Othman, N., & Halim, A. A. (2022). Indigenous Knowledge Documentation: Perspectives of Dusun and Bajau Communities in Kota Belud, Sabah, Malaysia. *International Journal of Academic Research in Business & Social Sciences, 12*(10), 162–175. DOI: 10.6007/IJARBSS/v12-i10/15205

Adu, K. K., & Ngulube, P. (2017). Key Threats and Challenges to the Preservation of Digital Records of Public Institutions in Ghana. *Information Communication and Society*, *20*(8), 1127–1145.

Ajani, Y. A., Oladokun, B. D., Olarongbe, S. A., Amaechi, M. N., Rabiu, N., & Bashrun, M. T. (2024). Revitalizing indigenous knowledge systems via digital media technologies for sustainability of indigenous languages. *PDT&C*, *53*(1), 35–44. DOI: 10.1515/pdtc-2023-0051

Akinwale, A. A. (2012) Digitization of indigenous knowledge for natural resources management in Africa. *Proceedings of the AERN Summit.* Paper presented at the 20th Anniversary Summit of the African Educational Research Network at North Carolina State University, Raleigh, USA, 19 May, pp.1-18.

Aremu, O. A., & Makunga, N.. (2022). Africa is a treasure of medicinal plants, here are the seven that are popular, In *The Conversation Africa, SAST*

Austin, P. K (2019). Digital Technology and Endangered Indigenous Languages: Challenges and New Opportunities. *Language Documentation and Conservation*, *13*, 356–382.

Balogun, T. (2023). Data management of digitized indigenous knowledge system in repositories. *Information Development*, *39*(3). Advance online publication. DOI: 10.1177/02666669231186777

Bandi, S., Angadi, M., & Shivarama, J. (2015). Best practices in digitization: Planning and workflow processes. In *International Conference on 'Emerging technologies and future of libraries: Issues and challenges'*, Gulbarga, Karnataka, India, 30–31 January 2015, pp. 332–339.

Biyela, N., Oyelude, A., & Haumba, E. (2016) Digital preservation of Indigenous Knowledge (IK) by cultural heritage institutions: a comparative study of Nigeria, South Africa and Uganda. *Paper presented at the SCESCAL Conference*, Ezulwini, Swaziland, 25 - 29 April, 2016.

Chakrabarty, S., & Hossain, K. (2024). Recognising indigenous traditional knowledge with medicinal value within a legal framework: An overview of the issues and challenges with special focus on India. *Man in India*, *104*(1&2).

Chigora, P., Masocha, R., & Mutenheri, F. (2007). The role of indigenous medicinal knowledge in the treatment of ailments in rural Zimbabwe: The case of Mutirikwi Communal lands. *Journal of Sustainable Development in Africa*, *9*(2).

Christian, G. E. (2009). Digitization, intellectual property rights and access to traditional medicine knowledge in developing countries: the Nigerian experience, *A research papers prepared for the International Development Research Centre (IDRC)* Ottawa, Canada. Available at http://idl-bnc.idrc.ca/dspace/handle/10625/41341

Dlamini, P., & Khanyile, N.C. (2021). Preservation of traditional medicinal knowledge: Initiatives and techniques in rural communities in KwaZulu-Natal. *Library Philosophy and Practice (e-journal)*. 4824.

Donaldson, D. R. (2020). Certification information on trustworthy digital repository websites: A content analysis. *PLoS One*, *15*(12), e0242525. PMID: 33296371

Essoungou, A. M. (2011). A social media boom begins in Africa: Using mobile phones, Africans join the global conversation. *Africa Renewal*, *3*, •••. https://www.un.org/africarenewal/magazine/december-2010/social-media-boom-begins-Africa

Ezema, I. J., & Ugwu, C. I. (2013). Electronic theses and dissertations in Nigeria university libraries: Status, challenges and strategies. *The Electronic Library*, *31*(4), 493–507.

Ghimire, (2021). Digitalization of indigenous knowledge in Nepal – Review article. *Acta Informatica Malaysia (AIM)*. 5(2). DOI: .DOI: 10.26480/aim.02.2021.42.47

Gorjestani, N. (2000). Indigenous knowledge for development. In *UNCTAD conference on traditional knowledge* (pp. 1–8). UNCTAD., DOI: 10.1111/an.1999.40.7.20

Greyling, E., & McNulty, N. (2011). *How to build an indigenous digital library through community participation: The case of the Ulwazi Programme.* Available at http://www.academia.edu/837039/How_to_build_an_Indigenous_Digital_Library_through_Community_Participation_the_case_of_the_Ulwazi_Programme

Gurib-Fakim, A. (2006). Medicinal plants: Traditions of yesterday and drugs of tomorrow. *Molecular Aspects of Medicine*, *27*(1), 1–93. DOI: 10.1016/j.mam.2005.07.008 PMID: 16105678

Harisanty, D., Variant Anna, N. E., Putri, T. E., Firdaus, A. A., & Azizi, N. A. A. (2023). Is adopting artificial intelligence in libraries urgency or a buzzword? A systematic literature review. *Journal of Information Science*, •••, 1–12.

Hippolyte, N., & Namaseb, L. (2011). Can IK be Communicated through Foreign Languages without Losing its Authenticity? *IKTC2011: Embracing Indigenous Knowledge Systems in a New Technology Design Paradigm.* ndigenous Knowledge Technology Conference. 2011. Windhoek, Namibia, 2 - 4 November 2011, pp. 116–117.

IFLA. (2002). *Guidelines for digitization projects for collections and holdings in the public domain, particularly those held in libraries and archives.* Available at https://www.ifla.org/VII/s19/sconsv.htm

Jennings, D., Johnson-Jennings, M., & Little, M. (2020). Utilizing webs to share ancestral and intergenerational teachings: The process of co-building an online digital repository in partnership with indigenous communities. *Genealogy*, *4*(70). Advance online publication. DOI: 10.3390/genealogy4030070

Mahomoodally, M. F. (2013). Traditional Medicines in Africa: An Appraisal of Ten Potent African Medicinal Plants. *Evidence-Based Complemetary and Altenative Medicine*, *2013*(1), 1–14. DOI: 10.1155/2013/617459 PMID: 24367388

Makonya, L., (2024). Safeguarding traditional Knowledge in Zimbabwe. *Manica Post*, 08 November 2024.

Maroyi, A. (2013). Traditional Use of Medicinal Plants in South-Central Zimbabwe: Review and Perspectives. *Journal of Ethnobiology and Ethnomedicine*, *9*(31). Advance online publication. DOI: 10.1186/1746-4269-9-31 PMID: 23642285

Masenya, T. M., & Ngulube, P. (2019). Digital preservation practices in academic libraries in South Africa in the wake of the digital revolution. *South African Journal of Information Management*, *21*(1), 1–9.

Masocha, M. (2002). *Identification of medicinal plants used in the treatment of abdominal diseases*. Prince Press.

Mawere, M. (2011). Ethical quandaries in spiritual healing and herbal medicine: A critical analysis of the morality of traditional medicine advertising in southern African urban societies. *The Pan African Medical Journal*, *10*, 1–10. DOI: 10.4314/pamj.v10i0.72212 PMID: 22187588

Mdhluli, T., Mokgoatsana, S., Kugara, S. L., & Vuma, L. (2021). Knowledge management: Preserving, managing and sharing indigenous knowledge through digital library. *HTS Theologies Studies.*, *77*(2), 1–7. DOI: 10.4102/hts.v77i2.6795

Mutula, M. S. (2014). Status of Digital Heritage Preservation Management in Eastern Africa. *The Electronic Library, 32*(3), 363–374.

Ngulube, P. (2012). Ghost in our machines: Preserving public digital information for the sustenance of electronic government in Sub-Sahara Africa. *Mousaion, 30*(2), 128–136.

Owiny, S., Mehta, K., & Maretzki, A. N. (2014). The use of social media technologies to create, preserve, and disseminate indigenous knowledge and skills to communities in East Africa. *International Journal of Communication, 8*, 234–247.

Parekh, H. (2000). *Digitization: An overview of issues.* Available at: https://ir.inflibnet.ac.in/server/api/core/bitstreams/81bdcf00-f3b9-4f63-b3d1-752d33869d4e/content

Perry, S. R. (2014) Digitization and digital preservation: a review of the literature. http://scholarworks.sjsu.edu/cgi/viewcontent.cgi?article=1170&context=slissrj

Sabbagh, K., Roman, F., Bahjat, E., Milind, S., & Sandeep, G. (2012). Maximizing the Impact of Digitization." *The Global Information Technology Report*, pp. 121–33. https://www3.weforum.org/docs/GITR/2012/GITR_Chapter1.11_2012.pdf

Sraku-Lartey, M., Acquah S.B., Samar, S.B., & Djagbletey, G.D. (2017). Digitization of indigenous knowledge on forest foods and medicines. *International Federation of Library Associations and Institutions*, 1(11). .

Thairu, W. (2007). A critical analysis of the impact of digitizing indigenous knowledge. *AAJIS*, 1(001).

Tora, A., & Heliso. (2017). Assessment of the indigenous knowledge and use of traditional medicinal plants in Wolaita Zone, Southern Ethiopia. *International Journal of Medicinal Plants and Natural Products, 3*(1). Advance online publication. DOI: 10.20431/2454-7999.0301003

Torri, M. C. (2010). Increasing knowledge and traditional use of medicinal plants by local communities in Tamil Nadu: Promoting self-reliance at the grassroots level through a community-based entrepreneurship initiative. *Complementary Health Practice Review, 15*(1).

Tredinnick, L. (2017). Artificial intelligence and professional roles. *Business Information Review, 34*(1), 37–41.

Tsekea, S. (2016). The position of intellectual property systems in the protection of indigenous knowledge. *Paper presented at SCECSAL XXII*, Ezulwini, Eswatini.

Twarog, S. (2004). Preserving, Protecting and Promoting Traditional Knowledge: National Actions and International Dimensions. In *Protecting And Promoting Traditional Knowledge: Systems, National Experiences And International Dimensions*, Twarog S, and Kapoor, P. United Nations Conference on Trade and Development: United Nations

WHO. (2008), *Fact sheet N°134.* Available at: https://www.who.int/mediacentre/factsheets/2003/fs134/en/

WHO. (2023), *Traditional Medicine,* Available at https://www.who.int/news-room/questions-and-answers/item/traditional-medicine

Yaga, D., Mell, P., Roby, N., & Scarfone, K. (2019). Blockchain technology overview. Crytography and Securitity. Available at: https://doi.org//arXiv.1906.11078 .DOI: 10.48550

Yirga, G. (2010). Assessment of indigenous knowledge of medicinal plants in central zone of Tigray, North Ethiopia. *African Journal of Plant Science*, *4*(1).

Chapter 7
Preservation of Indigenous Knowledge in Open Science:
Challenges and Opportunities

Elisha Mupaikwa
https://orcid.org/0000-0002-0313-7139
National University of Science and Technology, Bulawayo, Zimbabwe

ABSTRACT

In many areas, open science has created obstacles to indigenous knowledge. The peer-reviewed scholarly works published between 2000 and 2025 are reviewed in this chapter. The review's main conclusions demonstrated that intellectual property rights, knowledge appropriation, and loss of authority over indigenous knowledge are the main risks to indigenous knowledge. Documentation and preservation initiatives sponsored by the community may help overcome these issues. These tactics make use of online databases and tools.

INTRODUCTION

Historically, indigenous knowledge has been essential to communities' sustainable development. According to Ngulube (2023), a large number of academics, researchers, and researchers recognise the importance of IK in society and the necessity of preserving it, especially in fields related to natural resource management. Additionally, it has made it possible for communities to become resilient and adapt to changing environmental and climatic conditions, offered affordable means of obtaining and using locally available resources, empowered them to become

DOI: 10.4018/979-8-3373-0204-1.ch007

self-sufficient and enabled them to contribute to the advancement of scientific knowledge worldwide, guaranteed food security, and made it easier to preserve cultural diversity. These days, the open science movement incorporates this process of worldwide scientific knowledge growth (UNESCO, 2021). By encouraging the sharing of research data, coding, and publication, the open science movement aims to remove obstacles to scientific knowledge and make research more transparent, accessible, and collaborative. The tenets of open science promote cooperation, openness, and the joint creation and administration of knowledge. This has led to the FAIR principles being fulfilled, and in many developing nations where IK has not been thoroughly documented, the open science movement has had a variety of effects on IK, including expanding opportunities for IK accessibility (Ngulube and Chigwada, 2023). This is clear from a number of research and projects that show how open science can increase IK's accessibility to a wider audience, fostering cross-cultural dialogue and understanding (Ngulube, 2023). Furthermore, the internet and digital technical advancements have made it possible to document and preserve IK digitally, assisting indigenous groups in preserving their languages, cultural legacy, and IK for future generations. This open scientific movement has also made it possible for researchers, institutions, and indigenous people to work together and further facilitate knowledge sharing and co-creation (UNESCO, 2021). As a result, indigenous groups now have the chance to preserve and enhance their indigenous knowledge systems by utilising global knowledge systems. This approach promotes increased involvement in research, makes it easier to build on previous findings, and eventually speeds up the scientific method. However, this has led to issues with intellectual property rights, cultural appropriation, and a lack of control. When information, stories, and cultural heritage are shared publicly without the appropriate procedures or protections, indigenous communities may lose control over them. Additionally, political oppression, colonial legacies, and occasionally competing social and political interests have made it difficult for indigenous communities to manage their IK. Fernandez-Llamazares et al. (2023) claim that attempts to incorporate IK into global knowledge systems have led to a significant loss of IK. Such losses cause poverty, eviction, and continuous cultural deterioration by upsetting the base of the people's cultures and means of subsistence. Threats to IK have also surfaced from corporate and governmental interests in several areas (Lay, 2020). Although there have been a number of studies on IK and open science, this chapter aims to expand on previous findings by outlining general dangers to IK, threats arising from the open scientific movement, and recommendations for how different IK preservation techniques could mitigate these vulnerabilities. The chapter also looks at how open science may support initiatives to improve the calibre of IK research without jeopardising the values of data science, IK, research quality, and data quality. Therefore, the purpose of this chapter is to examine the difficulties that indigenous groups encountered when attempting to conserve their knowledge in open research. Researchers and policymakers also face these issues, in addition to indigenous communities. Additionally, the chapter looks at the prospects that the communities' efforts to preserve indigenous knowledge in open science present.

RELATED STUDIES

Threats to IK

The primary issue affecting oral cultures is the loss of traditional information and skills as a result of elders passing away or losing their memory, as well as the intentional or unintentional destruction of indigenous knowledge (Ngulube and Chigwada, 2023). The threats that indigenous groups have faced in their attempts to preserve indigenous knowledge are the main causes of these difficulties. Depending on the industries, these dangers have changed. But in the developing world, these difficulties have predominated. Indigenous cultures and their knowledge systems have been marginalised as a result of cultural homogenization brought about by modernisation and globalisation, which has led to the domination of Western life-styles. The basic erasure of the rich knowledge heritages of non-Western people was one of the effects of colonialism and apartheid on Indigenous knowledge systems in many nations. Indigenous knowledge was initially deemed invalid and non-scientific by colonialists and colonialism. The cultural and intellectual contributions of non-Western knowledge systems were methodically eradicated as part of this "conquest by naturalisation" approach (Catherine Odora Hoppers, 2021). These IKSs were concealed from the general public and left out of the core school curricula. The survival and use of IK have been endangered by the fact that the majority of public institutions, including universities across Africa, including South Africa, are still set up in this manner (Catherine Odora Hoppers, 2021). The loss of indigenous languages, which have traditionally been seen as trustworthy IK conduits, has made the effects worse. Furthermore, the exploitation of numerous indigenous tribes has resulted from the commercialisation of their natural resources. The deterioration of traditional beliefs, theft and misappropriation, and diminishing indigenous value are other issues that threaten IK's continued existence (Fernandez-LIamazares et al, 2021). Therefore, the digital revolution poses a threat to the survival of IK, a vital resource in modern society (Ngulube and Chigwada, 2023). The consequences of IK loss were also divided into three categories by Tang and Gavin (2016): alteration of traditional livelihood practices, alteration of traditional religion and beliefs, and loss of IK transmission pathways. Due to the involvement of foreign advancements and Western technical innovation concepts, these consequences are particularly evident in the decrease in IK transmission in indigenous communities (Malapane, Chanza, and Musakwa, 2024). IK techniques have been rapidly lost as a result of widespread usage of foreign practices, westernisation, and urbanisation (Neitzel et al., 2019; Ferguson et al., 2017). The primary cause of IK's demise, according to Malapane, Chanza, and Musakwa (2024), is oral transmission from one generation to the next; as the older generation dies, IK is more likely to vanish.

IK has been endangered by open science in a number of ways. First of all, cultural appropriation—the taking and use of intellectual property without due understanding, consent, or credit—can result from Open Science's emphasis on sharing and collaboration. When this occurs, indigenous tribes frequently lose control over their cultural history and knowledge, which can result in exploitation or misrepresentation (Ngulube, 2023). Since the open scientific movement emphasises data sharing and reuse rather than considering power imbalances and historical circumstances, it has been observed t,hat it ignores the rights and interests of indigenous people (Ngulube and Chigwada, 2023). Lack of cultural context is another danger facing IK in an open science setting. IK can be misunderstood and misused as a result of open science's tendency to ignore its cultural context and subtleties. Concerns over intellectual property rights have also surfaced. According to Ngulube and Chigwada (2023), the focus on open access in open science may be at odds with the rights and interests of indigenous peoples as well as those of intellectual property. This has a lot to do with the difficulties caused by power disparities that occasionally occur in open science IK research. When dominant groups profit from IK without properly acknowledging or compensating the communities, open research has the potential to maintain current power disparities. To ensure that IK is shared and preserved in a way that respects community rights and interests, it is possible to reduce these issues in a number of ways, including by creating and supporting community-led efforts or community-led open scientific initiatives. Furthermore, creating cultural norms for open research projects can assist in safeguarding IK and guarantee that it is communicated reasonably. Benefit-sharing is an additional mitigating technique. To guarantee that indigenous communities gain from the sharing of their knowledge, benefit-sharing procedures must be put in place.

Digitisation as a Benefit and a Threat to IK

Since digital technology offers creative ideas for recording, archiving, and disseminating this traditional knowledge, it has become crucial to the preservation and dissemination of Indigenous knowledge (IK). Digital technology can be utilised by indigenous groups who depend on oral traditions of their knowledge system to ensure that their cultural heritage is preserved for future generations (Gomez, 2012). Examples of technological tools that can be utilised to facilitate the process of information sharing and protection in a way that is easily accessible and culturally appropriate include digital storytelling platforms, mobile apps, and Geographic Information Systems (GIS).

These days, Indigenous peoples use radio broadcasts via the Internet, digitalised documents, and video conferencing technology. Indigenous history, culture, and human rights activism are primarily preserved and promoted through the employment of these technologies. According to Tesfa (2024), indigenous communities use the Internet for e-mail, chat rooms, radio stations, video conferencing, and basic information gathering through website browsing. Among the resources for representing Indigenous narratives and traditional knowledge are digital storytelling platforms (Shiri et al., 2021). In addition to sharing their tales with large audiences, the platforms enable Indigenous peoples to tell their stories to the world in a variety of media, including texts, audio, and videos.

The use of technology in Indigenous communities is currently the focus of numerous organisations, including educational initiatives that support and cater to the technological requirements of Indigenous peoples. While some institutions are collaborating with indigenous people to create collections, others already have indigenous materials that they are repatriating or attempting to make available to indigenous communities (Tesfa, 2024). Such information is being digitally captured using multimedia technologies. Digital papers are perfect for storing in databases for easy access and retrieval, as well as for sending via networks, such as the Internet. But they also make it easy for traditional knowledge to be misused and accessed illegally (Sen, 2005). In order to enforce traditional laws, protect indigenous heritage from being misused in ways that are culturally inappropriate or insensitive, and ensure that they receive fair compensation for their intellectual and cultural property, traditional owners must be able to define and control their rights and access to their resources (Sen, 2005).

Traditional knowledge (TK) preservation has major challenges in the digital age, especially due to the loss of Indigenous languages, which are the main carriers of oral traditions. Nearly half of the world's scripts, many of which are connected to minority languages, are still not supported by digital platforms, according to UNESCO. This is causing language extinction and intergenerational knowledge transmission to be disrupted (Septika Laily Anti and Ifan Awanda, 2025). These issues are made worse by technological constraints, especially in underdeveloped rural and Indigenous areas. While digital technologies provide preservation tools, TK holders are further isolated because of a “digital divide” caused by limited access to high-speed internet and sophisticated equipment (Septika Laily Anti and Ifan Awanda, 2025). Kugara and Makgoatsana (2022) are among those who have documented the difficulties in digitising IK. They claim that in South Africa, digitising the Vha Venda orakl tradition caused some cultural practices to seem outdated, while other cultural practices were characterised as unfeasible and impracticable, which ultimately resulted in the extinction of the cultural practices and values. According to Kugara and Makgoatsana's (2022) study, the digitalisation of Vha Venda IK was viewed as political and discriminatory due to the exclusion of indigenous people and the disrespect for their input. Incompatibilities between IK and other digital platforms have also arisen as a result of initiatives to digitise IK, which has slowed language extinction and interfered with the transmission of information between generations (Anti and Awanda, 2025).

The use of digital technologies for IK preservation has presented a number of difficulties, despite the quick development of these tools and the potential they present for IK preservation in an open research setting. The difficulties of digital exclusion are a few of them. Indigenous knowledge is still underrepresented in the mainstream discourse surrounding information technology (IT) and information systems (IS), even with the global growth of IT applications. Maintaining a rich cultural legacy while promoting digital fairness, social justice, and environmental sustainability can be achieved by bridging IT and Indigenous knowledge systems (Bawack et al., 2025).

Open Science Benefits to Indigenous Knowledge

Many of the open science practices are consistent with the transparent and cooperative methods employed in indigenous research and research on indigenous knowledge systems, even though open science is seen as a radical change in the ways that mainstream researchers create and share knowledge. Indigenous community members, for instance, can now improve their comprehension of their natural surroundings and fortify ties to Indigenous language, traditional knowledge, and customs through cooperative projects (Atalay, 2020). Furthermore, indigenous people are now more visible as a result of the use of open science in IK research and the field as a whole (Tapfuma and Hoskins, 2016). IK's visibility and recognition have increased thanks to the platform that open science has given it to reach a wider audience. IK can now be shared and used in new situations thanks to the worldwide sharing and collaboration made possible by this (Ocholla, 2025). Open Access articles are available to a larger audience, including researchers from developing nations who might not have access to subscription-based journals, which contributes to their greater visibility (Piwowar et al., 2018). Consequently, Open Access research helps to foster a more inclusive and varied scholarly discourse. By allowing academics from various institutional and geographic backgrounds to easily access and interact with one another's work, Open Access promotes increased collaboration (Lui et al., 2021). This free access encourages multidisciplinary research and collaborations that go beyond conventional lines, resulting in creative answers to global problems

(Gutam, 2024). Open access publishing is essential to traditional medicine since it makes study findings available to the public and other scholars (Pampel, 2023). By making research publicly available, open-access publications foster inclusion and transparency, which eventually benefits patients, medical professionals, and the larger scientific community. Posting manuscripts on preprint servers or in university repositories, as long as the authors make sure that everything included in the repository satisfies FAIR requirements, is another method to make research publicly accessible outside of publications (Ng et al., 2024).

Innovations in documentation and data preservation brought about by the open scientific movement have also benefited the larger field of IK and contributed to the protection of cultural assets. Communities' self-determination has been strengthened and made easier by the protection of cultural heritage (Pace, 2021). This promotes autonomy and self-determination by empowering communities to take charge of their knowledge. Innovations and applications in data and research management based on IK have been made possible by the adoption of open science principles in IK (Armeni et al., 2021). Increased accessibility of open-access platforms has allowed indigenous researchers and communities to publish their information in open-access journals, expanding their reach and influence, among other advantages of IK's adoption of open science. Additionally, this encourages a more informed and scientifically literate society by empowering educators, policymakers, practitioners, and the general public to interact with scientific research, supporting evidence-based practices, accountability, and transparency in research, and fostering informed decision-making (Suber, 2012; Willinsky, 2006).

For community-based research to have a greater impact, Open Access is essential. In their study on schistosomiasis communication in Uganda, Anyolitho et al. (2024) demonstrated how citizen science (CS) and participatory action research (PAR) enabled local communities to co-design health campaigns. In addition to improving knowledge dissemination, the effort promoted disease prevention behaviour modifications by directly engaging communities in the study process and making the results available. The outcomes of such research are guaranteed to reach a wider audience through open access, allowing comparable efforts in other areas to profit from the methods and conclusions. Similarly, by working together to document endangered languages, Gutam (2024) illustrated the value of empowering local communities in linguistic preservation. Both community members and scholars can profit from the project's results thanks to open access to linguistic materials like the Ekegusii dictionary, which guarantees the language's preservation and keeps it relevant in scholarly studies. Decolonising knowledge systems has been made possible by community empowerment, which has led to a more inclusive and varied worldview. The creation of community-led projects has also been made possible by this, giving indigenous groups the ability to direct the dissemination and use of their knowledge. Open access in engineering and technology makes it easier to convert scholarly research into workable solutions that have a direct positive impact on communities. The importance of research translation (RT) in engineering education is emphasised by Gutam (2024), who notes that Open Access to graduate research allows undergraduates and community members to gain access to state-of-the-art findings. This collaborative model places a strong emphasis on working together to co-create solutions for pressing problems, especially in community-driven initiatives like environmental risk assessment and waste recycling.

IK Preservation in Open Science

Indigenous knowledge systems (IKS) management and preservation have long been important issues for many societies worldwide. Various methods have been incorporated into IK preservation efforts, especially in the context of open science. According to the literature, community involvement, cultural sensitivity, monitoring and upholding the intellectual property rights of communities and IK holders, benefit sharing, and IK documentation and preservation are some of the tactics that can be employed to maintain IK in open research. Developing community-led projects, which are frequently started and run by members of the indigenous community, is one of the factors that must be taken into account when putting these techniques into practice. Establishing explicit guidelines for the exchange and application of indigenous knowledge is another important consideration. To ensure balanced power dynamics in indigenous communities during research, it is also essential to adopt open access methods that respect indigenous ownership and control of knowledge assets and to provide training and capacity building for indigenous people.

One of the strategies proposed by Christian (2009) for the successful and accessible preservation of IK is documentation and digital preservation; communities may gain the most from this once it is incorporated into open science. This is especially true when considering land use and medical expertise. Furthermore, according to Christian (2009), the digitisation and documentation of such traditional medicine knowledge is useful for defensive protection against expropriation by multinational pharmaceutical companies without remuneration and biopiracy. This is because documentation shows who owns and where information sources come from (Christian, 2009). Digital technology is the most modern method of conserving indigenous knowledge for future generations, even though oral tradition is still the main source of dissemination (Gomez, 2012). Ngulube (2023) contends that although digitisation is not a solution for all issues related to information access and storage, IK may gain a great deal from digitisation for preservation purposes, as demonstrated by a number of African efforts. Indigenous knowledge systems have been digitalised, opening up new avenues for their preservation, exchange, and application (Bologun,

2023). Digital preservation guarantees that IK is kept for as long as feasible and aids in its preservation and accessibility (Bologun, 2023). Information technology (IT) can be used to record, manage, and disseminate Indigenous knowledge, which is acknowledged alongside the growing appreciation of its importance and relevance to the modern world and the need to protect it (Hunter, 2005). In a setting where open science has put these essential resources in danger of going extinct, digital technologies can be extremely helpful in preserving IK. Digital archiving, or the creation of digital archives for IK, including text, photographs, and video, can potentially lower the risks of losing IK for many communities when resources allow. This has proven crucial for some nations, such as South Africa, due to the nation's numerous cultures and rich cultural legacy. In addition to digital preservation, online knowledge platforms for IK sharing and collaboration might be set up to facilitate IK development without data loss. In the current digital landscape of blockchain and cloud computing, collaborative research has become much easier. Metadata standards have helped many societies that have adopted digitisation as a preservation strategy by providing an intellectual structure and aiding in the interpretation of digitised content (Ngulube, 2023). To do this, more modern mobile documenting and sharing applications have been created. In a research on the management of digitised IK data in South African repositories, Bologun (2023) noted that community coordinators' involvement before data uploading guaranteed data integrity and correctness and prevented external intervention. Therefore, acquiring, conserving, and sharing indigenous knowledge in the digital age would aid in the preservation of cultural heritage because Indigenous knowledge is the wisdom and knowledge that generations of people have accumulated over time, which supports the preservation of cultural identity and heritage (Ngulube and Chigwada, 2023).

Numerous digital technologies are being investigated as a means of conserving indigenous heritage and knowledge. This is due to indigenous people's growing awareness of the potential advantages digital technologies might provide for their historical and cultural documentation and preservation. These groups want to employ new rights management technology as they become more aware of the potential for knowledge exploitation and infringement that comes with digitisation (Tesfa, 2024). For instance, stories, songs, dances, and other facets of oral-visual culture are being recorded on audio and video. In order to preserve or return the original artefacts to their native owners, museums are using 3D printers to duplicate them. A digital library interface can be used to access a compilation of these knowledge objects. Indigenous peoples are currently using technologies including online radio broadcasts, digitalised documentation, and video conferencing. Indigenous culture, custom, history, and human rights activism are primarily preserved and promoted through the employment of these technologies (Tesfa, 2024). The establishment of databases for conventional knowledge is another strategy that has gained popularity recently and should prove helpful in future intellectual property disputes (Adefila, Oladokun and Adewojo, 2024). However, due to the significant differences in the creation process, it is challenging to harmonise traditional knowledge with Western patent law (Tesfa, 2024). In addition to databases, indigenous tribes use the Internet for chat rooms, radio stations, video conferencing, email, and basic information gathering through website browsing. Additionally, user-generated material can be produced and shared by communities and individuals thanks to social media tools. Information, including different types of IK or skills, can be created, accessed, and shared inside social and geographic communities as well as among broader audiences using social media platforms like YouTube, Facebook, Google Docs, and Twitter. Users can upload, share, and watch videos on YouTube. Because YouTube is multimedia-based, both literate and illiterate persons can use it. In a library setting, local populations can see recorded recordings of indigenous knowledge (dancing, music, and agricultural practices) that have been published on YouTube. People can interact with others, discuss interests, publish videos, and join groups that share their interests on Facebook. They're all used to share IK and preserve it for future generations (Tesfa, 2024). According to Stevens and Stevens (2008), some institutions are already in possession of indigenous materials that they are repatriating or attempting to make available to indigenous people, while others are establishing collections in collaboration with indigenous groups. IK is also preserved via libraries, museums, archives, Facebook, and websites. When preserving IK using digital technologies in an open science environment, best practices are encouraged. From the literature, some of the best practices include prioritising community involvement, control and cultural sensitivity. Cultural sensitivity refers to the development or adoption of digital platforms to meet the specific cultural and linguistic needs of indigenous communities, incorporating features like culturally relevant metadata and access controls. It is also vital to ensure digital representation of IK capture in its richness and complexity, incorporating multimedia elements like videos, photographs and interactive maps (Siliutina et al., 2024). Furthermore, it is vital to establish platforms for data governance and sovereignty that recognise indigenous communities' rights to govern their data, including ownership, control, access and processing.

Numerous opportunities exist in the larger endeavours to maintain IK and cultural heritage in open science, as evidenced by all of this literature. Several studies have examined a range of topics, such as:

a) Digital preservation frameworks: Numerous frameworks for the sustainable digital preservation of cultural material have been developed by researchers, with a focus on eco-sufficiency and minimal environmental repercussions. A robust sustainability framework for digital preservation, for instance, has been created to minimise the environmental effects of ICT while optimising preservation potential.

b) Technologies for preservation: In order to protect and promote cultural heritage, open science programs have made use of technologies such as 3D visualisation, 3D modelling, augmented reality, and virtual reality. Documentation, study, and repair of cultural relics are made possible by these technologies (Siliutina et al., 2024).
c) Nanotechnology applications: In cultural heritage conservation, nanomaterials have been used to create novel, non-invasive methods of artefact preservation. Studies demonstrate how nanoparticles can be used to preserve and combine cultural heritage items.
d) Community engagement: In order to ensure that local knowledge and viewpoints are incorporated into preservation efforts, studies highlight the significance of community involvement in cultural preservation.
e) Community-led digital archives: Indigenous communities can create and manage their digital archives, ensuring control over their knowledge and cultural heritage.
f) Open access collaboration: Cultural heritage protection and promotion are made easier by open science projects that encourage cooperation and knowledge exchange among researchers, institutions, and communities.
g) Digital storytelling: Digital platforms can be used to share indigenous stories, traditions and knowledge, promoting cultural exchanges (Nappi, et al., 2024).
h) Language documentation: Digital technologies can help document and preserve indigenous languages, supporting language revitalisation efforts.
i) Collaborative research: Open science can facilitate collaborative research between indigenous communities and researchers, promoting co-creation and sharing of knowledge.
j) Digital repatriation: Digital technologies can facilitate the return of cultural heritage materials to indigenous communities, promoting cultural revitalisation and preservation(Yiyun, 2024).).

Ethics in IK Management in Open Science

Regarding current and future Indigenous cultural materials under their care, collecting institutions around the world have had to deal with new ethical conflicts and practical difficulties brought on by the quick development of digital technologies and the recent "digital turn" in archival science and information sciences (de Souza et al., 2016). For both present practice and future approaches, this presents a unique mix of legal, cultural, technical, institutional, and epistemological issues (de Souza et al., 2016). Humans' reliance on technology to process and store information abstractly has led to these issues, which have also been extended to information management

(IK). In addition to the well-known difficulties in managing IK, the open science movement has further raised ethical questions for IK research, endangering the legacy of indigenous communities. In their study on the ethical integration of indigenous and scientific knowledge systems for a sustainable mitigation of environmental and climate change challenges, Chanda and Alfred (2025) encountered many ethical issues, such as IKS proprietary rights, the loss and distortion of traditional, cultural, and religious beliefs, and unequal recognition of IKS.

From research, some of the ethical considerations that must be considered in IK research in open science include:

i) Ensuring communities are fully informed and voluntarily agree to any research and utilisation of their knowledge.
ii) Cultural respect and sensitivity: recognising the cultural background and refraining from misrepresentation or appropriation.
iii) Intellectual property rights: honouring community ownership and looking for defence strategies outside of the Western patent system (Okemwa, 2014).
iv) Equitable benefit sharing: putting in place an equitable system that allows communities to profit from the usage and commercialisation of their knowledge.
v) Power dynamics in research: establishing fair collaborations and guaranteeing community involvement in the study.
vi) Representation and voice: encouraging direct involvement in decision-making and self-representation.

These opinions are corroborated by Okemwa (2014), who noted that controversial topics like knowledge appropriation and collective vs individual ownership of information have persisted in Sub-Saharan Africa. According to Okwemwa (2014), the invasion of contemporary lifestyles and the destruction of traditional ways of life are two more concerns facing IK that affect the community's capacity to transmit knowledge to coming generations. The present IP system, which is intended to protect innovations and works by individuals or companies, particularly those relating to technology or pharmaceuticals, does not readily protect IK because of its living nature. IK-based innovations could be protected as trade secrets, patents, trademarks, or geographical indications (Paul, 2023).

CASE STUDIES

There are examples of cooperation between academics, archivists, and indigenous groups around the world to conserve and disseminate information. Powell (2016) provides instances of how the American Philosophical Society (APS) established the

Digital Knowledge Sharing program, which connected four indigenous groups: the Tuscarora Nation, the Penobscot Nation, the Anishinaabe (Ojibwe) communities in the US and Canada, and the Eastern Band of Cherokee Indians (EBCI). This project made it possible for these groups to exchange knowledge digitally.

Numerous nations have endeavoured to save their indigenous knowledge, leading to the emergence of various frameworks from these nations. For instance, the National Indigenous Knowledge System Policy is one of the most important policy initiatives in South Africa (Balogun, 2023). The duties of the government, communities, and other participants in this process are delineated in this policy framework, which offers a thorough approach to the management and preservation of Indigenous knowledge systems in South Africa. The Protection, Promotion, Development, and Management of Indigenous Knowledge Bill was also introduced by the South African government in 2016. It was enacted as the Protection, Promotion, Development, and Management of Indigenous Knowledge Act, 2019 by Act No. 6 of 2019 (Ncube, 2022; Bologun, 2023). In a report for the ARC LP, de Souza et al. (2016) documented additional instances of IK documentation using digital technologies among Aborigine communities in Victoria, Australia. They acknowledged the growing use of digital technologies to give access to a variety of information stored in collecting institutions that specifically pertains to Indigenous people, their culture, and heritage. The "Ara Irititja" initiative in Australia is one example of a project mentioned by Khan and Gope (2025) that demonstrates the effectiveness of digital archives in conserving and disseminating Indigenous Australian cultural heritage, rituals, and traditions while guaranteeing their availability for future generations. In a similar vein, Khan and Gope (2025) mention a few Kenyan programs that document Indigenous agricultural knowledge using mobile platforms, assisting farmers in increasing crop resistance to climate change.

Sen (2005) states that several initiatives are in place in India to promote and document IK. Government interest in indigenous medical systems, particularly Ayurveda, has been growing. One of the top priorities is the protection of medicinal plants that are used in agriculture and the medication manufacturing sector (Sen, 2005). The Traditional Knowledge Digital Library (TKDL), which houses publicly available data on significant medicinal plants, is another example from India. Several other projects on the digitisation of IK have been explored, and some of the examples are:

a) Māori knowledge in New Zealand: Numerous writers have examined various initiatives that have been conducted for the preservation and storage of Māori language, culture, and tradition through the use of digital platforms, including web archives (Ka'ai-Mahuta, 2012, 2019; Sharma et al., 2024).

b) Native American Language preservation: These programs aim to create digital preservation projects, such as the First People's Culture Council language (Lauter, 2023).
c) Australian Aboriginal Digital Archives: Aboriginal and Torres Strait Islander history and knowledge systems have been shared and preserved through these efforts (Wise and Kostelecky, 2018).
d) Inuit Digital Strategy: These projects look at how the Inuit people in Greenland and Canada use digital media to disseminate and preserve their cultural knowledge and customs (Chan et al., 2020).
e) The San Digital Archive (South Africa): The purpose of these activities was to communicate and preserve the San people's knowledge, customs, and stories (Ngulube and Chigwada, 2023; The Department of Science and Technology, nd).
f) Yoruba Ifa Digital Library (Nigeria): To preserve and disseminate Yoruba Ifa knowledge, customs, and cultural legacy, these projects created digital libraries (Elijah and Taiye, 2025).
g) Kenya's Indigenous knowledge digital repository: This project investigates the potential use of digital technology for the preservation and dissemination of IK from diverse Kenyan communities (Kiplagat, Chepken and Kiptoo, 2024).
h) African Storybook Project: This project encourages literacy and cultural preservation by producing and disseminating digital stories in African languages (Stranger-Johannessen et al., 2018).

CONCLUSION

According to the review, indigenous knowledge is essential to community development and the preservation of indigenous communities' well-being. However, the survival of indigenous knowledge and its transmission to future generations are currently threatened by modernisation, new scientific and technological advancements, and global research methodologies like the rise of open science movements. The assessment also demonstrated how digital technologies, particularly databases and online services, have enormous promise for preserving, documenting, storing, and sharing indigenous knowledge. However, the use of these technologies has once more led to issues with data quality, intellectual property rights, indigenous knowledge appropriation, and the loss of sovereignty over indigenous knowledge among indigenous communities. The assessment acknowledges these difficulties and suggests community-led efforts to preserve and document indigenous knowledge.

REFERENCES

Adefila, E., Oladokun, B. D., & Adewojo, A. A. (2024). Beginning of a new era: Leveraging digital twin technology in the preservation of indigenous knowledge systems in the disruptive era. *Library Hi Tech News*. Advance online publication. DOI: 10.1108/LHTN-05-2024-0088

Anti. S.L., and Awanda, I. (2025). Preserving traditional knowledge in the digital era: challenges and strategies. *Journal for Social Science and Society Studies.*

Armeni, K., Brinkman, L., Carlsson, R., Eerland, A., Fijten, R., Fondberg, R., Heininga, V. E., Heunis, S., Koh, W. Q., Masselink, M., & Moran, N. (2021). ´ Baoill, A, O.M., Sarafoglou, A., Schettino, Schwamm, H., Sjoerds, Z., Teperek, M., van den Akker, O.R., van't Veer, A., and Zurita-Milla, R. (2021). Towards wide-scale adoption of open science practices: The role of open science communities. *Science & Public Policy*, *48*(5), 605–611. DOI: 10.1093/scipol/scab039

Atalay, S. (2020). Indigenous Science for a World in Crisis. *Public Archaeology*, *19*(1–4), 37–52.

Balogun, T. (2023). Data management of digitised indigenous knowledge systems in repositories. *Information Development*, *39*(3), 425–438.

Bawack, R., Roderick, S., Badhrus, A. M., Dennehy, D., & Corbett, J. (2025). Indigenous knowledge and information technology for sustainable development. *Information Technology for Development*, *31*(2), 233–250. DOI: 10.1080/02681102.2025.2472495

Chan, E., Bourgeois-Doyle, D., Donaldson, M., & Haine-Bennett, E. (2020). *Toward a UNESCO Recommendation on Open Science: Canadian Perspectives*. UNESCO.

Chanda, A., & Alfred, K. (2025). Ethical Integration of Indigenous and Scientific Knowledge Systems for a Sustainable Mitigation of Environmental and Climate Change Challenges. *International Journal of Global Community*, *VIII*(1).

Christian, G. E. (2009). *Digitisation, intellectual property, rights and access to traditional medicine knowledge in developing countries-The Nigerian Experience. A research paper prepared for the International Development Research Centre*. IDRC.

De Souza, P., Edmonds, F., McQuire, S., Evans, M., & Chenhall, R. (2016). Aboriginal Knowledge, Digital Technologies and Cultural Collections Policy, Protocols, Practice. Melbourne Networked Society Institute Research Paper 4.

Elijah, I. S., & Mary Taiye, D. M. (2025). The Significance of The Digitalization Of Ifa in the Committee of Knowledge. *Cognizance Journal of Multidisciplinary Studies*, *5*(2), 208–216.

Ferguson, M., Brown, C., Georga, C., Miles, E., Wilson, A., & Brimblecombe, J. (2017). Traditional food availability and consumption in remote Aboriginal communities in the Northern Territory, Australia. *Australian and New Zealand Journal of Public Health*, *41*(3), 294–298. PMID: 28370899

Gomez, R. (2012). *Libraries, Telecentres, Cybercafés and Public Access to ICT: International Comparisons*. IGI Global., DOI: 10.4018/978-1-60960-771-5

Gutam, S. (2024). Empowering Research Communities: Open Access as a Catalyst for a Sustainable Knowledge Society. Researchgate.

Hoppers, C. O. (2021). Research on Indigenous knowledge systems: The search for cognitive justice. *International Journal of Lifelong Education*, *40*(4), 310–327. DOI: 10.1080/02601370.2021.1966109

Hunter, J. (2005). The Role of Information Technologies in Indigenous Knowledge Management. *Australian Academic and Research Libraries*, *36*(2), 109–124. DOI: 10.1080/00048623.2005.10721252

Ka'ai-Mahuta, R. (2012). Digital Technology: Contemporary methods for the Dissemination of Ancient Knowledge. The Use of Digital Technology in the Preservation of Māori Song. *Te Kaharoa. 5(2012), Special Edition, 2012, ISSN 1178-6035.*

Ka'ai-Mahuta, R. (2019). Preserving Indigenous voices: Web archiving in Aotearoa/New Zealand. *Interaction Design and Architecture(s) Journal - IxD&A, 41, (2019): 24 – 30.*

Khan, M. and Gope, L. (2025). Aligning Indigenous Knowledge with Digital Technology for Achieving Sustainable Development Goals.

Kiplagat, J., Chepken, C. and Kiptoo, C. (2024). Digital Repository for African History Culture and Heritage: A Case of Keiyo Community of Kenya. *iPRES 2024 Proceedings. iPRES 2024.*

Kugara, S. L. and Mokgoatsana, S.(2022). Challenges presented by the digitization of the Vha Venda oral tradition: An African Indigenous Knowledge perspective: HTS Teologies Studies.

Lauter, O. (2023). Challenges in combining Indigenous and scientific knowledge in the Arctic. *Polar Geography*, *46*(1), 62–74. DOI: 10.1080/1088937X.2023.2233578

Lui, P.O., Skewes, M.C., Gobrial, S., and Rollock, D. (2021). Advancing Transparency and Impact of Research and Impact of Research: Initiating ch: Initiating Crosstalk between Indigenous Research and Mainstream "Open Science". *Journal of Indigenous. .9(2021).*

Malapane, O. L., Chanza, N., & Musakwa, W. (2024). Transmission of indigenous knowledge systems under changing landscapes within the Vhavenda community, South Africa. *Environmental Science & Policy*, *161*, 103861.

Nappi, M. L., Buono, M., Chivăran, C., & Giusto, R. M. (2024). Models and tools for the digital organisation of knowledge: Accessible and adaptive narratives for cultural heritage. *Heritage Science*, *2024*(12), 112. DOI: 10.1186/s40494-024-01219-z

Ncube, B. (2022). Indigenous knowledge perspectives on water management and its challenges in South Africa. In *Current Directions in Water Scarcity Research (4)* (pp. 227–242). Elsevier.

Neitzel, A. L., Smalls, B. L., Walker, R. J., Dawson, A. Z., Campbell, J. A., & Egede, L. E. (2019). Examination of dietary habits among the indigenous Kuna Indians of Panama. *Nutrition Journal*, *18*(1), 44. DOI: 10.1186/s12937-019-0469-8 PMID: 31370836

Ng, J. Y., Wieland, L. S., Lee, M. S., Liu, J., Witt, C. M., Moher, D., & Cramer, H. (2024). Open science practices in traditional, complementary, and integrative medicine research: A path to enhanced transparency and collaboration. *Integrative Medicine Research*, *13*, 101047. PMID: 38799120

Ngulube, P. (2023). From Orality to Digital Assets: Managing Indigenous Knowledge in Africa in the Wake of the Open Science Movement. *The Serials Librarian*, *84*(5–8), 71–85. DOI: 10.1080/0361526X.2023.2296880

Ngulube, P., & Chigwada, J. (2023). Stakeholders in the acquisition, preservation and dissemination of indigenous knowledge projects. *Information Development*, •••, 1–18.

Ocholla, D. N. (2025). The complex marriage: interfacing open science and open indigenous knowledge. *Library Management. 46(3/4): 252-267. Emerald Publishing Limited e-ISSN: 1758-7921 p-ISSN: 0143-5124 DOI*DOI: 10.1108/LM-11-2024-0125

Ondari-Okemwa, E. (2014). Ethical Issues and Indigenous Knowledge Production and Use in Sub-Saharan Africa in the 21st Century. *Mediterranean Journal of Social Sciences*, *5*(23).

Pace, G. (2021). *Heritage Conservation and Community Empowerment. Tools for Living Labs. In book: Underground Built Heritage Valorisation: A Handbook.* CNR Edizioni.

Pampel, H. (2023). Promoting Open Access in Research-Performing Organisations: Spheres of Activity, Challenges, and Future Action Areas. *Publications*, *2023*(11), 44. DOI: 10.3390/ publications11030044

Paul, J. (2023). Traditional Knowledge Protection and Digitisation: A Critical Decolonial Discourse Analysis. *International Journal for the Semiotics of Law*, *2023*(36), 2133–2215. DOI: 10.1007/s11196-023-09989-8

Piwowar, H., Priem, J., Larivière, V., Alperin, J. P., Matthias, L., Norlander, B., & Haustein, S. (2018). The state of Open Access: A large-scale analysis of the prevalence and impact of Open Access articles. *PeerJ*, *6*, e4375. DOI: 10.7717/ peerj.4375 PMID: 29456894

Powell, T. B. (2016). Digital Knowledge Sharing: Forging Partnerships between Scholars, Archives, and Indigenous Communities. *Museum Anthropology*.

Sen, B. (2005). Indigenous knowledge for development: Bringing research and practice together. *The International Information & Library Review*, *2005*(37), 375–382.

Sharma, P., Martin, M., Swanlund, D., & Latham, C. (2024). Anderson, D and Waitangi Wood, W. (2023). A cloud-based solution for trustless indigenous data sovereignty: Protecting Māori biodiversity management data in Aotearoa New Zealand. *Transactions in GIS*, *28*, 836–857.

Shiri, A., Howard, D., & Farnel, S. (2021). Indigenous Digital Storytelling for Cultural Heritage Access and Preservation. *Proceedings of the Annual Conference of CAIS Actes Du Congrès Annuel De L'ACSI*. DOI: DOI: 10.29173/cais1202

Siliutina, I., Tytar, O., Barbash, M., Petrenko, N., & Yepyk, L. (2024). Cultural preservation and digital heritage: Challenges and opportunities. *Amazonia Investiga*, *13*(75), 262–273. DOI: 10.34069/AI/2024.75.03.22

Stevens, A. and Stevens, A. (2008). A Different Way of Knowing: Tools and Strategies for Managing Indigenous Knowledge. 58(2008): 25–33, 2008.

Stranger-Johannessen, E., Doherty, L., & Norton, B. (2018). The African Storybook and Storybooks Canada: Digital stories for linguistically diverse children.

Suber, P. (2012). *Open Access*. MIT Press., DOI: 10.7551/mitpress/9286.001.0001

Tang, R., & Gavin, M. C. (2016). A Classification of Threats to Traditional Ecological Knowledge and Conservation Responses. *Conservation & Society*, *14*(1), 57–70.

Tapfuma, M. and Hoskins, R. (2016). Visibility and accessibility of indigenous knowledge on open-access institutional repositories at universities in Africa. .DOI: 10.4018/978-1-5225-0833-5.ch011

Tesfa, S. (2024). *Preservation of Indigenous knowledge and appropriate technologies. ResearchGate. The Department of Science and Technology. (nd).* Indigenous Knowledge Systems. Republic of South Africa.

UNESCO. (2021). UNESCO Recommendation on Open Science. The General Conference of the United Nations Educational, Scientific and Cultural Organization (UNESCO), meeting in Paris, from 9 to 24 November 2021, at its 41st session. UNESCO.

Wise, M., & Kostelecky, S. R. (2018). Respecting the language: Digitising Native American language materials. *Digital Library Perspectives, 34*(3).

Yiyun, Z. B. A. (2024). Museum digital repatriation and case studies: exploring Guidelines for the future practice and digital bridge to cultural continuity. Master of Arts in Communication, Culture, and Technology Thesis. Graduate School of Arts and Sciences of Georgetown University. Washington, D.C.

KEY TERMS AND DEFINITIONS

Cultural heritage: This encompasses both tangible and intangible aspects of a group or society that are passed down through generations, representing their history, traditions, and values.

Digital archive: a system for storing and preserving digital information, similar to a physical archive but using electronic means.

Digital preservation: the set of processes and actions needed to ensure that digital information remains accessible and usable over the long term, despite technological changes and potential data loss.

Indigenous community: a group of people who are the original inhabitants of a particular geographic region, often predating colonisation or the establishment of current state boundaries.

Indigenous knowledge preservation: the active safeguarding and revitalisation of the unique, place-based knowledge, practices, and cultural heritage of indigenous peoples.

Indigenous knowledge: the understandings, skills, and philosophies developed by societies with long histories of interacting with their natural surroundings.

Open science: Open science is a movement that promotes transparency and accessibility in scientific research.

Traditional knowledge: a dynamic and evolving system of knowledge, know-how, practices, and innovations developed, sustained, and passed down through generations within a community, often deeply intertwined with its cultural and spiritual identity.

Chapter 8
A Wake-Up Call After a Disaster:
Initiating Archiving Household Documents Through a Digital Repository

Muhammad Rosyihan Hendrawan
https://orcid.org/0000-0003-0438-2571
University of Brawijaya, Malang, Indonesia

Aniesa Samira Bafadhal
https://orcid.org/0000-0002-5666-0660
University of Brawijaya, Malang, Indonesia

ABSTRACT

This chapter examines the development of a digital repository prototype by the Universitas Brawijaya to help effectively manage household records in Batu City, the Republic of Indonesia, which is prone to recurrent natural disasters. The 2021 flash flood that ravaged the area highlights the need for electronic records management to guarantee the accessibility and preservation of essential information during disasters. The digital repository enables people to securely save, access, and disseminate vital records, thereby improving disaster resilience and administrative efficacy. Notwithstanding its advantages, obstacles such as inadequate digital literacy and Internet accessibility endure, requiring extensive governmental programs to bolster infrastructure, augment digital literacy, and safeguard data security. The study promotes the extensive use of digital repositories to enhance community resilience to natural disasters, perhaps acting as a prototype for analogous projects in disaster-prone regions. A future study is advocated to enhance these initiatives and broaden their use.

DOI: 10.4018/979-8-3373-0204-1.ch008

INTRODUCTION

In disaster-prone regions, records are more than administrative tools—they are instruments of continuity, resilience, and legal identity. Household records such as birth certificates, family cards, property titles, and health documents are essential for families to access public services, prove ownership, and secure post-disaster assistance. The role of records management, therefore, extends beyond organizational efficiency to encompass disaster preparedness and social recovery. Effective records management regulates the lifecycle of records—creation, storage, access, and destruction—ensuring that information remains accessible, reliable, and legally compliant (Franks, 2013; Saffady, 2021). In this regard, robust systems mitigate the risk of data loss and ensure that information continuity can be maintained even under crisis conditions (Smallwood, 2019).

A well-functioning records management system also contributes to risk reduction by enabling quick retrieval and verification of vital information. It supports informed decision-making, enhances transparency, and fosters public trust, particularly during emergencies. Haddad et al. (2022) argue that strong recordkeeping infrastructures can expedite recovery operations by ensuring data integrity, reducing administrative friction, and securing legal identities. Furthermore, effective systems reduce storage costs, extend the lifespan of information, and provide audit trails for accountability (Dikopoulou & Mihiotis, 2012). From a disaster recovery perspective, the availability of secure, retrievable records is not merely operationally advantageous, it is essential for individual and community resilience.

Indonesia, a nation situated within the Pacific Ring of Fire, is frequently affected by natural disasters including floods, earthquakes, and volcanic eruptions. These recurrent events regularly result in the loss or degradation of physical records, impeding both individual recovery and state response. One significant example is the flash flood that struck Batu City in East Java Province on November 4, 2021. The disaster affected six villages, caused seven fatalities, displaced 89 households, damaged 35 homes, and submerged 33 others under thick mud. Notably, it compromised numerous family archives and administrative records, delaying the reissuance of essential documentation.

Figure 1. One of the areas affected by flash floods

Source: ANTARA News (2021)

The event severely impacted the Bumiaji and Junrejo districts, where improper land use and the conversion of catchment areas exacerbated the effects of extreme rainfall (Putra et al., 2022). As climate change intensifies and urban development continues without ecological planning, such flash floods are expected to increase in both frequency and intensity. The Batu City case highlights how physical archives are vulnerable to environmental hazards and underscores the need for digital solutions that ensure continuity of access and integrity of information.

In response, the Department of Library and Information Science and Universitas Brawijaya Archives developed SIMARTA, a digital repository prototype specifically designed to manage household records electronically. SIMARTA allows families to upload and access key documents, such as identification cards and land titles, via secure digital platforms. It is accessible through computers, tablets, or smartphones, and features encrypted storage, user-controlled access rights, and disaster-resilient recovery mechanisms.

This chapter posits that digital repositories, exemplified by SIMARTA, are essential components of disaster resilience infrastructure. By safeguarding household documentation, they not only preserve administrative functionality but also support emotional and psychological recovery in the wake of disasters. However, the effective implementation of such systems is not without challenges: digital literacy gaps, infrastructure constraints, and issues of data trust must be addressed for equitable access and community-wide benefit. Focusing on the Batu City flood, this chapter investigates how SIMARTA functioned during the crisis and analyzes the broader implications for integrating digital repositories into Indonesia's disaster governance framework.

In doing so, the chapter contributes to the discourse on digital resilience by demonstrating that digital repositories are not simply technical tools but socio-technical systems that intersect with issues of governance, community trust, and cultural adaptation. The chapter also advocates for inclusive, scalable, and secure digital recordkeeping policies that can withstand environmental disruptions. SIMARTA is presented as a prototype for future replication across disaster-prone regions in Indonesia and beyond, highlighting the evolving role of archival science in community resilience and disaster mitigation.

BACKGROUND

Records management is foundational to the continuity of administrative, legal, and social functions at the household level. Documents such as ownership certificates, family cards, birth certificates, and health records are not merely transactional artifacts; they are critical enablers of citizenship, access to rights, and institutional legitimacy (Smith, 2016). In contexts frequently impacted by natural disasters, such as Indonesia, these records play a particularly vital role. Effective management ensures that families can swiftly navigate post-disaster administrative processes, from insurance claims to identity verification and government relief applications.

The legal framework for records and archives in Indonesia is outlined in Act number 43 of 2009, which defines records as documentation of activities or events that possess administrative, legal, financial, scientific, or historical value. As Cole (2022) argues, when managed correctly, records transition from passive documentation to strategic assets with long-term benefits for both individuals and communities. This is especially relevant for personal and family archives, which may be required in the future for legal or evidentiary purposes.

Indonesia's exposure to a wide spectrum of natural hazards—including earthquakes, tsunamis, volcanic eruptions, floods, landslides, and wildfires—poses a persistent threat to physical records. The 2021 flash flood in Batu City stands as a sobering example of how these risks materialize. Water and mud infiltration severely damaged or destroyed a large number of personal records. Affected families reported delays in administrative processing, exacerbating the trauma of displacement and loss. Without valid documentation, many residents were unable to register for relief programs, proving the high stakes of resilient recordkeeping in disaster scenarios (Shepherd & Yeo, 2003).

Physical record recovery after a disaster is both technically complex and financially burdensome. Even in cases where damaged records can be restored, the procedures are typically time-consuming and involve high costs for digitization, validation, and re-issuance. Thus, proactive strategies that include digitization before disasters occur are significantly more effective and economical than post-crisis remediation.

In this context, SIMARTA emerges as a preemptive solution. Developed by the Department of Library and Information Science and Universitas Brawijaya Archives, SIMARTA is a digital repository system that enables the secure storage, categorization, and retrieval of household records. It is engineered with disaster resilience in mind, offering cloud-based access, encryption protocols, and customizable user permissions. Through SIMARTA, families in Batu City—and potentially other vulnerable regions—can protect their vital records from environmental threats and maintain administrative agency during crises.

Importantly, SIMARTA includes built-in privacy safeguards that restrict access to authorized users only, reinforcing the system's alignment with ethical information governance. The ability to limit access ensures that sensitive information, such as legal, financial, or health documents, remains confidential while still being available to trusted parties in times of need. In sum, this background establishes the rationale for digital records management as a core element of disaster preparedness. It also frames SIMARTA as a context-specific response to Indonesia's recurring natural hazards and legal recordkeeping obligations. The next sections will investigate its implementation and implications in greater detail, particularly through the lens of the Batu City flash flood.

METHODOLOGY

This study employs a qualitative case study methodology to examine the role of SIMARTA—a digital household record repository—in supporting disaster resilience in Batu City, East Java, Indonesia. Case studies are particularly suited for investigating complex, real-world interventions where contextual conditions are critical (Yin, 2018). By focusing on SIMARTA's deployment during and after the 2021 flash flood, this study seeks to understand how digital repositories function as socio-technical infrastructures in disaster-prone environments and to identify the practical barriers to their adoption and sustainability. The research design incorporated three data collection strategies:

a) Semi-structured interviews were conducted with 12 key stakeholders, including SIMARTA system developers from Universitas Brawijaya Archives, local government officials involved in disaster response, and community leaders in the affected sub-districts of Batu City. These interviews explored perceptions of the system's technical functionality, accessibility, and integration into disaster workflows.

b) Focus group discussions (FGDs) were held with 18 residents from three subdistricts in Batu City who were directly affected by the flash flood. Participants included both users and non-users of SIMARTA to provide balanced insights into public awareness, digital literacy levels, trust in the system, and practical constraints.
c) Document analysis included review of SIMARTA's technical documentation, user manuals, municipal disaster protocols, and policy frameworks from Universitas Brawijaya and local government archives. This triangulation ensured that the study was grounded in both user experience and institutional design.

Data collection was conducted between March and August 2023, with ethical oversight provided by the Universitas Brawijaya Research Ethics Committee. Informed consent was obtained from all participants, and identifying information was anonymized.

FINDINGS

SIMARTA was developed in early 2020 through a collaboration between the Department of Library and Information Science and Universitas Brawijaya Archives. The system was piloted in six neighborhood clusters (kelurahan) in Batu City. Its design includes a) encrypted cloud storage for secure data management; b) metadata-enabled search for rapid document retrieval; c) two-factor authentication and role-based access control to protect user privacy; d) mobile-friendly interface compatible with smartphones, tablets, and desktop devices. The platform was intended to serve both proactive (pre-disaster) and reactive (post-disaster) functions by allowing families to upload critical documents such as birth certificates, property titles, and family cards. To ensure adoption and community integration, SIMARTA's developers implemented a multi-pronged engagement approach.

Empirical evidence from the 2021 flash flood in Batu City provides strong support for the role of SIMARTA in enhancing disaster resilience through secure and rapid access to household records. According to post-disaster assessments conducted by Universitas Brawijaya, over 62% of households that had pre-registered with SIMARTA were able to access or recover essential documents within 48 hours of the flood. These records included birth certificates, family cards, and property ownership documents—materials critical for re-establishing identity, accessing relief services, and resuming administrative procedures.

For affected residents, SIMARTA offered a pathway to immediate documentation recovery. The ability to retrieve legal and identity records from cloud storage via smartphones or computers meant that survivors could begin post-disaster paperwork without delay. In contrast, households relying solely on physical records experienced significant administrative setbacks, with replacement processes taking several weeks and requiring multiple visits to municipal offices.

This digital advantage was particularly notable in education and healthcare. Respondents reported that SIMARTA-enabled document sharing expedited the re-enrollment of displaced children into schools and helped families access public health services more efficiently. Aid agencies also benefited, as they were able to verify family identity and eligibility status via SIMARTA, reducing dependency on temporary or paper-based affidavits.

Beyond administrative efficiency, SIMARTA contributed to the emotional well-being of users. Many participants likened the loss of physical records to the erasure of family history and identity. Knowing that their vital documents were safely stored in a secure digital repository offered a sense of reassurance during a period of immense vulnerability. This aligns with previous research that links record continuity to psychological stability in post-disaster settings (Shepherd & Yeo, 2003).

SIMARTA's access control features allowed users to share records securely with designated parties such as schools, healthcare providers, and local government officials. This reduced the risk of document falsification and minimized the logistical burden of providing physical copies. Users retained the ability to revoke access at any time, enhancing their control over personal data and increasing their trust in the system. Furthermore, records were categorized and tagged with metadata, allowing users to locate documents quickly based on type, title, or purpose. This facilitated more efficient administrative processing, particularly during high-stress situations when information clarity is paramount.

Despite these successes, the findings also revealed disparities in usage and access. Households in digitally marginalized areas, particularly those with poor internet connectivity or low digital literacy, were unable to fully benefit from SIMARTA's capabilities. Some families had never been informed about the platform, while others lacked the necessary devices or confidence to navigate it independently. These gaps underscore the importance of complementary strategies such as community outreach, mobile-friendly interfaces, and offline access features.

In summary, the Batu City case demonstrates that SIMARTA, when properly implemented, can offer substantial benefits in disaster contexts: administrative continuity, emotional resilience, and secure digital sovereignty. However, these outcomes are contingent on addressing infrastructural and socio-cultural barriers that influence equitable access and sustained use.

1) Electronic Records Management as a Strategy for Disaster-Resilient

The transition to electronic records management has become a central pillar in disaster resilience strategies, particularly for communities facing recurrent threats to physical infrastructure. In this context, SIMARTA functions not merely as a digital tool but as a strategic response to vulnerabilities inherent in analog recordkeeping systems. By digitizing vital household records, SIMARTA enables users to store, access, and protect essential documents in a secure and recoverable format—capabilities that are indispensable in high-risk environments.

Electronic records systems mitigate the risks associated with environmental damage by facilitating remote access and off-site storage, two features critical for continuity during crises. Bates et al. (2003) and Yeo (2018) emphasize that digital transformation in records management reduces the fragility of paper-based documentation, especially in emergency contexts where time-sensitive decisions hinge on access to legal, health, and property records.

Figure 2. The sign-up page in SIMARTA

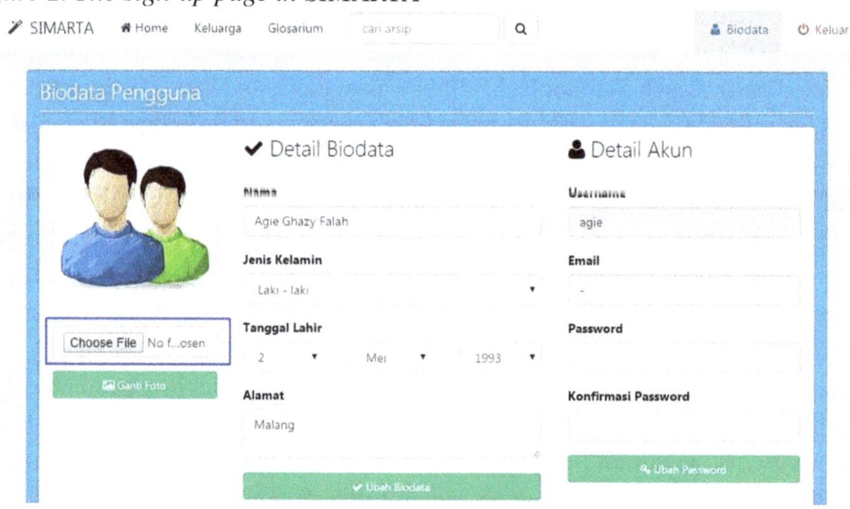

The design of SIMARTA reflects these principles. Through encrypted servers and user-level access control, the system provides centralized but secure data management for households. Each record stored is backed up regularly, reducing the chance of loss due to equipment failure or natural disaster. For example, following the 2021 flash flood in Batu City, SIMARTA enabled residents to access birth certificates and family cards via their smartphones within days of the disaster, an outcome that would have been unlikely with traditional records.

The system also addresses the evolving needs of modern families. According to Conway (2011) and Gardner and Davis (2017), digital family information requires new modes of preservation that are more adaptive and mobile. SIMARTA fulfills this by offering features such as mobile-friendly interfaces, metadata tagging for easy search, and a classification system that supports both structured and unstructured records. These design attributes align with user expectations in the digital age, especially in households that rely on flexible access points such as tablets and smartphones.

Developed under the research and community engagement initiatives of Universitas Brawijaya, SIMARTA has been positioned as a prototype for localized disaster mitigation through digital innovation. In early 2023, it was incorporated into the Universitas Brawijaya's community service portfolio to strengthen public resilience in Batu City, a region identified as highly vulnerable due to land-use changes and intensified rainfall (Putra et al., 2022). SIMARTA's proactive emphasis on record preservation—rather than post-disaster reconstruction—represents a significant shift in how communities prepare for environmental disruptions.

Moreover, SIMARTA supports data sovereignty at the household level, granting individuals the agency to manage, restrict, and share access to their digital documents. This decentralization of record control reduces dependency on centralized bureaucratic institutions during emergencies, allowing for more autonomous and responsive action.

In sum, SIMARTA exemplifies how electronic records management can evolve from a bureaucratic function into a strategic tool for disaster resilience. By safeguarding records in digital form, the system not only prevents loss but also empowers communities to recover more efficiently, preserve their identities, and protect their legal and social entitlements in times of crisis.

2) Obstacles and Opportunities in Establishing Electronic Record Management for Disaster Resilience

While the advantages of electronic recordkeeping in disaster contexts are increasingly recognized, the practical implementation of such systems, particularly in developing countries, remains fraught with challenges. SIMARTA's deployment in Batu City illustrates both the structural impediments and the transformative potential of digital repositories as tools for disaster mitigation. A primary obstacle to widespread adoption is digital literacy, especially among populations unfamiliar with information technologies. In rural and peri-urban parts of Batu City, many residents lacked the technical confidence to engage with platforms like SIMARTA. This finding is consistent with the broader literature, which identifies low digital literacy as a critical constraint in the Global South (Van Laar et al., 2017). Without adequate training, users are unlikely to adopt even well-designed digital systems, leaving significant segments of the population excluded from their benefits.

Infrastructure deficits also inhibit the effectiveness of electronic record management systems. Limited Internet coverage and frequent network disruptions were reported in several neighborhoods, particularly in mountainous and flood-prone areas. These limitations made it difficult for users to access their accounts or upload records, particularly during the flood emergency, when stable communication was essential. SIMARTA currently lacks an offline-access module or satellite sync capability, both of which would be necessary to serve communities with inconsistent connectivity.

Trust in digital systems poses another barrier. Despite SIMARTA's security features—such as two-factor authentication and encrypted storage—many users expressed concerns about data breaches and unauthorized access. These fears were exacerbated by limited awareness of the platform's privacy protocols. In contexts where public trust in institutional data governance is already fragile, introducing digital systems without robust public engagement can produce skepticism or outright resistance.

Finally, the system's limited outreach efforts constrained adoption. The SIMARTA team, operating with a modest university budget, was only able to reach a fraction of Batu City's population prior to the 2021 flood. As a result, many residents were unaware of the system's existence until after the disaster had occurred. This reactive model of dissemination undermines the preventive potential of digital repositories and suggests the need for long-term, integrated community engagement strategies.

Despite these challenges, several opportunities exist for expanding the impact and sustainability of electronic records management systems in disaster contexts. *First*, national policy alignment offers a powerful lever for institutionalizing digital repository practices. Indonesia's broader e-government and disaster risk reduction frameworks can incorporate SIMARTA-like systems as part of standard community preparedness strategies. The Republic of Indonesia's ongoing initiatives in digital transformation and smart cities provide fertile ground for expanding such platforms.

Second, technological innovation presents avenues for improvement. Enhancing SIMARTA with offline functionality, multi-language support, and lightweight mobile applications would improve its accessibility across diverse user groups. Partnerships with technology firms and mobile service providers could further subsidize data use or integrate repository access into zero-rated platforms for emergency use.

Third, community-based co-design holds promise for increasing relevance and adoption. As Ketelaar (2017) argues, participatory archival design that includes local knowledge systems fosters deeper engagement and trust. SIMARTA could incorporate oral histories, community maps, and indigenous risk narratives, transforming it from a static records platform into a living repository of community memory and resilience.

3) SIMARTA's Function in Mitigating Disaster Risks to Household Records

Universitas Brawijaya administers a server where SIMARTA enables users to securely store their household records in electronic format. In the case of disasters, SIMARTA is advantageous because of its enhanced accessibility and superior security relative to physical records. SIMARTA administers stringent security measures on a server to protect records from physical harm resulting from disasters. This server is designed for the long-term storage of users and incorporates frequent backups to mitigate the risk of data loss. One of the main reasons to complement the physical with digital records storage is protection against natural disasters. Physical records stored at home are vulnerable to damage from disasters. Natural disasters can permanently damage records, leading to loss and irretrievable information within them. SIMARTA offers electronic records management, which offers greater protection by storing records in electronic form on a centralized, secure server.

SIMARTA permits the storage of records on a server under the management of Universitas Brawijaya, a trusted institution. Periodic backups equip this server, enabling the restoration of stored records in the event of data damage or loss. The design of the SIMARTA storage server ensures its safe operation under various conditions, including emergencies. Users can feel more at ease with this data backup and recovery system, as it protects their records even in the event of an unexpected disaster. Data security plays a crucial role in electronic records management, particularly in Batu City, where sensitive information is contained in personal records. A secure storage system must be able to protect data from the risk of theft, unauthorized access, or data leakage.

SIMARTA uses encryption technology to achieve this security, preventing unauthorized parties from easily accessing stored data. The process of encryption secures data by transforming it into a format unreadable without a unique encryption key. SIMARTA incorporates a two-factor authentication feature in addition to encryption. This feature adds an extra layer of security by requiring users to enter a verification code sent to their device, in addition to a password. Dual authentication helps ensure that only the account owner can access stored records, minimizing the risk of unauthorized access.

Figure 3. The home page and the household records list in SIMARTA

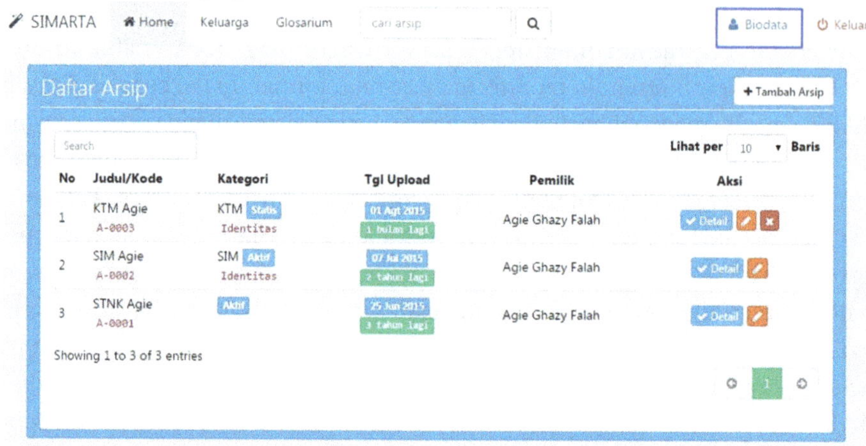

SIMARTA allows for greater flexibility in access compared to physical records. Users can access stored records anytime and anywhere, as long as they have an Internet connection. Emergencies, such as applying for assistance or filing post-disaster insurance claims, necessitate immediate access to records. With easy access, users do not need to worry about losing records, even if there is physical damage to the original storage location. SIMARTA also allows users to manage their access rights. Users can grant limited access to family members or other trusted parties. This feature allows users to share vital records without having to hand over the physical records, which increases their security from loss or damage. This access rights control provides greater flexibility in sharing records securely.

One important aspect of secure records storage is a data backup system. SIMARTA ensures that every record stored is periodically copied to a backup server. If the main server experiences disruption or damage, the backup server can still restore the stored data. Generally, this backup process occurs automatically at specific time intervals, alleviating users' concerns about record loss. Furthermore, data backup ensures users' peace of mind by distributing their data across multiple locations. Large-scale backup can also incorporate remote backup, storing data in various physical locations to mitigate the risk of data loss due to disasters that impact specific areas.

According to Duranti et al. (2013), electronic records storage is inseparable from the issue of personal data protection. Household records often contain personal and sensitive information, such as addresses, identity numbers, and financial information. Therefore, SIMARTA must comply with data protection standards following applicable regulations, such as Act number 27 of 2022 about the Personal Data Protection in the Republic of Indonesia. This regulation guarantees the protection of users' data from misuse or leakage to irresponsible parties. SIMARTA tries to implement a strict data protection policy to protect user privacy. Additionally, SIMARTA informs users about their data rights, which include the ability to access, correct, or delete their data as desired. With this data protection, users can feel safe storing sensitive records in electronic form. Data contingency and recovery are important parts of a secure records storage strategy.

In emergencies, such as when a disaster or disruption occurs to digital infrastructure, SIMARTA must have a contingency plan to ensure that data remains accessible and recoverable. Users can use this contingency in the form of temporary offline storage or alternative access. As a digital repository, SIMARTA incorporates a data recovery system that enables swift and secure records retrieval in the event of a disruption. This recovery process reduces the risk of data loss and ensures that users can continue to access their records even if there is a technical disruption. Secure electronic records storage requires policy support and public education. Government of the Republic of Indonesia policies, such as subsidies or incentives for digital application users, can encourage the community to use digital archiving more. In addition, digital education and literacy programs can increase public understanding of the importance of data security and how to use digital technology safely.

Public education also includes understanding data security risks and how to protect personal information in the digital world. Universitas Brawijaya, together with the Batu City Government or related institutions, can organize training and outreach aimed at building public trust in digital storage security. Secure records storage through SIMARTA provides a better protection solution from the risk of damage due to disasters, theft, or data leaks. Encryption technology, double authentication, and data backup are key components in secure storage. With the support of policies, public education, and contingency plans, digital archiving can be the main choice for managing household records safely and efficiently.

SIMARTA enables users to retrieve their electronic records from any Internet-enabled device. This feature is very beneficial in crises requiring rapid access to records for assistance or insurance claim processing. In the aftermath of a flood in Batu City, SIMARTA users retain access to their essential records, regardless of the loss or damage to the physical copies. In general, anytime and anywhere access refers to the ability of users to access records or data from various digital devices without time and location restrictions. Electronic records, unlike physical archives, enable users to open, view, or share records online through Internet-connected electronic devices (Vaarzon-Morel et al., 2021). Users with various needs can easily accomplish this using a computer, tablet, or smartphone.

Figure 4. Family cards as one of the records in SIMARTA

SIMARTA integrates this accessibility feature to simplify the management of household records, including birth certificates, family cards, land or house certificates, and health records, which are frequently required in administrative processes. This convenience means that users simply access the SIMARTA application, that has stored the records electronically. This feature is very helpful, especially in emergencies, such as during disasters or urgent needs, where users may need to immediately access vital records.

One of the main benefits of accessibility in SIMARTA is its ability to support emergency preparedness. In situations where quick access to records is essential, such as natural disasters or urgent health conditions, users can immediately access the records they need. For example, if a disaster occurs that causes the loss of physical records, users can still access digital copies of their archives through SIMARTA. This is very important, considering that disasters often result in irreversible damage or loss of physical records. The community in disaster-prone areas can feel more secure with anytime and anywhere access, as it allows them to access their vital records even in the event of physical damage to their original storage location. This also makes it easier to process insurance claims or apply for assistance, where identity records and proof of ownership are often required. Accessibility in SIMARTA also results in significant time efficiency. The process of searching for records in electronic format is much faster compared to searching for physical records, which often takes time because it is necessary to find the right storage location. SIMARTA organizes records into categories for easy search using a classification scheme.

Figure 5. SIMARTA organizes records into categories

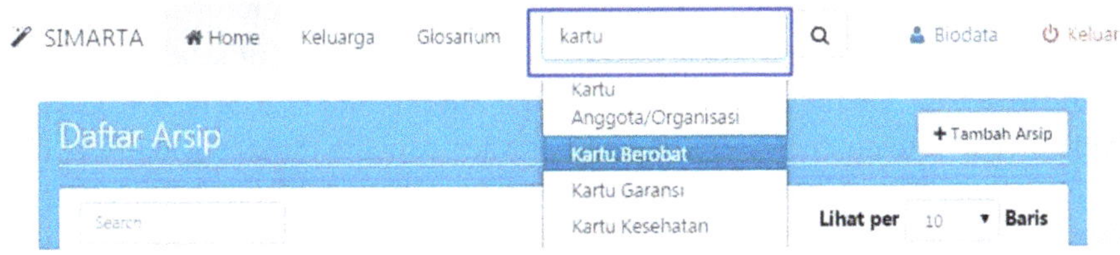

This efficiency is very useful in everyday life, where users may need records for various administrative or personal purposes. For example, when taking care of children's education or financial transactions that require supporting records, users do not need to return home or open a physical records storage place. This certainly speeds up the process and makes records management more effective. The anywhere access feature makes it easier for users who have high mobility or travel frequently because they do not need to carry physical copies of records that may be at risk of being lost or damaged.

SIMARTA allows users to access records from any location with an Internet connection. This flexibility is also very beneficial for communities living in rural or remote areas, where access to administrative services may be limited. With digital access, users can send copies of records directly from their devices to those in need without having to meet in person or send physical records. Several supporting technologies and infrastructure play a crucial role in achieving flexible access. Several key technologies enable accessibility in SIMARTA, including:

a) ***Cloud computing:*** A technology known as cloud computing enables the online storage of records on remote servers. Within SIMARTA, cloud computing guarantees the central storage of user-uploaded records on secure servers. Users can access these records via an Internet connection at any time, from anywhere, without having to rely on a specific device. Since cloud computing stores data in a location separate from the user, it also provides data security from physical interference or disasters.
b) ***A reliable Internet network:*** Accessibility in SIMARTA also requires a reliable Internet network, especially to enable rapid and smooth access to stored records. A stable Internet connection allows users to open records without long-term time constraints. According to Njogu (2022), the government and Internet service providers have an important role in ensuring adequate Internet infrastructure, especially in areas that are vulnerable to disasters or have limited access.
c) ***Multi-platform application:*** Various devices, including computers, tablets, and smartphones, can access SIMARTA. This multi-platform capability allows

users to open records from any device they own, eliminating the need to limit themselves to a single type. Mobile-friendly applications allow for easier and faster access, as smartphones are generally more accessible than desktop devices (Al-Sakran & Alsudairi, 2021). Although anytime and anywhere access provides great convenience, security and privacy aspects remain a major concern in electronic records management. SIMARTA maintains record access security through several steps, including double authentication and data encryption.

To ensure that only the account owner can access the records, double authentication necessitates additional verification, such as sending a one-time password (OTP) code to the user's device. In addition, the access control feature allows users to limit who can view or download certain records. This is crucial for maintaining privacy, particularly when the records hold sensitive information (Meyer & Fourie, 2021). This access control also allows users to share records with parties they trust, such as family members or colleagues, without having to worry about them falling into the wrong hands.

Despite the numerous advantages that come with anytime and anywhere access, users may encounter certain challenges when implementing it, including 1) Limited Internet infrastructure in remote areas, some areas in Batu City still have limited or unstable Internet access, which can affect users' ability to access records online; 2) Digital literacy, not all community in Batu City are familiar with the use of SIMARTA, so education and training are important to ensure optimal understanding and utilization of this accessibility feature; and 3) Trust in data security, some users may still doubt the security of their data when stored digitally, especially if the data can be accessed from anywhere.

The variety of features in SIMARTA provides significant convenience in managing household records. With technologies such as cloud computing and dual authentication, SIMARTA ensures that users can access their vital records safely from various locations and devices. This feature is very useful in emergencies, supports user mobility, and increases efficiency in records management. Implementing this accessibility feature is a step towards safer and more flexible archive management, even though there are still obstacles to overcome, particularly in terms of digital literacy and infrastructure.

SIMARTA enables users to provide access to family members or other trusted individuals, subject to certain access limitations. Ensuring data confidentiality is crucial for facilitating communication among family members during crises. In instances when a family member is unable to view a record, other accessible family members may facilitate the necessary administrative procedures. The restricted sharing feature allows users to share records with specific parties and provides control over the type of access allowed. Users can choose who can view the records, download them, or even grant temporary permissions that automatically expire after a certain period. This feature gives users the flexibility to share records for various administrative or personal needs without having to worry about the security and privacy of the shared records.

In SIMARTA, the limited access sharing feature allows users to share records with family members or third parties, such as lawyers or financial planners, who need temporary access to certain records. This feature streamlines the management of household records and lessens reliance on physically damaged or lost records. The limited access sharing feature in SIMARTA works through an access rights setting system that allows records owners to set certain restrictions, such as:

a) ***Access type selection*:** Users can determine whether invitees can only view records or also have the right to download them. In certain scenarios, users may opt for records reading only, without downloading, to maintain complete control over records distribution.
b) ***Limited-time access*:** SIMARTA offers the option to grant temporary access, with the automatic revocation of the records after a specified period. This feature is useful when users only need others to view the records for a certain period, such as during the administrative approval process.
c) ***Role or group-based access*:** Users can also set access based on roles or groups. For instance, users can restrict access to certain records to only family members designated as guardians or heirs.
d) ***Benefits of the limited access sharing feature*:** The limited access sharing feature on SIMARTA offers several key benefits that increase efficiency and security in managing household records, including:
    - ***Security and privacy:*** Security is a key factor in sharing electronic records, especially when they contain sensitive or personal information. With restricted access, users can ensure that only the right community can access the records. Users can also restrict the ability of third parties to download the records, minimizing the risk of unwanted data sharing. The restricted sharing feature provides flexibility in sharing records while maintaining data security and privacy.
    - ***Collaboration in records management:*** The restricted sharing feature also facilitates better collaboration in managing household records. For example, family members can access the same records without having to send physical copies. This is especially useful when multiple family members are involved in decision-making or need access to the same records for administrative or financial planning purposes. Each party can access the same records simultaneously, simplifying the need for physical records exchange.

- ***More efficient records management:*** Sharing records with restricted access allows users to maintain full control over the records they own. When the records owner wants to end access for a particular party, they can directly revoke access permissions. This is different from physical records, which, once handed over, are difficult to control again. This feature allows users to determine the access times and recipients of records, thereby enhancing the efficiency and flexibility of records management.
- ***Flexibility of access in emergency:*** In emergencies, the limited access sharing feature allows users to provide quick access to certain parties without having to meet in person or submit physical records. For example, in a medical emergency, users can provide access to their health records to doctors or medical personnel who need them. This flexibility ensures quick and accurate access to vital records without compromising privacy or data security.

The limited access sharing feature in SIMARTA provides greater flexibility and control in managing household records. This feature allows users to share records securely with certain parties and set access rights as needed. With security measures such as dual authentication, encryption, and activity logs, SIMARTA ensures that shared records remain protected. This feature is very useful for family administrative purposes, collaboration with external parties, and emergency needs, making it a more secure, efficient, and appropriate records management solution for today's digital society.

The deployment of SIMARTA offers reassurance to users residing in disaster-prone regions like Batu City, ensuring that their records remain accessible and safeguarded despite potential damage or loss of physical files. For several impacted individuals, SIMARTA offers a tangible option to ensure the continuation of home administration by reducing the risk of record loss. Despite SIMARTA providing an efficient solution for electronic records management, its application within the community encounters several substantial impediments and problems. The primary challenges encountered in the implementation of SIMARTA include:

a) ***Insufficient digital literacy:*** A primary barrier to the implementation of SIMARTA is insufficient digital literacy, particularly among elderly populations and in rural regions. Numerous communities lack comprehensive knowledge of the fundamental technologies necessary for the digitization of records. A deficiency in technological comprehension leads most individuals to choose conventional physical storage, despite its heightened hazards. Consequently, an effective and comprehensive digital literacy competency and training program is essential to promote the adoption of SIMARTA.

b) ***Restricted Internet connectivity:*** The SIMARTA necessitates Internet connectivity for best access and use. Nevertheless, inadequate Internet infrastructure in some areas, particularly in parts of Batu City, remains a significant impediment to the deployment of SIMARTA. Communities in Batu City with limited Internet connectivity face difficulties in accessing and uploading records, thereby reducing the benefits of SIMARTA. This dilemma underscores the need for enhanced telecommunications infrastructure in disaster-prone regions like Batu City.
c) ***The dissemination and promotion*:** Despite SIMARTA presenting a viable solution, its restricted socializing scope results in widespread unfamiliarity with the application's advantages among many individuals. Universitas Brawijaya has socialized several communities, but budget constraints have limited this initiative to specific areas in Batu City. Numerous communities acknowledged that they had not received information about SIMARTA from family or friends for an extended period and had never encountered direct advertisements or outreach efforts. This underscores the need for a more comprehensive and persistent socialization campaign to enhance public knowledge about the significance of SIMARTA and effective records management.
d) ***Perceptions of security and trust:*** Certain individuals remain apprehensive over the protection of personal information and the risk of data breaches in SIMARTA. Uncertainty about data storage and access control shapes this view. Despite SIMARTA's implementation of security elements, including user authentication and a restricted access system, the communication regarding the data protection mechanisms in place needs enhancement to foster public trust. These challenges suggest that SIMARTA within the community requires a holistic strategy, including the enhancement of digital literacy, the establishment of infrastructure, and continuous social engagement. Batu City Government and Universitas Brawijaya support is essential to enhance the use of SIMARTA as a component of disaster mitigation strategies in Batu City.

Finally, educational institutions like Universitas Brawijaya are well-positioned to sustain and scale such systems. Through academic research, technical refinement, and policy advocacy, universities can play a critical role in bridging digital archives with real-world governance needs. In summary, while implementation of electronic record management systems like SIMARTA is not without obstacles, these challenges are surmountable. With the right mix of policy support, technical upgrades, and community engagement, digital repositories can evolve into cornerstone infrastructures for disaster-resilient societies.

DISCUSSION

The implementation of SIMARTA in Batu City offers a concrete example of how digital repositories can enhance community resilience in disaster-prone environments. The evidence collected through field interviews, focus groups, and usage reports indicates that SIMARTA successfully facilitated the secure recovery of vital household records, expedited administrative response, and provided psychological reassurance to disaster survivors. However, the broader implications of this deployment also highlight the complexity of embedding digital infrastructure in socially diverse and infrastructurally uneven contexts.

1.) Validation of SIMARTA's Core Strengths

One of the most salient outcomes of SIMARTA's use was its ability to mitigate bureaucratic paralysis following the 2021 flood. Households that had enrolled in the system before the disaster were able to retrieve digital versions of identity documents, property titles, and family records within 48 hours, compared to weeks or even months for those reliant on paper documentation. This efficiency gap had tangible effects: users were able to resume school enrollments, apply for government aid, and verify health entitlements with minimal disruption.

The system's dual authentication and encrypted cloud storage also ensured that personal data remained protected during and after the event. From a design perspective, SIMARTA reflects best practices in secure electronic recordkeeping, including metadata indexing, access control, and mobile accessibility. These features enabled decentralized access without sacrificing data integrity—a critical advantage in post-disaster scenarios where centralized systems are often impaired.

Furthermore, the psychological benefits of SIMARTA were clearly articulated by users. The ability to "see" one's documents—even in the aftermath of physical loss—served as a symbolic anchor of stability. This aligns with archival science literature suggesting that personal records serve not only legal but also emotional functions (Shepherd & Yeo, 2003).

2.) Limitations Revealed through Deployment

Despite its technical strengths, SIMARTA's deployment exposed several limitations that speak to broader challenges in disaster informatics. *First*, uneven digital literacy across user demographics hindered equitable access. Older adults, rural households, and those with limited exposure to smartphones found the interface difficult to navigate. This echoes findings in digital divide literature, which emphasize that technological solutions alone cannot bridge socio-educational gaps (Van Laar et al., 2017).

Second, infrastructure constraints, including poor Internet connectivity in remote areas, and reduced system reliability during peak demand periods. SIMARTA's lack of offline functionality or caching options rendered it temporarily inaccessible to users in the most affected zones. This undermines the core objective of digital resilience and signals the need for hybrid models that combine digital and physical safeguards.

Third, data governance transparency was identified as a weak point. Although the system employed advanced security protocols, many users were unaware of how their data was managed, who had access, or what recourse existed in the event of a breach. Trust is foundational to any digital repository system, and future implementations must prioritize user-centered design that communicates data policies and rights.

3.) Toward an Inclusive Socio-Technical Model

The SIMARTA case underscores that disaster-resilient digital repositories are not merely technical solutions; they are socio-technical systems embedded in cultural, political, and infrastructural realities. A key lesson is that user engagement cannot be an afterthought. Outreach must be continuous, multilingual, and adapted to the needs of diverse users. Community co-design processes, including participatory training, localized content, and integration of indigenous knowledge, can increase both relevance and legitimacy (Gilliland & Caswell, 2015; Ketelaar, 2017).

From a systems perspective, SIMARTA's impact points to the need for resilience by design, not just in data storage but in user access, redundancy, and institutional interoperability. Governments and universities must view such repositories not as one-off innovations but as evolving infrastructures requiring policy alignment, budgetary support, and technical stewardship.

In short, SIMARTA demonstrates the promise and limitations of digital records systems in disaster contexts. It validates the potential of information systems to deliver timely, equitable, and secure responses. Yet it also warns that without sustained investment in inclusion, outreach, and infrastructure, such systems risk becoming tools of fragmentation rather than integration.

IMPLICATIONS FOR POLICY

The SIMARTA initiative in Batu City presents valuable insights for policymakers seeking to integrate digital repositories into broader disaster mitigation frameworks. Its deployment demonstrates that secure and accessible digital recordkeeping can play a central role in post-disaster response and recovery. However, the efficacy of such systems is heavily dependent on enabling policies in four critical domains: infrastructure, education, regulation, and institutional collaboration.

1.) Enhancing Digital Infrastructure

A significant barrier to the implementation of digital archiving in the Republic of Indonesia is the restricted Internet connection in some regions, particularly in disaster-prone locales like Batu City. This constraint impedes the use of digital repositories like SIMARTA and obstructs communication and swift action during a crisis. To facilitate the execution of more equitable electronic records management, the government might develop a strategy to enhance information and communication technology infrastructure, particularly in disaster-prone regions. Establishing Internet networks in rural regions or subsidizing Internet access in disaster-prone areas might enhance community resilience to disasters.

Digital infrastructure refers to networks and devices that enable access to Internet-based services, such as telecommunications networks, Internet connections, data centers, and digital devices. In disaster-prone areas, digital infrastructure is critical because it provides rapid and secure access to information, coordinates with rescue teams, and reports emergencies. Additionally, even if the disaster damages or loses their physical records, the community can still access vital records electronically. Digital access that is available before, during, and after a disaster also facilitates the recovery process and reduces the economic and social impacts of damage to physical records. Electronic records management through SIMARTA is becoming increasingly relevant for communities in disaster-prone areas because this application allows users to store their vital records in the cloud or a secure data center. Adequate digital infrastructure supports SIMARTA in providing fast and reliable access for the community so they can obtain records that may be needed in the post-disaster relief process.

Improving digital infrastructure in disaster-prone areas includes several key components that support the accessibility, security, and sustainability of digital services. The availability of a stable Internet connection is the basis of digital infrastructure that enables continuous access to digital services, including SIMARTA. In disaster-prone areas, reliable Internet allows the community to upload and access vital records in real-time, communicate with relevant parties, and obtain the latest information on the situation. To achieve this, the government and Internet service providers can work together to expand the Internet network to areas that do not yet have access, as well as increase network capacity in areas that are already accessible but still have limited connections. In areas with inadequate Internet infrastructure, initiatives such as installing satellite-based Internet networks can be a solution to ensure accessibility. Satellite connections do not rely on physical infrastructure, making them more resilient to disruptions caused by natural disasters.

Data centers and cloud computing facilitate secure and centralized data storage, enabling access from multiple locations (Shobaruddin et al., 2024). In the context of disaster mitigation, Universitas Brawijaya managed data centers to provide regular backups so that stored data remains safe even when a disaster occurs. SIMARTA application users, for example, can store their records in the cloud, which is connected to a secure data center. The data center's secure location ensures data access even in the event of a breakdown in the user's area. A disaster recovery system also equips the data center, enabling swift data recovery following a disaster. This ensures that services such as SIMARTA continue to operate even if there is a technical disruption or breakdown in a particular physical location.

Affordable digital devices allow the community to access digital services anytime and anywhere. However, in disaster-prone areas, the devices used must be durable enough to withstand emergency conditions. Providing digital devices specifically designed for disaster situations, equipped with long-lasting batteries, is also very useful. In addition, the government can provide incentives or subsidies to increase the accessibility of digital devices for communities in vulnerable areas. In disaster-prone areas where electricity supplies can be disrupted, digital infrastructure necessitates a stable energy source. Backup power sources, such as generators or solar panels, ensure that Internet networks and digital devices continue to function during a disaster. With a reliable source of energy, the community can continue to access digital services and communicate with authorities for emergency assistance. Additionally, backup power sources ensure the operation of data centers, enabling users to access records storage services like SIMARTA.

Emergency communication systems, such as two-way radios, satellite telephone networks, or other emergency communication networks, serve as a backup for communication, complementing the Internet itself. These networks facilitate communication between individuals and search and rescue teams or the government, particularly in the event of Internet network disruption. The community in need can still receive information by using this emergency communication system as a complement to digital infrastructure.

However, improving digital infrastructure in disaster-prone areas is not without challenges. Some common obstacles include 1) Budget constraints and high costs, the cost of building and expanding digital infrastructure is very high, especially in remote or hard-to-reach areas. Internet infrastructure, data centers, and backup energy resources require large investments, which are often prohibitive, especially for local governments with limited budgets; 2) Challenging geography, many disaster-prone areas in the Republic of Indonesia have challenging geographic conditions, such as mountains or small islands, which make the development of digital infrastructure more difficult. These areas require special solutions, such as satellite connections or telecommunications infrastructure designed for extreme conditions; 3) Lack of digital literacy, although digital infrastructure is available, the lack of digital literacy among communities in disaster-prone areas is a challenge. Efforts are necessary to improve digital understanding and skills, as a community may not be accustomed to using digital services or applications like SIMARTA; 4) Network stability and reliability, in some disaster-prone areas, the available Internet network may be less stable or prone to disruption. Network reliability is critical to ensuring continuous access during a disaster, so regular network maintenance and quality improvement are required.

Despite the significant challenges, improving digital infrastructure in disaster-prone areas brings several important benefits. Adequate digital infrastructure allows communities to receive up-to-date information on disasters, such as early warnings or evacuation guides. With quick access to emergency information, communities can take faster and more effective steps to protect themselves and their families. With digital infrastructure, communities can access SIMARTA to store household records in electronic format. This is very important in protecting records from the risk of physical damage caused by disasters, so that communities do not lose vital records needed to process post-disaster assistance or recovery.

Digital infrastructure facilitates communication and coordination between communities and the government, rescue teams, and aid agencies. During a disaster, communities can report on their conditions, ask for help, or provide important information through digital channels. This facilitates the process of distributing aid and recovery. With adequate digital infrastructure, communities can resume their economic activities more quickly after a disaster occurs. Access to digital services allows them to stay in touch with users and partners who help accelerate post-disaster economic recovery. With the right digital infrastructure improvements, disaster-prone areas can become more prepared and resilient in the face of disaster risks, while ensuring that communities can still access critical services and information to support their daily lives.

2.) Formulating and Encompassing a Digital Literacy Initiative

The community's inadequate digital literacy poses a substantial barrier to the implementation of SIMARTA. Digital literacy is crucial for the community to comprehend the advantages of SIMARTA and to use it efficiently and securely. The government and associated entities, including the Records and Archives Professional Association, Universitas Brawijaya, the National Archives, and the Ministry of Communication and Digital of the Republic of Indonesia, must develop a more comprehensive digital literacy initiative that encompasses fundamental technological knowledge, electronic records management, and the significance of personal data security.

Disaster-prone areas, like Batu City, necessitate heightened preparedness due to the potential loss of physical records or access to crucial information. In an emergency, information about evacuation locations, early warnings, and safety guidelines is essential. Digital literacy allows the community to get this information through digital channels such as social media, government websites, or special disaster mitigation applications. SIMARTA enables individuals to digitally manage their records, which is crucial for disaster recovery.

Digital literacy ensures that the community not only understands how to upload records but also how to protect their data safely. With adequate digital literacy, the community can use technology to communicate with rescue teams, access government assistance, and manage other administrative processes. This accelerates the process of social and economic recovery after a disaster. According to Lee (2024), the development of a comprehensive digital literacy program in disaster-prone areas is an important investment in building community resilience and preparedness. With key components such as basic technology education, data security, and electronic records management, this program can equip the community to access digital information and services in emergencies. By implementing effective strategies, collaboration, ongoing training, and government support, the community can enhance digital literacy, fostering a society that is more prepared, resilient, and capable of leveraging technology for its well-being.

3.) Funding and Subsidization Policy

Establishing and sustaining electronic records incurs expenses related to infrastructure, server storage, and the development of data security measures. Consequently, the government must contemplate sustainable financing strategies to maintain SIMARTA, enabling its continued development and provision of optimum services to the public. The government may implement subsidy schemes to encourage the public to transition to effective electronic records management without incurring substantial expenses.

Electronic records management is not only a need for large organizations or institutions but also for the public, especially those living in disaster-prone areas such as Batu City. It provides superior security and accessibility compared to physical archives. However, its management incurs costs for infrastructure, application development, and public education on its use. Funding and subsidy policies for electronic records management are important long-term investments in facing the digital era and increasing disaster risks. With the right funding, communities can more easily access digital repositories, protect their vital records, and improve digital literacy.

This policy will strengthen community resilience to disasters, reduce the burden on government administration, and accelerate sustainable digital transformation across the region in the Republic of Indonesia.

4.) Enhancing Data Protection Policies

Data security is a crucial element of electronic records management since household records often include sensitive personal information. In this context, rules that safeguard the confidentiality and security of user data must be reinforced, for instance, by referencing the Personal Data Protection Law in the Republic of Indonesia, which has commenced implementation. This policy must include the protection of user data and the management of privacy breaches.

Managing records in electronic form brings great benefits, especially in terms of accessibility and efficiency. However, SIMARTA also increases data security risks because records stored in this system often contain sensitive information, such as personal identity, health data, and financial information. Adequate data protection is very important to prevent unauthorized access, maintain user privacy, build public trust, and comply with laws and regulations. For data protection policies in SIMARTA to run effectively, several main components must be implemented, namely:

a) ***Data encryption***: The process of converting data into a code that no one without the encryption key can read is known as encryption. In the context of electronic records management, encryption ensures that stored data is safe from unauthorized access. Encryption also protects data in transit between the server and the user's device. SIMARTA can implement end-to-end encryption to ensure that only authorized users can read data stored in the system.
b) ***Two-factor authentication:*** Double authentication is an additional security measure that requires users to enter a second verification code in addition to their password in SIMARTA. The purpose of this measure is to restrict data access to only the genuine account owner. Double authentication aids in preventing unauthorized access, particularly when a third party may have discovered the user's password.
c) ***Access rights and permissions control:*** Access control gives users the ability to control who can view, change, or download certain records. For example, in SIMARTA, users can grant limited access to family members or third parties. This policy allows users to have full control over their data and reduces the risk of it falling into the wrong hands.
d) ***Activity log and monitoring system:*** This system logs every action on the data in the system, detailing who accessed the records, when they did it, and what they did. This monitoring helps detect suspicious activity and provides clear

information about each access in SIMARTA. In the event of a data breach, the activity log can serve as evidence for further investigation.

e) ***Data recovery and disaster recovery procedures*:** SIMARTA should have procedures in place to quickly recover data in the event of a disaster or technical disruption. This includes backing up data regularly and storing copies in a separate location. Disaster recovery procedures guarantee the prompt recovery of data without any loss of critical information.
f) ***Policy transparency and user education*:** The providers need to be transparent about their data protection policies. Users should be aware of their rights, such as accessing, correcting, or deleting their data. Educating users about data security practices, such as the importance of strong passwords, is also part of a strong policy in SIMARTA.

While data protection policies are essential, their implementation often faces several challenges, including costs, low digital literacy among users, evolving security threats, and compliance with varying regulations. To overcome the above challenges and ensure that data protection policies in an archives digital repository are effective, several strategies can be implemented, including:

a) ***Mutual collaboration:*** *The* Batu City Government can collaborate with Universitas Brawijaya to ensure strong data protection. Universitas Brawijaya can assist in the development of better encryption systems, double authentication, and access monitoring. This collaboration also allows SIMARTA to improve security technology as new threats develop.
b) ***Digital literacy programs on data protection*:** It is important to improve the digital literacy of the community about data protection. Educational programs that focus on data security can help users understand how to protect their data, such as recognizing phishing, using strong passwords, and protecting their devices from unauthorized access. These programs can be conducted through workshops, online campaigns, or written guides from SIMARTA.
c) ***Regular policy and technology updates*:** Data security requires continuous updates. Data protection policies must be evaluated and updated regularly to ensure that systems remain secure against evolving threats. Security technologies such as encryption, authentication, and access monitoring also need to be improved as technology advances for SIMARTA.
d) ***Regular security audits and evaluations:*** Regular security audits and evaluations are important steps to ensure that data protection policies are functioning properly. A competent internal or external party can conduct this audit to identify security gaps and provide recommendations for improvement. Periodic

evaluations also help measure the effectiveness of data protection policies and ensure that protection measures remain relevant.

e) ***Increased transparency with users***: Transparency is key to building user trust in SIMARTA. Service providers must be open about their data protection policies and provide clear information about user rights. Regulations on data storage, user access, and independent data protection are all part of this transparency.

A strong data protection policy in SIMARTA has many long-term benefits, including increasing user trust, reducing legal risks, securing the community's personal information, and driving digital transformation. SIMARTA should strengthen its data protection policies to safeguard user data against evolving security threats. With policies that include data encryption, dual authentication, access control, and monitoring systems, SIMARTA can be more secure and reliable. Through the implementation of the right strategy, governments and service providers can build public trust in electronic records management while ensuring compliance with applicable regulations. This policy not only protects data but also strengthens digital resilience and encourages public participation in the increasingly advanced digital era.

5.) Institutional Collaboration to Enhance the Resilience of Electronic Records

Effectively managing electronic records for disaster mitigation requires cooperation across many institutions, including governmental, academic, private sector, and community groups. This partnership is essential for the integration of resources, information, and assistance required for the management of household records. All the entities might collaborate to provide more comprehensive services. They must collaborate to address the various challenges in electronic records management, including the need for reliable infrastructure, the development of data security technology, and the need for digital literacy programs for the community. By working together, all entities can provide more optimal resources, improve technical expertise and data security, expand digital access and literacy, and reduce the risk of data loss. Each entity involved in the collaboration to support the resilience of electronic records has a different role according to its expertise and function, such as:

a) ***Government institutions*** have an important role in providing regulations, policies, and resources to support SIMARTA. The government can also provide basic infrastructure, such as data centers and reliable Internet networks. In addition, the government can set data security standards and protect user rights in electronic records management. The government can also work with national and international institutions to bring in additional technological assistance or funding as needed.

b) ***Educational and research institutions*:** Play a crucial role in the research and development of electronic records management standard, guidelines, and technology. In addition, universities can also provide training to the community to improve digital literacy. Universities can also practically apply their research results through collaboration with other institutions, bolstering the resilience of electronic records management, data encryption technology, backup systems, or the development of a more efficient digital repository.
c) ***Private sector and technology companies*:** This sector has a role in providing sophisticated software, hardware, and data security services. With their expertise in information technology, companies can contribute to the development of secure and efficient storage solutions, including cloud storage services that are accessible anytime and anywhere. Additionally, companies can offer the necessary expertise for system maintenance and security monitoring of the digital repository.
d) ***Nonprofit and community organizations*:** Often have access to funding, technology, and experience in disaster management in various regions. Nonprofit and community organizations can also be involved in organizing digital literacy programs, especially in areas that still have limited access to technology. Institutions can realize collaboration in various forms, such as the establishment of shared data centers, collaborative digital literacy and training programs, the creation of integrated security and technology standards, joint funding and grants, and joint monitoring and evaluation.

Collaboration between entities to support electronic records management resilience is a strategic step that allows for safer and more effective programs. By utilizing the resources, technology, and expertise of various entities, this collaboration strengthens infrastructure, improves data security, and expands digital literacy in the community. Through the right strategies, such as the formation of consortia, standardization of protocols, and efficiency of resource management, collaboration between entities can provide long-term benefits for electronic records management resilience and help communities manage vital records more securely in the digital era.

6.) Contingency Plan During Emergencies

The government must establish a contingency plan to maintain electronic records during emergencies, anticipating potential Internet disruptions or damage to digital equipment. This contingency plan may include temporary offline storage protocols or alternate access ways for those in remote areas. Contingency plans for electronic records management are important for several key reasons, including ensuring continued accessibility, reducing the risk of data loss, minimizing economic and administrative impacts, and addressing ongoing threats.

An effective contingency plan includes several key components designed to minimize the impact of a disaster on electronic records. A contingency plan for electronic records management during emergencies is an essential step to ensure that critical data and records remain secure and accessible even if a disruption occurs. With key components such as data backup, disaster recovery, infrastructure redundancy, and staff training, a successful contingency plan can reduce the impact of a disaster and expedite the recovery process. Despite the challenges, the community ensures the effective implementation of these contingency plans through strategies like crowdfunding, developing disaster-resilient infrastructure, and increasing digital literacy.

Moreover, digital repositories like SIMARTA elucidate and safeguard history for posterity, guaranteeing authenticity, preservation, and societal consistency. Digital repositories function as both storage for records and community organizations that aggregate historical data, serving as tools for the construction of collective memory. Digital repositories influence historical and scientific viewpoints and enhance the national identity of communities (Hendrawan et al., 2024).

In the digital age, repositories face significant challenges, especially with the introduction of electronic materials that require different management and preservation methods compared to physical records. Digitization has emerged as a priority to guarantee the accessibility and preservation of information, yet it requires sufficient technical resources, established standards, and skilled personnel. This problem entails preserving the legitimacy of records while promoting public access and ensuring that digital repositories achieve their strategic objectives.

CONCLUSION AND FUTURE RESEARCH

This chapter demonstrates that electronic records management via a digital repository called SIMARTA plays a crucial role in assisting communities in managing the repercussions of disasters, particularly in safeguarding and preserving vital records susceptible to harm from natural disasters. The flash flood that struck Batu City in November 2021 underscored the pressing need for a safe and readily available household electronic records management system. The destruction and loss of physical archives due to floods highlight the inadequacy of conventional documenting methods in mitigating the increasing risk of disasters caused by climate change and other natural disaster phenomena.

The Universitas Brawijaya developed SIMARTA, which provides a digital solution to manage household records in electronic format. Even if calamities compromise or destroy their physical records, this method allows individuals to digitally save, access, and share their records. SIMARTA's functionalities, which provide remote access, ensure data security, and promote sharing with family members, instill a feeling of security for users, particularly those residing in disaster-prone regions.

This chapter identified several substantial challenges in the implementation of electronic records management across the community. Lack of digital literacy restricts Internet access in some regions, and concerns over data security are barriers that must be addressed to enhance the community's readiness and confidence in transitioning to electronic records. To ensure the successful implementation of the electronic records management program in disaster mitigation, the government must consider comprehensive policies that enhance digital infrastructure, promote digital literacy, ensure data safety, and provide financial assistance to the community. Given the rising frequency and severity of disasters in the Republic of Indonesia, the integration of electronic records into domestic life is an unavoidable progression. The government or other entities may adopt and enhance SIMARTA as a prototype to expand this service to diverse regions. With appropriate legislative support, electronic records management via repositories may enhance community resilience and save essential records and information for the future.

Numerous avenues for advanced research exist that can enhance comprehension of electronic records management within the framework of disaster mitigation. These include developing a records management model, assessing the effectiveness of digital literacy programs for electronic records management, investigating data security vulnerabilities in digital repositories, devising strategies for electronic records management in remote areas, comparing the use of electronic records management in disaster prevention across various regions and countries, particularly developing ones, and analyzing the social and economic impacts of implementing electronic records management. The future proliferation of research may facilitate the steady enhancement of electronic records management implementation in the Republic of Indonesia, rendering it a more effective and adaptable option for addressing more complex disaster risks.

REFERENCES

Al-Sakran, H. O., & Alsudairi, M. A. (2021). Usability and accessibility assessment of Saudi Arabia mobile e-government websites. *IEEE Access : Practical Innovations, Open Solutions*, *9*, 48254–48275. DOI: 10.1109/access.2021.3068917

ANTARA News. (2021, November 6th). Rumah Rusak Akibat Banjir Bandang di Kota Batu Bertambah. Antara News. Retrieved from https://www.antaranews.com/berita/2506153/rumah-rusak-akibat-banjir-bandang-di-kota-batu-bertambah Access on October 11th 2024.

Bates, D. W., Ebell, M., Gotlieb, E., Zapp, J., & Mullins, H. C. (2003). A proposal for electronic medical records in US primary care. *Journal of the American Medical Informatics Association : JAMIA*, *10*(1), 1–10. DOI: 10.1197/jamia.m1097 PMID: 12509352

Cole, T. (2022). Developing and Maintaining Practical Archives: A How-To-Do-It Manual. 3rd ed. By Gregory S. Hunter. *Archival Issues, 41*(2). https://doi.org/DOI: 10.31274/archivalissues.15645

Conway, P. (2011). Archival quality and long-term preservation: A research framework for validating the usefulness of digital surrogates. *Archival Science*, *11*, 293–309. DOI: 10.1007/s10502-011-9155-0

Dikopoulou, A., & Mihiotis, A. (2012). The contribution of records management to good governance. *The TQM Journal*, *24*(2), 123–141. DOI: 10.1108/17542731211215071

Duranti, L., Eastwood, T., & MacNeil, H. (2013). *Preservation of the integrity of electronic records* (Vol. 2). Springer Science & Business Media., DOI: 10.1007/978-94-015-9892-7

Franks, P. C. (2013). *Records and information management.* American Library Association.

Gardner, H., & Davis, K. (2017). *The app generation: How today's youth navigate identity, intimacy, and imagination in a digital world.* Yale University Press., DOI: 10.12987/9780300199185

Gilliland, A. J., & Caswell, M. (2015). Records and their imaginaries: Imagining the impossible, making possible the imagined. *Archival Science*, *16*, 53–75.

Haddad, A., Habaebi, M. H., Islam, M. R., Hasbullah, N. F., & Zabidi, S. A. (2022). Systematic review on ai-blockchain based e-healthcare records management systems. *IEEE Access : Practical Innovations, Open Solutions, 10*, 94583–94615. DOI: 10.1109/access.2022.3201878

Hendrawan, M. R., Isa, A. M., & Samsudin, A. Z. H. (2024). Initiating Memory Institutions Convergence Through Digital Convergence in Indonesian World Heritage Sites. In *Multidisciplinary Approach to Information Technology in Library and Information Science* (pp. 231–251). IGI Global., DOI: 10.4018/979-8-3693-2841-5.ch013

Ketelaar, E. (2017). Archival turns and returns: Studies of the archive. In *Research in the Archival Multiverse* (pp. 228–268). Monash University Publishing.

Lee, P. C. (2024). Unveiling community resilience: The integral role of public libraries. *Journal of Library Administration, 64*(2), 194–216. DOI: 10.1080/01930826.2024.2305072

Meyer, A., & Fourie, I. (2021). Ethical Challenges and Protection of Privacy. In *Autoethnography for Librarians and Information Scientists* (pp. 92–111). Routledge., DOI: 10.4324/9781003014775-7

Njogu, H. W. (2022). Effects of droughts on the delivery of infrastructure services in Kenya. *In Natural Resources Forum* (Vol. 46, No. 2, pp. 221-244). UK: Blackwell Publishing Ltd. https://doi.org/DOI: 10.1111/1477-8947.12251

Putra, D., Sofwany, B., Zulfinanda, H., & Kamaruddin, I. (2022). Flash Flood (Δ) Risk and Damage Assessment in Batu, East Java. *Jurnal Teknosains, 12*(1), 72–84. DOI: 10.22146/teknosains.78483

Saffady, W. (2021). *Records and information management: fundamentals of professional practice*. Rowman & Littlefield.

Shepherd, E., & Yeo, G. (2003). *Managing records: A handbook of principles and practice*. Facet Publishing.

Shobaruddin, M., Hendrawan, M. R., & Bafadhal, A. S. (2024). Multi-Disaster Risk Mitigation through Augmented Reality Conservation Map for Enhancing Cultural Heritage Attitude. *Environment-Behaviour Proceedings Journal, 9*(SI18), 205–213. DOI: 10.21834/e-bpj.v9isi18.5484

Smallwood, R. F. (2019). *Information governance: Concepts, strategies and best practices*. John Wiley & Sons., DOI: 10.1002/9781119491422

Smith, K. (2016). The importance of records in the right to information. In *Integrity in Government through Records Management* (pp. 149–159). Routledge.

Vaarzon-Morel, P., Barwick, L., & Green, J. (2021). Sharing and storing digital cultural records in Central Australian Indigenous communities. *New Media & Society*, *23*(4), 692–714. DOI: 10.1177/1461444820954201

Van Laar, E., Van Deursen, A. J., Van Dijk, J. A., & De Haan, J. (2017). The relation between 21st-century skills and digital skills: A systematic literature review. *Computers in Human Behavior*, *72*, 577–588. DOI: 10.1016/j.chb.2017.03.010

Yeo, G. (2018). *Records, information, and data: exploring the role of record-keeping in an information culture*. Facet Publishing., DOI: 10.29085/9781783302284

Yin, R. K. (2018). *Case study research and applications: design and methods* (6th ed.). Sage.

KEY TERMS AND DEFINITIONS

Community Preparedness: The proactive measures taken by a community to anticipate, plan for, and reduce the impact of disasters. This includes public education, training programs, emergency simulations, and strategic planning to ensure that communities can respond swiftly and recover efficiently when a disaster occurs.

Community Resilience: The capacity of a community to absorb, adapt to, and recover from the adverse effects of disasters or disruptions. Digital technologies enhance this resilience by enabling the secure storage of critical information and ensuring rapid access to essential records during emergencies.

Digital Literacy: The ability of individuals to effectively understand, navigate, and use digital technologies. Digital literacy is fundamental for communities to utilize digital platforms safely and efficiently, especially for managing and protecting personal information in a digital environment.

Digital Repository: A structured and secure system designed for the storage, management, and long-term preservation of digital documents and records. Digital repositories often include advanced search, access control, and metadata features to facilitate efficient information retrieval and sharing.

Disaster Mitigation: Preventive actions aimed at reducing the severity and consequences of disasters on people, property, and the environment. In the context of digital recordkeeping, this includes protecting household records through electronic storage to prevent their loss or destruction during natural disasters.

Disaster Resilience: The ability of a system, community, or society to withstand, adapt to, and recover from the impacts of disasters. This encompasses restoring social, economic, and institutional functions, often with the help of preparedness strategies and technological support.

Household Records: Vital personal and family documents such as birth certificates, marriage licenses, property deeds, health records, and identification cards. These records are essential for accessing public services, proving legal identity, and managing family affairs before, during, and after emergencies.

Natural Disasters: Severe and often unpredictable natural events, such as floods, earthquakes, volcanic eruptions, landslides, and tsunamis, that can result in loss of life, property damage, and significant disruption to society and the environment.

Records Management: A professional discipline focused on the systematic control of records throughout their life cycle—from creation and use to final disposition. The objective is to ensure that records are accessible, reliable, and properly maintained to support legal, administrative, and historical needs.

Technology Adoption: The process by which individuals, organizations, or communities accept and integrate new technologies into their daily practices. In this context, it refers to the uptake of digital repositories as a preferred method for managing and protecting vital records in disaster-prone settings.

Chapter 9
Challenges in the Digital Preservation of Indigenous Knowledge in Southern African Countries

Kwanisai Mukwerete
https://orcid.org/0009-0000-6761-5665
University of South Africa, Bulawayo, Zimbabwe

Shadreck Ndinde
https://orcid.org/0000-0001-8678-9265
National University of Science and Technology, Bulawayo, Zimbabwe

ABSTRACT

This chapter explores the challenges faced in digitally preserving Indigenous knowledge concerning Southern African countries. It examined the significance of digitising Indigenous Knowledge, emphasising the processes involved in its digitisation. However, even though the digitalisation of Indigenous knowledge is a malleable idea, it has opened a 'Pandora's box' of challenges. The study used a desk study in which the researchers gathered data from the existing literature using literature reviews from books, journal articles, internet sources, and newspapers; this chapter will explore the need to preserve indigenous knowledge, although most of it is not preserved. This chapter identifies the barriers to the practical digital preservation of Indigenous Knowledge and offers insights and recommendations for creating sustainable and culturally sensitive digitisation strategies. The findings

DOI: 10.4018/979-8-3373-0204-1.ch009

aim to contribute to academic discourse and inform policymakers, researchers, and practitioners seeking to bridge the gap between traditional knowledge systems and modern technological advancements.

INTRODUCTION

The digital preservation of Indigenous Knowledge (IK) is a critical and timely subject, especially in Southern African countries where rich cultural heritage and traditional knowledge are at risk of being lost due to modernisation, lack of resources, and other challenges (Masenya, 2023). IK represents the rich traditions, practices, and wisdom developed by Indigenous communities over generations, often rooted in unique interactions with their environment, culture, and heritage (Bihari, 2023). Several attempts have been made to address the universal preservation challenges of digital materials. The United Nations Educational, Scientific, and Cultural Organisation (UNESCO) and the International Federation of Library Associations and Institutions (IFLA) are prominent international organisations active in these global projects. The 2003 UNESCO General Conference's "draft charter" on preserving digital heritage was one such attempt. The charter serves as a statement of principles to help Member Nations create national policies for the preservation and accessibility of digital heritage (Osedo, 2013).

In Southern African countries, this knowledge is vital in agriculture, healthcare, environmental conservation, and sustainable living (Lwoga, Ngulube and Stilwell, 2010). However, as the world increasingly relies on digital technology, preserving this invaluable knowledge has become more urgent (Ngulube, 2025). Digitisation offers a promising solution, enabling the documentation, storage, and dissemination of IK to ensure its longevity and accessibility for future generations. Despite its potential, numerous obstacles jeopardise the sustainability and efficacy of digitally preserving IK in Southern Africa. Social dynamics, moral challenges, intellectual property issues, and insufficient technological infrastructure complicate digitisation. Furthermore, much of this knowledge is undocumented, making it challenging to gather and retain without running the danger of loss or misinterpretation.

This study focuses on African countries that are part of the Southern African Development Community (SADC), which are usually the focus of digital preservation efforts for IK in Southern Africa. These include nations in the region, including Mozambique, Angola, Lesotho, Namibia, Zambia, Malawi, South Africa, Zimbabwe, Botswana, and Swaziland (Eswatini). Every nation's distinctive cultural heritage and Indigenous Knowledge Systems (IKS) are essential to their sustainability and sense of self. For instance, with programs like the National Recordal System, South Africa has made considerable progress in digitising IK. Balogun and Kalusopa

(2021) buttress that South Africa is among the select African nations that have dedicated significant resources to developing and digitalising IKS. Other nations may encounter differing difficulties because of disparities in their finances, policy frameworks, and technological infrastructure. According to Greyling and McNulty (2012), an example of digital preservation of Indigenous Knowledge is the Ulwazi Programme and the Digital Innovation South Africa (DISA), which made IK available online. The Ulwazi Programme is a community-driven digital library that preserves and shares IK through a wiki-style database. It enables digital literacy, preserves cultural heritage, and fosters collaboration between libraries, researchers, and locals, promoting digital literacy. If adopted by Southern African nations, such innovative ideas and methods will advance the preservation of IK.

This chapter examines these challenges and categorises them into four key domains: technological, ethical, intellectual, and social. The study aims to illuminate pathways toward sustainable and culturally sensitive digital preservation strategies by exploring the processes involved in digitisation and offering practical insights. Ultimately, this work seeks to bridge the gap between traditional knowledge systems and modern technological advancements, safeguarding the cultural heritage of Southern African communities in an ever-changing digital era.

METHODOLOGY

This study employed a desk research approach, gathering data from existing literature. Desk research relies on secondary sources for data collection, such as books, journals, websites, reports, and other published materials (Kiely, 2024). Cheong et al. (2023) highlight that this method involves synthesising information from various resources to understand the research topic comprehensively. Morgan (2022) notes that desk research is beneficial when researchers do not have the time or money to undertake field research. The study analysed current literature and case studies to explore challenges and provide insights into real-world applications. The analysis focused on organising, compiling, and interpreting data to draw valid conclusions. Desk research is a practical methodology for understanding concepts thoroughly by utilising already available information. It offers researchers opportunities to enhance their understanding of a topic, identify gaps in knowledge, and complement original findings with context and background details.

Secondary research relies heavily on existing materials, such as academic papers, reports, surveys, and online content, often accessible through libraries and digital platforms. This approach enables researchers to evaluate information critically, build upon established knowledge, and inform subsequent phases of their work. Desk research is faster, more cost-effective, and more accessible than primary research methods. It requires less time and resources, eliminating the need for recruitment or surveys. Desk research can be combined with other sources to provide a comprehensive overview. Desk Research Methods have limitations such as data quality, less control, potential bias, and limited control. To mitigate bias, users should rely on credible sources or corroborate evidence with multiple sources. Secondary data analysis based on published data has been successful. The original data may also serve as the basis for secondary data analysis if preserved in an archive (Church, 2002). Desk research primarily employs qualitative and quantitative methodologies. On the one hand, the qualitative methodology focuses on gathering descriptive information to understand concepts, trends, or phenomena deeply (Largan and Morris, 2019). This involves analysing textual data from books, articles, reports, and websites to identify patterns and interpret meanings. On the other hand, the quantitative methodology centres on numerical data, relying on statistical analysis to conclude. It involves reviewing datasets, surveys, and reports with measurable figures, percentages, or frequencies. By integrating both methods, desk research provides a balanced perspective, combining numerical precision with contextual depth to comprehensively address research objectives. Examples of desk research methods include literature review, competitor analysis, social listening, and consumer intelligence. Qualitative methodology and literature review methods were used in this study. With literature review methods, research findings can be synthesised in a literature review to demonstrate meta-level evidence and identify areas requiring more investigation (Snyder, 2019). This is a crucial step in developing theoretical frameworks and conceptual models.

CONCEPTUAL FRAMEWORK

Imenda (2014) defined a conceptual framework as the outcome of combining multiple linked ideas to provide a more comprehensive knowledge of a research challenge, a phenomenon of interest, or an explanation or prediction of a particular event. To successfully apply a conceptual framework to the difficulties in the digital preservation of IK in Southern African nations, this study used one since it views comprehension of important concepts or components of theories and models as essential. The authors built their conceptual framework for this chapter.

Figure 1. Conceptual framework

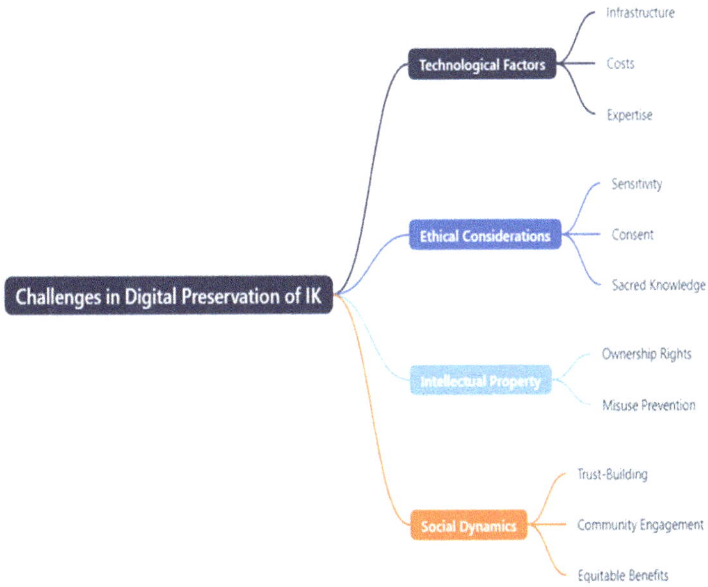

Source: Authors' contribution (Mukwerete & Ndinde, 2025).

The conceptual framework avails the digital preservation of IK's unique challenges across four key domains:

Technological: Limited access to advanced tools and infrastructure, especially in rural areas, can hinder digitisation efforts. Rapidly evolving technology also risks the obsolescence of stored data. The digital preservation of IK in Southern African nations is beset by technological difficulties (Malapane, Chanza and Musakwa, 2024). For example, rural Zimbabwe frequently lacks electricity and dependable internet connectivity, which makes it challenging to digitise and store IK efficiently (Nherera and Mukora, 2024). Therefore, limited access to digital instruments puts traditional farming methods and herbal medicines at risk of disappearing. In South Africa, rural villages struggle with issues like antiquated technology and software, whereas urban areas enjoy superior infrastructure (Seretse, Chukwuere, Lubbe, and Klopper, 2018). The necessity for sustainable technological solutions is demonstrated by initiatives such as the Taung community's digitisation of cultural traditions (Masenya, 2023). Although Botswana has taken steps to preserve IK, such as recording customary rituals like Bogwera (Aiseng, 2023). It has difficulties because it lacks financing for innovative digital technologies. These illustrations highlight how critical it is to close infrastructure gaps and guarantee that protected data is accessible for a long time.

Ethical: It is essential to make sure that IK is conserved with dignity and with their permission. It is, nevertheless, overly concerning when sensitive cultural information is misused or exploited. One top priority is appropriately preserving IK (Balogun and Kalusopa, 2021). Heritage loss, mistrust, and cultural insensitivity can result from misuse or exploitation. Preservation of cultural heritage is a shared responsibility that involves individuals, communities, governments, and international organisations. To prevent the cultural erosion from occurring, consider the following crucial elements (Stephenson, 2023). Complete participation and decision-making power over who owns what knowledge and what is retained. Thus, their originality and cultural integrity are ensured. Explicit consent processes that enable communities to restrict IK sharing while protecting personal information should be in place. Governments and organisations must implement legal frameworks that acknowledge and safeguard IK owned by the community, avoiding its abuse for profit. According to community norms, IK data should be stored securely, and only authorised personnel should have access.

Intellectual: Ownership and intellectual property rights over IK are often contested. Balancing community rights with broader access can be complex. Handique and Dubey's 2022 article examines the legal mechanisms for protecting IK, a valuable resource for Indigenous communities. The article highlights the tension between traditional intellectual property laws and IK's collective nature. It discusses the limitations of existing legal instruments and emphasises the need for innovative solutions, international cooperation, technological advancements, and capacity building. The article calls for a paradigm shift towards holistic approaches, acknowledging IK's value in addressing global challenges and honouring Indigenous peoples' rights. It calls for a collaborative dialogue among stakeholders to ensure adequate protection and sustainable use of IK.

Social: Resistance to digitisation due to fears of cultural erosion or loss of control over heritage is common. Building trust and collaboration between communities and institutions is essential. Boamah and Liew's (2017) conference paper explores the challenges faced by memory institutions in digitising cultural heritage resources, focusing on developed and developing countries. They highlight the need for addressing attitudinal issues, such as fear of permanent loss of heritage, lack of interest from stakeholders, and conflicting ideas on the digitisation process. They argue that these issues should be explicitly discussed in discussions about cultural heritage digitisation and preservation, aiming to eliminate gaps in future memory.

Importance of Digitising IK

There are several reasons why digitising IK is essential. Digitising IK is crucial in preserving the knowledge for future use (Masenya, 2024). According to Nettleton (2020), preserving IK in digital format helps many people digitally protect IK, allowing them to access it globally. For example, the University of Cape Town's Digital Heritage Project persevered in music, oral history and other digital stories, making them globally accessible. On the other hand, in Masenya (2024), digital preservation has become a popular method for revitalisation and safeguarding IK in recent years. Plockey (2014: 21), revealed that "African nations are going through many changes resulting in a loss of traditional, cultural and customary knowledge as a result of a lack of preservation and digitisation of African Indigenous Knowledge (AIK)". To avoid losing IK, Bawack, Roderick, Badhrus, Dennehy, and Corbett (2025) revealed that documenting IK can help to preserve the knowledge for future generations. Amunkete (2020) indicated that, "Indigenous knowledge is mostly present within older generations, and if it is not preserved, this knowledge will die with its custodians." Various formats are used in the digital preservation of IK as put forward by Chigwada and Ngulube (2023) that IK can be in the form of digital texts, audio recordings, video recordings, photographs, digital artefacts and objects, interactive multimedia, maps, digital archives and repositories, and digital exhibitions and online collections. Balogun (2023) revealed that the initiative in digital preservation in Africa is the African Indigenous Science and Knowledge Systems, which is available online at (https://africahistory.net/). Further, Balogun and Kalosopa (2023) revealed that, in South Africa, the University of KwaZulu-Natal (UKZN) is responsible for maintaining the Indigenous Knowledge Systems website (www.iks.ukzn.ac.za.)

Further, digitising IK is especially important because it helps to guard against cultural erosion in a society. Mbigi (2019) indicate that IK can be lost since most of the knowledge is transmitted orally; their IK can be lost since most of the knowledge is transmitted orally; there are issues with globalisation, modernisation, and generational shifts. Biyela, Oyelude and Haumba (2016) opine that digitisation is seen to ensure the long-term preservation and worldwide accessibility of Indigenous Knowledge (IK). Mbigi (2019) provided an example of the KwaZulu-Natal Museum, which has digitised its materials, including medicinal plants, Zulu oral traditions, and traditional crafts, that can be used in the future even if the current generation passes on. Digitising IK is especially important, as it helps in cultural revitalisation, as indicated by Louw and King (2021) in Namibia. Under the Khoisan Language Project, the Khoisan people are recording their language with translations, thereby preserving their heritage. Digitising IK can influence the research and documentation of knowledge. According to Chikozho and Mabhena (2020), the digital preservation of IK is critical in documenting the knowledge. For example, the African Traditional and Indigenous Knowledge System project in Zimbabwe focuses on documenting IK used for farming and biodiversity, among other things. Further, digitising IK can foster collaboration in a community. Pottier (2021) reveals that digitising IK under the Ubuntu Net Alliance in Southern Africa triggered cooperation and dissemination of traditional farming methods among the communities in South Africa. Digitising IK has economic benefits that can benefit the local communities. For example, Mapungubwe Heritage Site in Limpopo, South Africa, shows how digital preservation of IK has manifested economic benefits by participating in tourism and educational enterprises. Mdhluli, Mokgoatšana (2022) opine that, looking at the extinction of VhaVenda oral history, the importance of digitising oral history cannot be overemphasised. The VhaVenda (Venda or Vhangona) are Southern African Bantu people residing close to the South African-Zimbabwean border.

Challenges with Digitisation

The available literature indicated that many challenges emanate from digitally preserving IK. Masenya and Ngulube (2020) revealed that organizations encounter numerous obstacles in their efforts to safeguard digital content, such as insufficient institutional dedication and engagement, no established standards, policies, or procedures for digital preservation, limited resources, a shortage of skills and training, inadequate funding, minimal collaboration and partnership initiatives, and issues related to technological obsolescence. Further, Bishoff and Smith (2015) pointed out that the obstacles to establishing sustainable digital preservation programs include insufficient funding, a shortage of expertise, inadequate institutional backing, and uncertainty about how to begin. Chigwada and Ngulube (2023) cited several authors indicating that, the obstacles encompass issues in the digital preservation of IK include cultural sensitivity and respect, gaining access to indigenous communities, concerns over intellectual property and ownership, the digital divide, language differences, ethical considerations, copyright regulations, data privacy and security, sustainability and long-term preservation, insufficient funding and resources, a shortage of expertise, and the suitability of various technologies. Moreover, Adu and Ngulube (2016) and Biyela et al. (2016) revealed that digital preservation of IK includes technological obsolescence, lack of awareness, lack of finance to sustain the project, security, legislation and piracy.

Biyela, Oyelude, and Haumba (2016) investigated the level of digitisation of IK in South Africa, Uganda, and Nigeria. Most of the world's heritage treasures, even those created digitally, have been lost, and some cannot be restored because of neglect. The authors cite numerous studies to support their claim. Semi-structured interviews were used to collect the data addressed and provided in this research. The study suggested the need for intensifying digitisation projects of IK found in rural communities, a collaborative approach, increased funding, and capacitating information professionals in digitising heritage resources.

1. Technological Challenges and Digital Divide

Rapid technological changes can pose serious challenges in digitising IK since, once digitised, technology can change. Once technology changes, recovering the data saved will be challenging. Masenya (2021) indicated that technological obsolescence is a significant challenge that complicates the long-term preservation of digital records. The same sentiments were also put forward by De la Porte and Higgs (2019), who stated that poor technological infrastructure can hamper the digitalisation of cultural knowledge. Further, in Namibia, Amunkete (2020) in a study titled "A model for the digital preservation of Indigenous Knowledge on medicinal plants in Namibia via an e-learning platform" indicated that there is still a lack of technology in the digital preservation of IK.

The digital preservation of IK in Southern African nations encounters difficulties due to the digital divide. Dutta (2019), Mwanza (2022), and the United Nations (2021) highlight the problem of the digital divide, indicating that numerous indigenous communities often do not have consistent access to reliable internet and digital tools. This lack of connectivity hampers their ability to participate in digital preservation efforts and access digital resources. These challenges include restricted access to technology, inadequate skills and infrastructure, intellectual property and cultural sensitivity concerns, and insufficient funding and policy support. Most Southern African countries, such as Zimbabwe, Zambia, and Mozambique, have limited access to technology such as computers, the internet, scanners, and voice recorders, the main tools for digitising IK. Some rural areas do not have access to electricity or the Internet, which is very bad since most IK systems are found in rural areas. Some of the countries have limited infrastructure that is needed for the digitisation of IK, which includes unreliable power supply and a shortage of storage.

There is also a severe shortage of qualified staff who can work on the digitisation of IK in Southern Africa. Asogwa (2011) also believes that a lack of technical expertise is a common problem in many African countries, leading to a shortage of human capital. Balogun (2023) revealed that IK recorders need to be trained on how to use the equipment effectively, proper interviewing skills, taking photos and recording videos to capture top-quality data. However, most countries in Southern Africa, such as Zimbabwe and Zambia, have suffered due to a 'brain drain' where qualified personnel of crucial importance in digitising IK look for greener pastures in neighbouring countries such as South Africa. Another challenge that might affect the digitisation of the IKS in Southern Africa is the digital divide. According to Osei and Moyo (2020), the digital divide affects most Southern African countries. Most people in rural areas do not have access to the internet and the technology needed to digitise IK. Mthembu (2021) reiterates that most African countries have no electricity limited or unreliable internet access in rural areas where IK is shared.

This hampers the digital preservation of IK since technological skills are also scarce in rural areas.

Dela Porte and Higgs's (2019) paper, “Challenges in the digitisation of cultural heritage material in the Western Cape, South Africa,” revealed that 76% of the participants recognised that the staff's inadequate technical skills and insufficient training posed a challenge, whereas 70% noted that the lack of knowledge affected digitisation efforts. Kala (2023) found that many African nations, such as Zambia, struggle with restricted access to electricity and internet services. The lack of reliable power and a strong telecommunications infrastructure impedes these countries' ability to fully adopt and utilise technology.

Further, the rapid changes in technologies can also have a detrimental effect on the digital preservation of IK in Southern Africa. Chigwada and Ngulube (2023) buttress the point that technological obsolescence is a challenge in the management of IK, of which they recommend the need to keep the hardware and software. The issue of technological obsolescence is a challenge in IK management. Sraku-Lartey (2017) notes that modern technologies, including the hardware, software, and data formats for creating and storing digital collections, can be costly because they quickly become obsolete. This rapid obsolescence complicates access to preserved IK. Carrado (2019) suggests that while technology can become outdated, it is essential to establish sustainable funding sources and policies to support ongoing digital preservation efforts.

2. Intellectual Property Infringements

Digital preservation of IK in Southern African countries faces significant challenges, particularly concerning intellectual property (IP) infringements. One major issue is the lack of legal frameworks that specifically protect IK from appropriation by external entities. For instance, the case of the Hoodia plant, traditionally used by the San people of Southern Africa for its appetite-suppressing properties, illustrates this challenge. In the early 2000s, a South African company patented a formulation derived from Hoodia without adequately compensating or recognising the San communities. This led to widespread controversy over the exploitation of IK. This situation highlights the vulnerability of indigenous communities, which often lack the resources to enforce their rights or engage in legal battles against well-funded corporations. Further, digitising IK can have some complicated intellectual property issues. According to Natural Justice (2023) in South Africa, in 2009, Nestle filed a patent application on the uses of rooibos and honeybush for treating skin disorders, inflammation. Balogun (2023) revealed that the government of South Africa, through the National Recordal System, to prevent issues related to do with intellectual property and biopiracy, led to the formation of the Indigenous Knowledge

Systems Documentation Centres (IKSDCs) in South Africa across the nine provinces to cater for the fifty indigenous communities. However, Sraker-Lartey et al. (2017) indicated that, currently, there is no global agreement on how to legally protect the rights of indigenous communities over their cultural knowledge systems, nor on how to ethically promote and responsibly utilise this knowledge. Ntsebeza (2022) indicates that no legal intellectual frameworks govern the use of IK. In Southern Africa, no specific laws of IK govern the use of IK, which is communally owned. IK is impeded in an ethical setting. Therefore, digital preservation can change the meaning. This can lead to misinterpretation of IK systems.

3. Ethical Challenges

Ethical challenges hamper the digitisation of IK in Southern Africa. Awuor (2023) put forward the same sentiments, revealing that digitising IK has ethical concerns such as control, access and consent. Chigwada (2024) buttresses that preservation and sharing of IK in the digital era pose distinct ethical and technical challenges. During the first session of the seminar series on the second day, Chigwada stressed the significance of integrating the CARE (Collective Benefit, Authority to Control, Responsibility, and Ethics) and FAIR (Findable, Accessible, Interoperable, Reusable) principles for the digital preservation of IKS. The CARE principles focus on the ethical management of IK, ensuring that the rights and benefits of Indigenous communities are prioritised. Conversely, the FAIR principles aim to improve data accessibility and usability, but they might compromise community control and the safeguarding of sensitive information. Therefore, it is essential to secure ethical approval before collecting IK from its rightful owners. Amunkete (2020) found that preserving IK of medicinal plants using an e-learning platform in Namibia required proper engagement with knowledge owners.

Awuor (2023) indicated a need to address ethical issues in collecting, preserving and managing IK. Therefore, Žukauskas et al. (2018) suggested that for researchers to work legally and respectfully, they should follow ethical considerations. Similar sentiments were put forward by Kamal and Amani (2020), who indicated that the challenge is ensuring that the digitisation of IK respects and reflects Indigenous communities' values, traditions, and rights. Ritchie (2014:16) advises that the 'new technology should be approached with vigilance for oral tradition'. Meanwhile, Murphy (2021) revealed that to digitise IK without the consent and involvement of the communities is to perpetuate colonial legacies of appropriation. Special advice from Mishra, A. (2019), digitisation can empower Indigenous communities, but it must be approached with a framework prioritising their rights and voices. According to Okeke (2023), digitisation must not be seen merely as a technological process;

it is also a cultural and political one, requiring deep engagement with indigenous communities.

According to Loupis and Saffy (2020), IK is kept in a society and is transmitted by word of mouth (orally). Therefore, digitising IK can pose serious challenges since it can be misinterpreted after being it. The same sentiments were opined by Hernandez (2025) that the digital era in preserving IK could lead to the misappropriation and abuse of traditions and practices, exacerbating the injustice they have endured.

Further, Loupis and Saffy (2020) were concerned with the ownership of IK since it could be accessed and used without permission from the founders of the knowledge. Some IK cannot be conveyed in digital format, which challenges future knowledge users. In Kenya, Kanyabuhinya and Juma (2022) indicated that there are policies such as the Traditional Knowledge in the Protection of Traditional Knowledge and Cultural Expression Act 2016 (No.33 of 2016) for protecting IK.

4. Lack of Policies

Another significant problem, especially in Southern African nations, is the absence of regulations about the digital preservation of IK. Masenya (2023) observed that the communities and owners of IK lack the procedures and guidelines necessary to properly preserve their IK digitally. The study suggests that anticolonial policies and tactics for safeguarding IKS should be implemented, and that community members should focus on genuine training and awareness-raising opportunities. Another significant obstacle is the lack of laws that enable digital preservation in Africa and enable policies, standards, and strategies to be implemented (Osede, 2013). This is because legislators and policymakers typically do not know or understand the requirements for digital preservation; as a result, they create laws that either neglect or cover digital preservation issues insufficiently (Wamukoya & Mutula, 2005). The National Library of Australia (2003) states that one major obstacle to digital documents is the absence of a law that supports them. The issue of internet links concerning copyright laws is also connected to this. The problem is that African national legislation has not adequately stated the right to copy for preservation, and the software copyright is needed to access digital assets. For example, to access the same information whenever needed, a subscriber to an internet-based information service must constantly renew the access license due to copyright regulations, even for things that have been paid for a long time. Once more, African nations, governments, and institutions cannot create timely and well-informed preservation policies due to the speed and cost of the digital revolution.

IK digitisation and preservation initiatives frequently transgress cultural norms and customs without defined and thorough frameworks. There is a great need for strong policies to implement the preservation of IK. Balogun and Kalusopa's (2021) study on *A framework for the digital preservation of the IKS in repositories in South Africa* indicates a need to have policies that cover the management of IK that an organisation preserves. The purpose of the study was to assess the digital preservation policies and plans for long-term digital preservation in selected repositories in South Africa, with a view to developing a digital preservation framework for the preservation of the IKS in South Africa. This study also reveals that digital curation, policy formulation and disaster preparedness plans, to some extent, are measures said to be in place for the digital preservation of IKS. The study was, however, limited to the NRS initiative in South Africa. The study calls for governments in Africa to develop policies that will ensure the digital preservation of IKS. Balogun (2023) opines that the South African government developed a policy framework to promote the preservation of IK.

In 2004, recognising the importance of IK, the South African government adopted the National Indigenous Knowledge Systems Policy, a platform for recognising, affirming, developing and protecting IKS in South Africa (Pretorius and Bezuidenhout, 2011). The Department of Science and Technology (DST) initiated the National Recordal System (NRS) through the National indigenous knowledge systems policy framework to coordinate and standardise the capture, storage, maintenance and dissemination of science and technology-related data on IKS in South Africa. In 2014, the DST promulgated an additional IKS Bill, which aims to improve IK management in South Africa further. Such initiatives should be developed in other nations to promote the preservation of IKS. Additionally, there is a need for disaster planning, policy development, and digital curation for the digital preservation of IKS. Governments and stakeholders in Southern Africa must work together to create regulations that safeguard intellectual property, accommodate cultural sensitivities, and provide fair access to knowledge that has been conserved. A well-thought-out plan is also necessary for organisations to have a durable and well-planned digital legacy (De la Porte and Huggins, 2019).

5. Financial challenges

The challenges associated with the digital preservation of IK in Southern African countries are intricately linked to inadequate funding. Many nations in this region are grappling with fundamental socio-economic challenges, such as providing food security, adequate housing, and essential healthcare for their populations. Mdhluli, Mokgoatsana and Kugara (2022) opine that digitising IK can bring challenges, such as a lack of resources to manage the process. Ajani et al. (2024) revealed that preserving IK in the digital age typically demands significant resources, such as financial support, technological infrastructure, and qualified personnel. Chowdhury et al. (2023) found that many Indigenous communities lack the financial capacity for a large-scale digitisation project. Sraku-Lartey, Acquah, Samar and Djagbletey (2017) highlighted the vital role of funding in digitising IK, noting that expenses such as transportation, accommodation, and research instruments. Meanwhile, Mdhluli et al. (2021) revealed that the digital preservation of IK involves processes such as acquisition, preservation, and dissemination, all of which demand substantial financial and human resources.

As a result, allocating financial resources to digital preservation initiatives is often viewed as a secondary concern, if not an outright luxury. For example, in Zimbabwe, the National Archives has struggled to digitise its extensive collections due to a lack of financial support, prioritising immediate humanitarian needs over long-term preservation efforts. This prioritisation reflects a broader trend across Southern Africa, where governments frequently face competing demands for limited resources. This leads to the perception that investment in the digital preservation of IK misallocates funds. According to Jaya (2006), several governments in the Sub-Saharan Africa region have respected the role of IKS and have developed mechanisms to integrate it for sustainable development. Governments in Southern Africa seem to be shunning IK.

According to De la Porte and Higgs (2019), a challenge arises from insufficient financial resources and technological infrastructure, including software, hardware, and connectivity. Digitisation initiatives can be costly, and it's essential to account for staff training expenses alongside the necessary technical infrastructure. Mdhluli, Mokgoatsana and Kugara (2022) state that capturing, managing and preserving digital information is expensive and requires skilled personnel and the right equipment. Additionally, the same scholars found that digitising VhaVenda oral tradition has raised challenges concerning the methodologies, funding and politics in preserving digital heritage materials for archivists and librarians. Moreover, the absence of dedicated budgets for the digital preservation of IKS has resulted in several failed digitisation projects throughout the region.

Another notable case is the South African Heritage Resources Agency's (SAHRA) initiative to digitise historical documents and oral histories, which faced significant setbacks due to insufficient funding and inadequate infrastructure. Initially hailed as a vital step towards safeguarding cultural heritage, the project faltered, revealing the stark realities of bureaucratic inefficiency and the misalignment of priorities. Similarly, the Namibian government's efforts to preserve Indigenous languages through digitisation have been hampered by a lack of financial commitment, leaving many valuable oral traditions at risk of being lost to future generations. These examples highlight the critical need to re-evaluate funding priorities, emphasising the importance of digital preservation as integral to Southern African societies' cultural and historical fabric. However, Pickover (2008) revealed that the Digital Innovation South Africa (DISA) project in South Africa, which was funded by the

Andrew W. Mellon Foundation which attracted many academic institutions, research and heritage sites.

6. Social Challenges

The digital preservation of IK faces significant social challenges that stem from the complex interplay between technology, culture, and power dynamics. One of the primary challenges is the risk of cultural appropriation and misrepresentation. Indigenous communities often have unique knowledge systems that are deeply intertwined with their cultural identities. When this knowledge is digitised without proper context or consent, there is a threat of it being extracted and commodified by external entities, leading to a loss of authenticity and control over their narratives. In a study that was carried out by Chigwada and Ngulube (2023) titled, "Librarians' role in the preservation and dissemination of indigenous knowledge" found that during the acquisition process, the librarians reported encountering opposition from certain IK holders, who were reluctant to share their IK, particularly when they were aware that it was being documented.

Furthermore, the ethics of consent and ownership in digital preservation present a complex social challenge. Many indigenous cultures emphasise communal ownership of knowledge, contrasting sharply with individualistic approaches prevalent in Western frameworks. According to Senanayake (2006), IK is regarded as a social capital of people living in poverty as their main source for survival in terms of food production, shelter and even control over their own lives. This raises questions about who has the authority to digitise and disseminate IK and how to navigate the intricacies of consent. Chigwada and Ngulube (2023) indicated that libraries need to engage with IK in a culturally sensitive manner that honours the beliefs, practices, and protocols of Indigenous communities. Neglecting this approach can result in misunderstandings and conflicts, obstructing collaboration and knowledge exchange. Bagley (2018) in a study titled, "Toward an Effective Indigenous Knowledge Protection Regime Case Study of South Africa" reported that a Xhosa group in South Africa entered a benefit-sharing with a multinational corporation Schwabe Pharmaceuticals on the use of Pelargonium sidoides, a plant used for treating respiratory problems of which a rival group also went to claim the benefits and they won. Engaging indigenous communities in the decision-making process is crucial yet often overlooked in the rush to digitise. Establishing ethical guidelines and collaborative frameworks that respect indigenous sovereignty and cultural protocols is essential for fostering trust and ensuring that digital preservation efforts are genuinely beneficial to the communities they aim to serve.

7. Language Challenge

Language can be another challenge in the digital preservation of IK in Southern Africa since many indigenous knowledge systems are expressed in local languages that may not have standardised written forms. Many indigenous languages are spoken in Southern African countries. The region is home to a rich diversity of indigenous languages, including Zulu and Xhosa in South Africa, Venda in Zimbabwe, and languages such as Bemba in Zambia, Makhuwa in Mozambique, Setswana in Botswana, and Swahili in Tanzania. These languages are vital carriers of cultural heritage and traditional knowledge, yet their oral and unwritten nature complicates efforts to document and preserve this information digitally.

According to Settee (2008), language is a key factor in collecting and preserving IK. In their research, Sraku-Lartey et al. (2017) discovered that language plays a crucial role in the gathering and safeguarding of IK. It is central to culture and the retention of knowledge, serving as either a unifying force or an obstacle in preserving IK. For instance, during their study on IK in Ghana, the Akan people were identifying trees using their native language, which translated into English, both costly and challenging. Cosijn et al. (2002) indicated that in preserving IK, language can be a barrier since most IK is stored in native languages. In Namibia, Gallert, Winschiers-Theophilus, Kapuire and Stanley (2016) in their study titled "Indigenous Knowledge for Wikipedia: A Case Study with an OvaHerero Community in Eastern Namibia" found that translating the findings from local language to English might distort the meaning of the message or narrative. Addressing these linguistic challenges is essential to ensure that IK is accurately preserved and accessible for future generations.

8. Lack of Standards

The importance of establishing standards in the digital preservation of IK cannot be overstated. The study by Balogun and Kalusopa (2021) highlights critical issues surrounding the web archiving of indigenous knowledge systems in South Africa, particularly the absence of standardised policies for digitisation. Meanwhile, according to Amechi (2015) and Bagley (2019), the initiative to safeguard IK in South Africa resulted in the development of the National Recordal System (NRS) by the Department of Science and Technology (DST). The project aims to organise and standardise the collection, storage, management, and distribution of science and technology-related Indigenous Knowledge System data within the country. According to Chigwada and Ngulube (2023), there was a need for Zimbabwe to revive its Zimbabwe Resource Centre for Indigenous Knowledge or start a new organisation in charge of the management of IK in Zimbabwe. The organisation should play a pivotal role in the acquisition, preservation and dissemination of IK in the digital era. Most of these projects remain on paper due to a lack of implementation, due to a lack of funding from the government. Chigwanda and Ngulube (2023) buttress the same point that most of the IK preservation programs fail because of a lack of resources to implement the projects successfully.

This lack of standardisation poses significant challenges for the effective preservation of IK, which is essential for maintaining its accessibility, authenticity, and integrity over time. The researchers found out that there were no standards in the digital preservation of knowledge in many countries in Southern Africa besides South Africa, which had come up with Web digital preservation. Without established guidelines, digital preservation teams may struggle to determine what knowledge should be preserved, leading to potential oversights or biases in the selection process. Moreover, the absence of standards regarding digital formats can lead to long-term issues such as format obsolescence, where the technology to access certain formats becomes outdated, or data corruption, which could render valuable knowledge irretrievable.

9. Data Security

Data security represents a significant challenge in the digital preservation of IK in Southern Africa. As information is converted into digital formats, it becomes vulnerable to a range of threats, including viruses, hacking, data theft, and unauthorised access. Researchers such as Chigwada and Ngulube (2023) have highlighted that security issues are pervasive across various types of libraries and information repositories managing IK in the digital age. They warn that digital systems are susceptible to malware attacks, system crashes, and theft, all of which can lead to catastrophic content loss. The risks are compounded by the fact that indigenous communities often regard their cultural knowledge as sacred, making the protection of this data not just a technical necessity but an ethical imperative. These risks not only threaten the integrity and availability of valuable cultural heritage but also raise ethical concerns regarding the protection of sensitive and sacred information. If such data falls into the wrong hands or is lost due to technical failures, it could result in the irreversible loss of important cultural expressions and traditional knowledge that communities seek to preserve for future generations. Scholars like Masenya (2022) and Mdhuli et al. (2021) emphasise the importance of addressing data privacy and security concerns proactively. They argue that digital preservation efforts must incorporate comprehensive security measures, such as encryption, strong password policies, firewalls, and regular data backups, to shield sensitive IK from theft, misuse, or accidental loss.

10. In active role of Librarians in Africa

Librarians in Southern Africa are guided by a mandate from the International Federation of Library Associations and Institutions (IFLA), which emphasises the significant role libraries and archives play in collecting, preserving, and disseminating IK. According to IFLA (2002), libraries and archives are encouraged to implement programs that focus on the collection and preservation of indigenous and local traditional knowledge resources. Further, Biela, Oyelude and Haumba (2016), as a result, in October 2003, the thirty-second session of the United Nations Educational, Scientific and Cultural Organization (UNESCO) approved a Charter focused on the Preservation of Digital Heritage. The purpose of this Charter was to enhance initiatives aimed at protecting documentary heritage resources. Unfortunately, much of this IK is transmitted through oral traditions, songs, and community elders, practices that may not ensure its safety for future generations. Research by Masenya (2022) indicates that rural communities continue to rely heavily on traditional methods, such as storytelling and community practices, to share their IK, even in the context of today's digital age.

In Southern Africa, many library schools tend to pay mere lip service to the importance of preserving IK without adequately equipping librarians with the necessary skills and tools for digital preservation. This gap hinders the ability of libraries to fulfil their mandate and adequately support the preservation of IK for future generations.

RECOMMENDATIONS

1. Governments should establish laws protecting IK from exploitation, ensuring communities have legal recourse against unauthorised use. For instance, create frameworks similar to Kenya's Traditional Knowledge Protection Act. Indigenous communities must be involved in drafting these laws to ensure their interests and rights are adequately represented.
2. Governments and NGOs should prioritise investments in technological infrastructure, particularly in rural areas, to ensure access to the internet and electricity. There is a need to facilitate access to necessary tools for digitisation, such as computers, scanners, and software, particularly in communities where IK is most prevalent.
3. Training is essential for effectively digitising IK, as failures in this process can often stem from inadequate training. By providing training, individuals digitising IK can acquire the necessary skills to collect and manage this valuable information. There is a greater need to implement training programs targeting local communities to develop skills in digital preservation and technological literacy that can empower them to manage their knowledge. There is a need to partner

with universities and technical schools to offer courses on digital preservation, focusing on IK.

4. Librarians, archivists, and other information professionals should develop effective strategies for digitising IK while adhering to all ethical guidelines and obtaining the necessary clearances.
5. Integrate the CARE principles (Collective Benefit, Authority to Control, Responsibility, and Ethics) alongside FAIR principles (Findable, Accessible, Interoperable, Reusable) in digitisation projects to ensure ethical management of IK. Before digitisation, obtain consent from knowledge holders and engage with communities to understand their needs and expectations regarding preserving and sharing their knowledge.
6. There is a need to ensure that digitisation projects reflect the cultural context of IK, avoiding commodification and misrepresentation. Develop guidelines that outline best practices for handling and digitising IK that honour the traditions and values of Indigenous communities.
7. Develop detailed ethical standards that emphasise community approval, respect for cultural traditions, and rights to ownership during the digitisation process. These guidelines should be created collaboratively with indigenous communities to honour their knowledge systems and cultural protocols, helping prevent misrepresentation, misuse, and cultural disrespect.
8. Allocate resources toward infrastructure, technology, and training designed for indigenous populations to reduce the digital gap. Improving digital literacy and providing affordable connectivity will enable communities to actively safeguard and manage their cultural knowledge, promoting inclusion and reducing marginalisation.
9. Develop national and regional standards for the digital preservation of IK, covering aspects such as data formats, metadata, language considerations, and long-term accessibility. These standards should adhere to global best practices while remaining adaptable to local needs, ensuring data authenticity, consistency, and sustainability.
10. Support initiatives aimed at recording and digitising indigenous languages, including oral traditions and regional dialects, with proper transcription and translation. This effort will help preserve linguistic diversity, retain cultural subtleties, and strengthen the authenticity of digital repositories of IK.
11. Use advanced cybersecurity measures, such as encryption, controlled access, and regular data backups, to safeguard sensitive cultural information from theft, tampering, or accidental loss. Establish transparent data ownership and access rights policies to ensure indigenous communities maintain control over their cultural assets.

CONCLUSION

Digital preservation of IK is of crucial significance because of the oral nature of IK. Despite the numerous benefits obtained from digitising IK, such as the need to document the knowledge, collaboration benefits, and economic benefits, among other things, innumerable challenges emanate from digitalising IK, standards, such as ethical challenges and a lack of skills. However, it will be a smooth process with dedicated support from donor funding.

REFERENCES

Adu, K. K., & Ngulube, P. (2016). Key threats and challenges to the preservation of digital records of public institutions in Ghana. *Information Communication and Society*, *20*(8), 1127–1145. DOI: 10.1080/1369118X.2016.1218527

Aiseng, K. (2023). Challenges and opportunities of preserving African indigenous knowledge using digital technologies: The case of Bogwera. Digital Preservation and Documentation of Global Indigenous Knowledge Systems, 123–141.

Ajani, Y., Oladokun, B., Olarongbe, S., Amaechi, M., Rabiu, N., & Bashorun, M. (2024). revitalizing Indigenous knowledge systems via digital media technologies for sustainability of Indigenous languages. Preservation. *Digital Technology & Culture*, *53*(1), 35–44. DOI: 10.1515/pdtc-2023-0051

Amechi, E.P. (2015). Leveraging traditional knowledge on the medicinal uses of plants within the patent system: The digitisation and disclosure of knowledge in South Africa. Potchefstroom Electronic Law Journal/Potchefstroomse Elektroniese Regsblad 18(1): 3071–3101.

Amunkete, K. N. (2020). A model for the digital preservation of indigenous knowledge on medicinal plants in Namibia via an e-learning platform. https://uir.unisa.ac.za/items/78076e2a-b80e-46fa-a3b6-ee2a9333e874

Awuor, A.S., Kamau, G.W. & Owano, A. (2023). The role of digital humanities in the preservation of indigenous knowledge at the National Museums of Kenya.

Bagley, M. A. (2019). The fallacy of defensive protection for traditional knowledge. *Washburn Law Journal*, *58*, 323.

Balogun, T. (2023). Digitization of Indigenous knowledge systems in Africa: The case of South Africa's National Recorded System (NRS). *Records Management Journal*, *33*(1), 88–102.

Balogun, T., & Kalusopa, T. (2021). A framework for digital preservation of Indigenous knowledge system (IKS) in repositories in South Africa. *Records Management Journal*, *31*(2), 176–196. DOI: 10.1108/RMJ-12-2020-0042

Balogun, T. & Kalusopa, T. (2021). Web archiving of indigenous knowledge systems in South Africa. Information Development, 38(4), 658-671. (Original work published 2022)DOI: 10.1177/02666669211005522

Bawack, R., Roderick, S., Badhrus, A., Dennehy, D., & Corbett, J. (2025). Indigenous knowledge and information technology for sustainable development. *Information Technology for Development*, •••, 1–18. DOI: 10.1080/02681102.2025.2472495

Bihari, S. (2023). Cultural heritage and indigenous knowledge: Reviving traditions for future generations. Sustainable Development Goals in SAARC Countries: Key Issues, Opportunities and Challenges, 1, 24-32.102.

Bishoff, L., & Smith, C. (2015). Managing digital collections survey results. *D-Lib Magazine : the Magazine of the Digital Library Forum*, *21*(3), 1–11.

Biyela, N., Oyelude, A., & Haumba, E. (2016) Digital preservation of indigenous knowledge (IK) by cultural heritage institutions: a comparative study of Nigeria, South Africa and Uganda. (Paper presented at the SCESCAL Conference, Ezulwini, Swaziland, 25 - 29 April). http://hdl.handle.net/20.500.11910/10196 http://hdl .handle.net/20.500.11910/10196

Boamah, E., & Liew, C. L. (2017). Conceptualising the Digitisation and Preservation of Indigenous Knowledge: The Importance of Attitudes. Lecture Notes in Computer Science, *International Conference on Asian Digital Libraries*. DOI: 10.1007/978-3-319-70232-2_6

Chigwada, J. (2024). Harnessing indigenous knowledge in the digital era. https://www.unisa.ac.za/sites/myunisa/default/News/Articles/Harnessing-indigenous-knowledge-in-the-digital-era (10 April 2025).

Chigwada, J. & Ngulube, P. (2023). Librarians' role in the preservation and dissemination of indigenous knowledge. IFLA Journal, 50(2), 242-256. (Original work published 2024)DOI: 10.1177/03400352231217270

Chikozho, C., & Mabhena, R. (2020). Harnessing Traditional Ecological Knowledge for Sustainable Environmental Management: Digital Preservation in Southern Africa. *The International Journal of Environmental Studies*, *35*(4), 280–294.

Chowdhury, G., McLeod, J., Lihoma, P., Teferra, S., & Wato, R. (2023). Promoting access to Indigenous information in Africa: Challenges and requirements. *Information Development*, *39*(3), 611–623. DOI: 10.1177/02666669211048488

Church, R. M. (2002). The effective use of secondary data. *Learning and Motivation*, *33*(1), 32–45.

Cosijn, E., Pirkola, A., & Bothma, T.. (2002). Information access in indigenous languages: A case study in Zulu. *South African Journal of Library and Information Science*, *8*(2), 94–103.

De la Porte, B., & Higgs, R. (2019). Challenges in digitization of cultural heritage material in the de la PorteI, B and Higgs, R. (2019). Challenges in digitisation of cultural heritage material in the Western Cape, South Africa. https://scielo.org.za/scielo.php?script=sci_arttext&pid=S1560-683X2019000100030 (Accessed 07 April 2025).

Digital Preservation Coalition. (2025). What is digital preservation? https://www.dpconline.org/digipres/what-is-digipres

Dutta, U. (2019). Digital preservation of indigenous culture and narratives from the global south: In search of an approach. *Humanities (Washington)*, *8*(2), 68.

Gallert, P, Winschiers-Theophilus, H, Kapuire, G & Stanley, C. (2016). Indigenous Knowledge for Wikipedia: A Case Study with an OvaHerero Community in Eastern Namibia.

Greyling, E., & McNulty, N. (2012), June). "How to build an indigenous digital library through community participation: the case of the ulwazi programme", Information for Sustainable Development in a Digital Environment. *XXth Standing Conference of Eastern, Central and South Africa Library and Information Associations (SCECSAL).*

Handique, P., & Dubey, A. (2022). Legal mechanisms for protecting Indigenous knowledge. *Interdisciplinary Studies in Society, Law, and Politics*, *1*(1), 4–13.

Hernandez, A. (2025). Preserving and transferring Indigenous knowledge in a digital world. https://www.giga-hamburg.de/en/events/conferences-and-workshops/co-la-borate-2-preserving-and-transferring-indigenous

IFLA. (2002). IFLA Statement on Indigenous Traditional Knowledge. https://www.ifla.org/publications/ifla-statement-on-indigenous-traditional-knowledge/

Imenda, S. (2014). Is there a conceptual difference between theoretical and conceptual frameworks? *Journal of Social Sciences*, *38*(2), 185–195.

Jaya, E. (2006). The role of university in promoting indigenous knowledge systems in Zimbabwe with reference to traditional practices in rural areas", *2nd International Appropriate Technology Conference*, Bulawayo, Zimbabwe, 12–15 July 2006, pp. 1-7.

Kamal, S. & Amani M. (2020). Digitization and Indigenous Knowledge: Challenges and Opportunities in Africa

Kiely, T. J. (2024). What Is Desk Research? Meaning, Methodology, Examples https://www.meltwater.com/en/blog/desk-research (Accessed 18 November 2024).

Kumar, A. (2022). Protection of Traditional Knowledge. Issue 2 Indian JL & Legal Rsch., 4, 1.

Loupis, N., & Saffy, L. (2020). Indigenous Knowledge and Digital Repositories: Ethical and Legal Considerations. *International Journal of Cultural Heritage, 15*(2), 134–146.

Louw, E., & King, G. (2021). Digital Preservation and the Revitalization of Endangered Indigenous Languages in Southern Africa. *Journal of Language and Cultural Revitalization, 8*(1), 67–82.

Lwoga, E. T., Ngulube, P., & Stilwell, C. (2010). Managing indigenous knowledge for sustainable agricultural development in developing countries: Knowledge management approaches in the social context. *The International Information & Library Review, 42*(3), 174–185.

Malapane, O. L., Chanza, N., & Musakwa, W. (2024). Transmission of indigenous knowledge systems under changing landscapes within the Vhavenda community, South Africa. *Environmental Science & Policy, 161*, 103861.

Masenya, T., & Ngulube, P. (2020). Factors that influence digital preservation sustainability in academic libraries in South Africa. *South African Journal of Library and Information Science, 86*. Advance online publication. DOI: 10.7553/86-1-1860

Masenya, T. M. (2022). Digital preservation of indigenous knowledge in South African rural communities. In Tshifhumulo, R., & Makhanikhe, T. (Eds.), *Handbook of Research on Protecting and Managing Global Indigenous Knowledge Systems* (pp. 317–340). IGI Global.

Masenya, T. M. (Ed.). (2023). *Digital Preservation and Documentation of Global Indigenous Knowledge Systems*. IGI Global.

Masenya, T. M. (2023). Revitalization and digital preservation of Indigenous Knowledge Systems for sustainable development of Indigenous communities in South Africa. *The Serials Librarian, 84*(5-8), 86–102.

Masenya, T. M. (2024). Revitalization and Digital Preservation of Indigenous Knowledge Systems for Sustainable Development of Indigenous Communities in South Africa. https://www.tandfonline.com/doi/abs/10.1080/0361526X.2023.2277962 (Accessed 21 November 2024).

Mbigi, L. (2019). Digital Tools for Safeguarding Endangered Indigenous Knowledge in Southern Africa. *International Journal of Cultural Heritage Studies, 14*(3), 142–156.

Mdhluli, T. D., Mokgoatšana, S., & Kugara, S. L.. (2021). Knowledge management: Preserving, managing and sharing indigenous knowledge through digital library. *Hervormde Teologiese Studies*, *77*(2), a6795.

Morgan, H. (2022). Conducting a qualitative document analysis. *The qualitative report*, *27*(1), 64–77.

Mthembu, P. (2021). Challenges in Digital Archiving and Preservation of Indigenous Knowledge in Southern Africa. *Journal of Digital Preservation*, *12*(4), 45–59.

Mukasa, G., & Kamusiime, N. (2012). The value of managing library information resources in digital form in Uganda. *African Research & Documentation*, *118*, 69–78.

Mulalap, C. Y., Frere, T., Huffer, E., Hviding, E., Paul, K., Smith, A., & Vierros, M. K. (2020). Traditional knowledge and the BBNJ instrument. *Marine Policy*, *122*, 104103. DOI: 10.1016/j.marpol.2020.104103

Murphy, L. (2021). The Ethics of Digitizing Indigenous Knowledge: An African Perspective, Mwanza, A.J. (2022) Can digital technology be used for addressing digital literacy in the dissemination of indigenous knowledge in rural areas? *South African Journal of Library and Information Science*, *88*(1), 1–11.

National Library of Australia. (2003). Guidelines for the Preservation of Digital Heritage.

Natural Justice. (2023). Rooibos robbery: a story of bioprospecting in South Africa", [video], available at: https://naturaljustice.org/video/rooibosrobbery-a-story-of-bioprospecting-in-south-africa/

Nettleton, K. (2020). The Role of Digital Repositories in Enhancing the Accessibility of Indigenous Knowledge. *Journal of Information Science*, *46*(2), 231–245.

Ngulube, P. (2025). Leveraging information and communication technologies for sustainable agriculture and environmental protection among smallholder farmers in tropical Africa. *Discover Environment*, *3*(1), 1–17.

Nherera, C., & Mukora, F. N. (2024). Digitalisation of Higher Education in Zimbabwe: A Challenging Necessity and Emerging Solutions. *Journal of Comparative & International Higher Education*, *16*(2).

Ntsebeza, L. (2022). Traditional Knowledge and Intellectual Property Law in Southern Africa: The Case for Digital Protection. *Journal of African Law*, *66*(1), 58–79.

Okeke, C. (2023). Reclaiming Digital Spaces: Indigenous Knowledge and Technology in Africa.

Osedo, O. A. (2013). Challenges of digital preservation in Africa. Available at SSRN 3496885. Paris: UNESCO.

Pickover, M. (2008). The DISA project. Packaging South African heritage as a continuing resource: Content, access, ownership and ideology. *IFLA Journal*, *34*(2), 192–197.

Plockey, F. (2014). The role of Ghana public libraries in the digitization of indigenous knowledge: Issues and prospects. *The Journal of Pan African Studies*, *6*(10), 20–36.

Pretorius, R. I. Ë. T. T. E., & Bezuidenhout, H. E. N. N. I. E. (2011). National Recordal System IK holder catalogue process.

Ritchie, D. A. (2014). *Doing oral history*. Oxford University Press.

Senanayake, S. G. J. N. (2006). Indigenous Knowledge as a key to sustainable development. Journal of Agricultural Sciences – Sri Lanka 2(1). https://www.researchgate.net/publication/265197993_Indigenous_knowledge_as_a_key_to_sustainable_development

Seretse, M., Chukwuere, J., Lubbe, S., & Klopper, R. (2018). Problems around accessing information in rural communities. *Alternation Journal*, *25*(1), 214–244.

Settee, P. (2008). *Native Languages Supporting Indigenous Knowledge*. United Nations, Department of Economic and Social Affairs.

Snyder, H. (2019). Literature review as a research methodology: An overview and guidelines. *Journal of Business Research*, *104*, 333–339.

Sraku-Lartey, M., Acquah, S. B., Samar, S. B., & Djagbletey, G. D. (2017). Digitization of indigenous knowledge on forest foods and medicines. IFLA Journal, 43(2), 187-197. (Original work published 2017).DOI: 10.1177/0340035216681326

Stephenson, J. (2023). *Culture and sustainability: Exploring stability and transformation with the cultures framework*. Springer Nature.

Sullivan (2016). The John Marshall review of intellectual property law. https://repository.law.uic.edu/cgi/viewcontent.cgi?article=1392&context=ripl

UNESCO Institute for Statistics. (2009). *UNESCO Framework for Cultural Statistics*. UNESCO.

United Nations. (2021). Indigenous peoples and connectivity: Bridging the digital divide. Available at: https://www.un.org/development/desa/dspd/2021/04/indigenous-peoples-digital-divide/

Wamukoya, J., & Mutula, S. M. (2005). Capacity-building Requirements for E-records Western Cape, South Africa. *South African Journal of Information Management, 21*(1).

KEY TERMS AND DEFINITIONS

Cultural heritage: The Merriam-Webster dictionary (2025) defined culture as "the customary beliefs, social forms, and material traits of a racial, religious, or social group." UNESCO Institute for Statistics (2009) defined cultural heritage as artefacts, monuments, groups of buildings and sites, as well as museums that possess a variety of values such as symbolic, historical, artistic, aesthetic, ethnological or anthropological, scientific, and social importance. Sullivan (2016) indicated that cultural heritage covers physical aspects like structures, monuments, scenery, archival documents, books, artworks, and artefacts; intangible elements such as folklore, customs, language, and knowledge; and natural aspects, including landscapes of cultural importance and biodiversity.

Digital Preservation: According to the Digital Preservation Coalition (2025) digital preservation involves a series of managed activities aimed at ensuring ongoing access to digital materials for as long as needed. It encompasses all the measures necessary to retain access to digital content despite media deterioration, technological advancements, or organisational changes. Mukasa and Kamusiime (2012) defined digital preservation as a process and all activities and management of information over an extended period to be accessed in future.

Ethical Considerations: According to Bhandari (2024), ethical considerations in research refer to a group of principles that direct how research is planned and conducted. Researchers and scientists are required to follow a specific code of conduct when gathering data from individuals. Mirza, Bellalem and Mirza (2023) indicated that ethics concern the obligations that researchers have toward their participants, audience, society, and academic communities. To ensure they follow the principles of good research practices, researchers should consult established ethical guidelines.

Indigenous Knowledge: Ajani et al. (2024) conceptualise IK as the collective wisdom, customs, and cultural practices passed down through generations within Indigenous communities. Similarly, Grey (2014) describes IK as encompassing the expressions, practices, beliefs, understandings, insights, and experiences of Indigenous groups, shaped over centuries of deep interaction with their specific territorial environments. Sraku-Lartey, Acquah, Brefo, and Djagbletey (2017) define IK as traditional knowledge orally transmitted across generations and primarily preserved in the minds of local individuals. Aisega (2023) extends this understanding, emphasising that communities worldwide have shared IK over generations, addressing diverse aspects such as agriculture, architecture, human behaviour, marriage, childbirth, healthcare, natural disasters, climate, natural resources, and interpersonal relationships.

Technological advancements: According to Burton (2023), technological advancement is an ongoing cycle of innovation, refinement, and the adoption of new tools, methods, and systems aimed at boosting efficiency, productivity, and convenience across different sectors. It has transformed our lifestyles, work environments, and means of communication, while also paving the way for new opportunities in exploration and discovery. Nonetheless, it presents notable challenges such as ethical concerns, job losses, and information saturation, which demand careful reflection and proactive approaches.

Traditional Knowledge: Traditional knowledge (TK) is knowledge, know-how, skills and practices that are developed, sustained and passed on from generation to generation within a community, often forming part of its cultural or spiritual identity (Kumar, 2022). TK is transmitted between generations through stories, songs, dances, carvings, paintings and performances. Traditional knowledge can be found in a wide variety of contexts, including agricultural, scientific, technical, ecological and medicinal knowledge as well as biodiversity-related knowledge (Mulalap et al., 2020).

Chapter 10
Challenges in Digital Preservation of Indigenous Knowledge within the Open Science Framework

Sibongile Chituku
https://orcid.org/0009-0005-2026-5764
Africa University, Mutare, Zimbabwe

ABSTRACT

This chapter explores the challenges of digitally preserving Indigenous Knowledge (IK) within the Open Science Framework, highlighting conflicts between IK's communal nature and principles of openness. Key challenges include ethical dilemmas (consent, ownership, cultural misrepresentation), technological barriers (limited infrastructure, rapid obsolescence), and legal issues (intellectual property laws vs. traditional knowledge). It addresses cultural sensitivities (respecting protocols, community resistance) and resource constraints (financial limitations, shortage of trained personnel). The 7C model (Co-design, Conceptualisation, Collection, Correction, Curation, Circulation, Creation) is proposed to navigate these complexities. The chapter advocates for community-led, culturally respectful approaches to ensure IK preservation while aligning with Open Science ideals.

DOI: 10.4018/979-8-3373-0204-1.ch010

INTRODUCTION

Indigenous knowledge (IK) represents a vast repository of wisdom developed over centuries, deeply rooted in the traditions, practices, and cultural heritage of Indigenous communities. It encompasses knowledge systems related to agriculture, biodiversity, health, ecology, and cultural practices (UNESCO, 2017; First Nations Information Governance Centre, 2014). As globalization and technological changes continue to accelerate, digitally preserving Indigenous knowledge ensures its survival and accessibility for future generations (Biyela et al., 2016)

Efforts to digitally preserve Indigenous knowledge have gained traction worldwide. For instance, Indigenous communities in Canada have adopted digital platforms to document oral traditions and ecological practices. These projects aim to protect cultural heritage while bridging traditional knowledge and modern science (First Nations Information Governance Centre, 2014). Similarly, audio and video recordings are used to safeguard endangered languages, ensuring their transmission to younger generations and supporting cultural revitalization initiatives (UNESCO, 2017).

By employing Digitisation, Indigenous communities preserve ecological practices and traditional insights, thus contributing to sustainability research. For example, traditional ecological knowledge related to land management, biodiversity, and water conservation has been recorded and shared through digital archives (Biyela et al., 2016). This not only fosters cultural pride within communities but also informs global initiatives in sustainability and environmental conservation.

This chapter provides a comprehensive overview of the various obstacles faced in preserving Indigenous knowledge in the digital age. It emphasizes the importance of Indigenous knowledge, which is often deeply rooted in cultural practices and traditions, and highlights the need for its preservation to ensure that it is not lost to future generations (Chiwanza, Musingafi, & Mupa, 2012).

One of the primary challenges discussed is the issue of intellectual property rights. Indigenous knowledge is often not formally documented, making it difficult to protect under conventional intellectual property laws. This can lead to bio-piracy and exploitation, where external entities may use this knowledge without proper consent or compensation (Chiwanza et al., 2012).

The chapter also addresses the cultural sensitivity required in digitizing Indigenous knowledge. It is crucial to understand and respect the cultural contexts and sensitivities associated with this knowledge to avoid misrepresentation or misuse, which can lead to cultural erosion and loss of trust (Boamah & Liew 2017).

Technological barriers are another significant challenge. Many Indigenous communities have limited access to digital technologies and the internet, which hampers efforts to digitize and preserve their knowledge. This digital divide needs to be addressed to ensure that these communities can actively participate in the preservation process (Balogun & Kalusopa, 2021).

Additionally, the chapter highlights the importance of community engagement. There can be resistance from within the communities themselves, as some fear the permanent loss of their heritage, while others may lack interest in Digitisation efforts. It is essential to involve Indigenous communities in the Digitisation process and address their concerns to ensure successful preservation (Chiwanza et al., 2012). Resource constraints are also discussed, as Digitisation projects require significant financial and human resources. Many memory institutions in developing countries struggle with limited budgets and expertise, making it challenging to undertake such projects (Balogun & Kalusopa, 2021).

The need for sustainable data management practices to ensure the long-term preservation and accessibility of digital records is emphasized in this chapter. This involves not only digitizing the information but also maintaining and updating digital archives to prevent data loss or degradation (Boamah & Liew 2017).

Overview of Open Science Frame work

According to Foster (2017), the Open Science Framework (OSF) is a versatile tool that fosters open and centralized workflows by facilitating the capture of various aspects and outputs throughout the research lifecycle. This includes conceptualizing research ideas, designing studies, storing and analyzing collected data, as well as writing and publishing reports or papers. The OSF is developed and maintained by the Center for Open Science (COS), a nonprofit organization founded in 2013 that conducts research into scientific practices, supports scientific research communities, and develops research tools and infrastructure to manage and archive research. The COS advocates for openness, integrity, and reproducibility in research across various scientific disciplines (Center for Open Science, 2017). While the OSF offers a range of tools and services to assist researchers, this review primarily focuses on its core functionality, with brief mentions of some other available tools and services (ParaFoster, 2017).

The Importance of Integrating Indigenous Knowledge into Open Science Frame work

Indigenous Knowledge (IK) is vital for preserving the cultural heritage, traditions, and environmental wisdom of Indigenous communities. It encompasses a vast array of knowledge passed down through generations, including practices related to agriculture, medicine, sustainable resource management, and community life. Preserving IK ensures the continuation of this valuable information, which can contribute to global efforts in biodiversity conservation, climate change adaptation, and sustainable development (Boamah & Liew, 2017; Maasz et al., 2018; le Roux, 2017).

Biodiversity Conservation

Indigenous knowledge plays a crucial role in biodiversity conservation. Over generations, Indigenous communities have developed a profound understanding of their local ecosystems, enabling them to implement practices that maintain ecological balance. These include sustainable methods of hunting, fishing, and agriculture, which safeguard biodiversity hotspots and promote ecological harmony (Mirza, 2020; Chukwuma & Mpofu, 2019).

Eduardo Brondízio's extensive research in the Amazon underscores the critical role of Indigenous knowledge in global conservation efforts. His work has revealed how local communities actively combat deforestation, fires, and other environmental challenges, showcasing their resilience and commitment to sustainability. Brondízio has documented numerous grassroots initiatives led by Indigenous and rural communities, such as agroforestry and community-led governance of land and biodiversity. These efforts not only preserve the Amazon's rich biodiversity but also provide valuable insights for global climate resilience strategies (Brondízio, 2025; Karjalainen & Okere, 2018).

Globally, the contributions of Indigenous knowledge have been increasingly recognized in policymaking. Brondízio's work with the Intergovernmental Science-Policy Platform on Biodiversity and Ecosystem Services (IPBES) has emphasized the integration of Indigenous practices into global conservation frameworks, such as the Kunming-Montreal Global Biodiversity Framework. This integration highlights how Indigenous knowledge can guide sustainable land stewardship and ecological restoration (Brondízio, 2025; Mirza, 2020).

Furthermore, Brondízio's contributions to the Intergovernmental Science-Policy Platform on Biodiversity and Ecosystem Services (IPBES) Global Assessment have highlighted the indispensable role of Indigenous and local knowledge in addressing biodiversity loss. This work has influenced international frameworks like the Kunming-Montreal Global Biodiversity Framework, emphasizing the integration of Indigenous practices into global conservation policies (Mirza, 2020; Brondízio, 2025; Karjalainen & Okere, 2018). By recognizing and integrating Indigenous knowledge into broader scientific and policy frameworks, we can enhance biodiversity conservation and ensure a sustainable future for all.

The United Nations Scientific Advisory Board (UNSAB) has acknowledged the importance of Indigenous and local knowledge systems in sustainable development, emphasizing their role in food security, climate change mitigation, and biodiversity conservation (UNESCO, 2016). By recognizing and integrating Indigenous knowledge into global scientific and policy frameworks, we can enhance sustainable development efforts and ensure that Indigenous communities are active participants in shaping a sustainable future.

Indigenous knowledge (IK) holds immense significance in the global scientific community and is integral to sustainable development. It encompasses centuries-old traditions, ecological practices, and cultural heritage that reflect a deep, intuitive understanding of local environments. By integrating Indigenous knowledge with modern science, we can address complex global challenges like biodiversity loss, climate change, and food insecurity while fostering cultural preservation and intergenerational knowledge transmission (Mirza, 2020; Smith, 2012).

Climate Change Adaptation

Indigenous knowledge provides invaluable strategies for adapting to climate change. Having observed environmental patterns over centuries, Indigenous communities have developed techniques for managing and mitigating the impacts of climate variability. These methods, when paired with scientific approaches, can enhance climate resilience at both local and global levels (Karjalainen & Okere, 2018).

An example of this is seen in Vanuatu, where Indigenous ecological indicators—such as changes in fish behavior—have been combined with modern scientific data to develop an early-warning system for ciguatera fish poisoning. This innovative collaboration illustrates the practical value of integrating traditional knowledge with scientific methods to address localized climate challenges (MSN, 2025).

More broadly, Indigenous communities' focus on long-term sustainability and harmony with nature offers critical lessons for global climate policies. Their practices prioritize maintaining environmental resilience over generations rather than maximizing short-term gains, which serves as a model for international climate strategies (Karjalainen & Okere, 2018; UNESCO, 2016).

Sustainable Land and Resource Management

Indigenous communities have long practiced sustainable land and resource management, demonstrating methods that emphasize ecological balance and intergenerational responsibility. These traditional practices—such as crop rotation, agroforestry, and water conservation—are vital for ensuring food security and mitigating environmental degradation (Chukwuma & Mpofu, 2019).

The United Nations Scientific Advisory Board (UNSAB) has emphasized the importance of Indigenous and local knowledge systems in sustainable development. These practices contribute not only to climate change mitigation and biodiversity conservation but also to the preservation of cultural heritage (UNESCO, 2016). For example, traditional irrigation techniques in arid regions maximize water efficiency while minimizing ecological damage. Similarly, Indigenous forest management practices, such as selective logging and natural regeneration, ensure the long-term sustainability of ecosystems (Mirza, 2020).

Collaborative Integration

Recognizing and integrating Indigenous knowledge into broader scientific and policy frameworks is essential for addressing global challenges. Collaboration between Indigenous communities, researchers, and policymakers ensures that this knowledge is preserved, respected, and actively utilized in efforts to build a sustainable future (Smith, 2012; Meertens, 2018).

To foster such collaboration, it is critical to establish trust and transparency. This involves engaging Indigenous communities as active participants in decision-making, respecting cultural protocols, and acknowledging the value of their contributions (Meertens, 2018). Efforts such as the United Nations Declaration on the Rights of Indigenous Peoples (UNDRIP) and the Nagoya Protocol reinforce the importance of Indigenous participation in shaping sustainable development strategies (Nagoya Protocol, 2010; UNESCO, 2016).

The various challenges faced in the digital preservation of Indigenous knowledge, including technological, cultural, legal, and resource-related issues. Identifying the challenges in the digital preservation of Indigenous knowledge is crucial to understanding the complexities involved.

TECHNOLOGICAL CHALLENGES IN DIGITIZING INDIGENOUS KNOWLEDGE

The preservation of Indigenous knowledge in digital formats is imperative to safeguarding the rich cultural heritage, environmental wisdom, and unique practices of Indigenous communities. However, numerous technological challenges impede efforts to digitize and preserve this invaluable repository of information. These obstacles encompass limited access to digital technologies, inadequate digital infrastructure, lack of expertise, and the digital divide that exists between Indigenous communities and more developed regions. Addressing these challenges is critical to ensuring that Indigenous knowledge is not lost in the era of technological advancement (Boamah & Liew, 2017; Maasz et al., 2018; le Roux, 2017).

Limited Access to Digital Technologies

Indigenous communities often reside in remote or marginalized areas where access to digital tools and resources remains scarce. Without the necessary technological infrastructure, these communities face significant barriers to digitizing their knowledge systems. For example, Biyela et al. (2016) emphasize that the digital divide—the gap between those with access to modern technology and those without—limits the ability of Indigenous groups to engage in Digitisation initiatives. This divide manifests in various ways, including a lack of computers, mobile devices, or advanced digital tools necessary for recording and preserving oral histories, cultural practices, and ecological knowledge.

Moreover, the absence of basic technological resources restricts the opportunities for Indigenous peoples to participate actively in global discussions and projects related to digital preservation. Even when external organizations initiate Digitisation efforts, the lack of local participation due to limited access often results in the marginalization of Indigenous voices in how their knowledge is represented and stored (Borrero, 2013; EcoLiterateLaw, 2024).

Digital Infrastructure Deficits

In many Indigenous regions, digital infrastructure such as stable electricity and internet connectivity is either underdeveloped or entirely absent. The First Nations Information Governance Centre (2014) highlights that inadequate infrastructure poses a substantial obstacle to Digitisation processes. For instance, some areas face intermittent or unreliable electricity, making it difficult to operate computers and other Digitisation equipment consistently. Additionally, limited access to the internet hinders the ability to store knowledge in cloud-based systems, which are increasingly used for digital archiving.

The lack of infrastructure extends beyond electricity and internet access to include facilities such as community-based Digitisation centers. Many developing countries lack the resources to establish these centers in Indigenous regions, further complicating efforts to ensure that Indigenous knowledge is preserved securely and accurately. Without addressing these infrastructure deficits, the goal of creating equitable access to digital preservation tools remains unfulfilled.

Lack of Technical Expertise

Digitisation projects demand specialized skills, including proficiency in recording, cataloging, and preserving information in digital formats. Indigenous communities, especially in developing countries, often lack the technical know-how to undertake such initiatives independently (UNESCO, 2017). This shortage of expertise is particularly apparent in rural areas, where access to education and training in digital technology is limited.

The absence of skilled personnel also affects the quality of Digitisation projects. Even when external organizations assist Indigenous communities, there is often a mismatch between the priorities of the communities and the methodologies employed by these organizations. For instance, external entities may prioritize efficiency or cost-effectiveness over cultural sensitivity, leading to the loss or misrepresentation of important nuances in Indigenous knowledge (Boamah & Liew, 2017). Empowering Indigenous people with the necessary technical expertise through training programs and workshops is essential for ensuring that they maintain control over their knowledge systems.

THE DIGITAL DIVIDE

The digital divide remains a pervasive issue, deepening the inequities faced by Indigenous communities in accessing modern technology. This gap not only prevents these communities from digitizing their knowledge but also from benefiting from advancements in science, technology, and education. As Biyela et al. (2016) argue, the digital divide perpetuates cycles of marginalization, as communities without access to technology are excluded from decision-making processes, funding opportunities, and global networks.

his divide is exacerbated by factors such as economic inequality and geographic isolation. Many Indigenous communities struggle to afford the high costs associated with acquiring, maintaining, and upgrading digital equipment (Biyela et al., 2016). Geographic isolation further limits their ability to access resources or technical support, creating a vicious cycle where the lack of access to technology perpetuates underdevelopment (First Nations Information Governance Centre, 2014; Karjalainen & Okere, 2018)..

Sustainability of Digitisation Projects

Another major challenge in the Digitisation of Indigenous knowledge lies in ensuring the long-term sustainability of projects. Many Digitisation initiatives are launched as short-term projects funded by external organizations or governments. However, once the funding ends, the projects are often abandoned, leaving Indigenous communities without the means to continue the preservation process. Karjalainen and Okere (2018) note that sustainability requires not only financial support but also the establishment of local institutions and infrastructure capable of maintaining and updating digital archives over time.

Additionally, sustainability is hindered by the rapid pace of technological advancement. As digital tools and formats evolve, previously digitized content may become obsolete, requiring regular updates and migrations to new platforms. Indigenous communities that lack the resources to adapt to these changes risk losing access to their own knowledge, defeating the purpose of preservation efforts (Karjalainen & Okere, 2018; Biyela et al., 2016).

ETHICAL AND CULTURAL CONSIDERATIONS

While not strictly technological, ethical and cultural concerns intersect with the technical challenges of Digitisation. Indigenous knowledge is deeply tied to cultural practices, spiritual beliefs, and community identities, making its preservation a sensitive issue. Digitizing sacred or proprietary knowledge without the consent of the community can lead to ethical dilemmas and potential misuse of the information (Boamah & Liew, 2017). For example, the commercialization of Indigenous medicinal knowledge by pharmaceutical companies is a recurring issue, highlighting the need for strict ethical frameworks and intellectual property protections (Chukwuma & Mpofu, 2019; Boamah & Liew, 2017). Indigenous communities must be actively involved in decisions regarding what knowledge to digitize, how it should be stored, and who should have access to it. Ensuring community ownership and control over digital archives is critical to addressing these ethical and cultural challenges (Boamah & Liew, 2017).Addressing the Challenges

To overcome the technological challenges in preserving Indigenous knowledge, a multifaceted approach is required. First, governments and international organizations must invest in improving digital infrastructure in Indigenous regions. This includes providing reliable electricity, affordable internet access, and community-based Digitisation facilities (First Nations Information Governance Centre, 2014; Biyela et al., 2016). Second, training programs should be developed to equip Indigenous peoples with the technical skills needed for Digitisation. These programs should be culturally sensitive and designed in collaboration with Indigenous communities to ensure alignment with their values and priorities (UNESCO, 2017).

Third, partnerships between Indigenous communities, governments, and academic institutions can foster sustainable Digitisation efforts. These partnerships should prioritize long-term funding, capacity building, and the use of technologies that are adaptable to the unique needs of Indigenous regions (Karjalainen & Okere, 2018). Finally, ethical guidelines and legal protections must be strengthened to ensure that

Indigenous knowledge is not misappropriated or exploited (Boamah & Liew, 2017; Chukwuma & Mpofu, 2019).

Cultural Challenges in Digitizing Indigenous Knowledge

The Digitisation of Indigenous knowledge is a complex yet essential endeavor that seeks to preserve the rich traditions, practices, and histories of Indigenous communities. However, the process is fraught with cultural challenges that require careful consideration and sensitivity. These challenges encompass issues such as misrepresentation, loss of trust, resistance from within the communities, and the need to respect traditional practices and protocols. Addressing these challenges is crucial to ensuring that Digitisation efforts are both effective and culturally respectful.

Understanding the Context of Indigenous Knowledge

Indigenous knowledge is deeply rooted in the cultural, spiritual, and social fabric of communities. It encompasses a wide range of practices, from ecological management and medicinal knowledge to storytelling and spiritual rituals (Smith, 2012). This knowledge has been passed down through generations, often orally or through practices, and is inherently tied to the cultural identity of its custodians.

Digitizing this knowledge requires a profound understanding of the cultural contexts in which it exists. Failure to recognize these contexts can lead to misrepresentation, where knowledge is taken out of its cultural and spiritual framework, stripping it of its deeper meanings and significance. For instance, the commercialization of Indigenous medicinal knowledge by pharmaceutical companies often ignores the cultural practices and beliefs associated with these remedies, reducing them to mere commodities (Smith, 2012; Boamah & Liew, 2017). This can erode the trust that Indigenous communities place in external organizations and Digitisation efforts.

The Risk of Cultural Erosion

One of the most significant cultural challenges in digitizing Indigenous knowledge is the potential for cultural erosion. By transferring knowledge from its traditional mediums—such as oral storytelling, rituals, or carvings—into digital formats, there is a risk of detaching it from its cultural context. This detachment can result in the loss of intangible elements, such as the emotions, gestures, and sacredness that accompany traditional knowledge-sharing practices (Chiwanza, Musingafi, & Mupa, 2012).

For instance, a story that has been passed down orally may lose its richness and depth when recorded in a purely textual format. The act of storytelling often involves intricate gestures, vocal intonations, and a deep connection between the storyteller and the audience, which are difficult to capture in digital formats (Meertens, 2018). As a result, Digitisation may inadvertently contribute to the dilution of cultural heritage.

Resistance from Indigenous Communities

Another significant challenge is resistance from within the communities themselves. Indigenous people may fear that digitizing their knowledge will lead to its exploitation or loss. According to Chiwanza, Musingafi, and Mupa (2012), there is a widespread concern among Indigenous communities that their knowledge, once digitized, could be misused by external entities or commodified without their consent. These fears are not unfounded, given the history of exploitation and marginalization that many Indigenous groups have faced.

Additionally, some community members may lack interest in or understanding of Digitisation efforts. For those who view traditional practices and oral transmission as sacred, the idea of digitizing knowledge may seem alien or even disrespectful. This resistance highlights the importance of engaging with communities in a meaningful way to address their concerns and ensure their active participation in the process (Smith, 2012).

Respecting Traditional Practices and Protocols

Respecting the traditional practices and protocols of Indigenous communities is a cornerstone of successful Digitisation efforts. Indigenous knowledge is often governed by specific cultural norms and rules regarding who can access, share, and use it. For example, certain spiritual or medicinal knowledge may be restricted to elders or specific groups within the community (Meertens, 2018). Digitizing such knowledge without adhering to these protocols can lead to cultural offenses and loss of trust.

Consulting with community elders and leaders is essential to navigating these cultural norms effectively. Elders are often regarded as the custodians of knowledge and play a crucial role in determining what can be digitized and how. According to Meertens (2018), collaboration with elders ensures that Digitisation efforts align with the values and priorities of the community. This collaborative approach not only respects cultural norms but also fosters trust and ownership among community members.

Balancing Access and Control

One of the core tensions in digitizing Indigenous knowledge is balancing access and control. While Digitisation aims to make knowledge more accessible, unrestricted access can lead to its misuse or misappropriation. For instance, open-access digital archives may unintentionally expose sensitive cultural information to inappropriate uses, such as commercialization or cultural appropriation (Boamah & Liew, 2017). Ensuring that Indigenous communities retain control over their digital archives is critical to addressing this challenge.

To achieve this balance, Digitisation projects must incorporate mechanisms that allow communities to control who can access their knowledge and for what purposes. This may include the use of restricted-access digital platforms, where certain knowledge can only be viewed by authorized users, as determined by the community (Chiwanza, Musingafi, & Mupa, 2012).

Building Trust Through Collaborative Approaches

Trust is a fundamental component of successful Digitisation efforts. Building trust requires transparent and collaborative approaches that involve Indigenous communities at every stage of the process. According to Smith (2012), fostering trust begins with recognizing the sovereignty and agency of Indigenous peoples over their knowledge. This means that external organizations and researchers must act as facilitators rather than controllers, ensuring that the community retains ownership of the knowledge being digitized.

Collaboration also extends to the design and implementation of Digitisation projects. By involving Indigenous communities in decisions about what knowledge to digitize, how it should be stored, and who should have access to it, Digitisation efforts can align with the values and priorities of the community. For example, some communities may prioritize the Digitisation of endangered languages or ecological knowledge as a way to preserve them for future generations (Meertens, 2018).

Education and Awareness

Education and awareness are critical to overcoming resistance and fostering community participation. Many Indigenous communities may lack awareness of the potential benefits of Digitisation, such as preserving knowledge for future generations or sharing it with broader audiences to promote cultural understanding. Conducting workshops and awareness campaigns can help bridge this gap by demonstrating how Digitisation can complement traditional practices rather than replace them (Chiwanza, Musingafi, & Mupa, 2012).

Moreover, education should be a two-way process. External organizations and researchers must take the time to learn about the cultural norms, values, and practices of the communities they work with. This cultural competency is essential to ensuring that Digitisation efforts are respectful, effective, and aligned with community needs.

LEGAL CHALLENGES IN PRESERVING INDIGENOUS KNOWLEDGE

Indigenous knowledge (IK) is a treasure trove of cultural, environmental, and historical wisdom that has been passed down through generations of Indigenous communities. Despite its significance, preserving and protecting this knowledge faces numerous legal challenges that highlight the inadequacies of conventional intellectual property (IP) laws. These legal systems often fail to accommodate the communal, oral, and intergenerational nature of Indigenous knowledge, leaving it vulnerable to exploitation, misappropriation, and bio-piracy (Chiwanza, Musingafi, & Mupa, 2012; UNESCO, 2017). Below is an exploration of these challenges, as well as potential frameworks for ensuring the fair and equitable treatment of Indigenous knowledge.

The Unique Nature of Indigenous Knowledge

Indigenous knowledge is fundamentally different from the forms of knowledge that most modern legal systems are designed to protect. IP laws, such as patents and copyrights, are structured to safeguard individually-authored and tangible creations, often with a defined date of origin. In contrast, Indigenous knowledge is typically oral, collectively owned, and developed over centuries (Chiwanza, Musingafi, & Mupa, 2012). For instance, traditional medicinal practices or agricultural methods are often the result of contributions from entire communities over generations, making it impossible to attribute ownership to a single person.

This mismatch between the communal characteristics of Indigenous knowledge and the individualistic nature of IP laws creates significant gaps in legal protection. Indigenous communities may struggle to claim ownership or control over their knowledge within these frameworks, rendering it susceptible to exploitation by external entities. This includes corporations, researchers, and other parties seeking to capitalize on Indigenous knowledge without proper acknowledgment or compensation (UNESCO, 2017).

The Threat of Bio-Piracy

One of the most pressing legal challenges faced by Indigenous communities is the threat of bio-piracy. This occurs when corporations or researchers exploit Indigenous knowledge—often related to medicinal plants, traditional healing methods, or ecological practices—to develop commercial products without obtaining consent or sharing benefits with the knowledge holders. This practice not only undermines the rights of Indigenous communities but also commodifies their cultural heritage in ways that are often contrary to their values (Chiwanza, Musingafi, & Mupa, 2012).

The pharmaceutical industry provides numerous examples of bio-piracy. Many drugs and treatments have been developed based on Indigenous medicinal knowledge, but these communities rarely receive credit or financial compensation. The lack of formal documentation or patents for such knowledge makes it difficult for Indigenous peoples to contest these cases in court (UNESCO, 2017). As a result, bio-piracy continues to erode the cultural and intellectual sovereignty of Indigenous communities around the world.

Legal Frameworks and Their Limitations

Existing legal frameworks, including international agreements like the World Intellectual Property Organization (WIPO) treaties, offer limited protection for Indigenous knowledge. Most of these frameworks are designed to address the needs of modern industries and fail to account for the unique characteristics of IK. For example, the patent system requires detailed documentation and evidence of innovation, which is incompatible with the oral and communal traditions of Indigenous communities (Nagoya Protocol, 2010).

Similarly, copyright laws focus on protecting creative works in tangible forms, such as books, music, or art. While these laws may offer some protection for Indigenous cultural expressions, they are ill-equipped to address the broader scope of Indigenous knowledge, such as traditional ecological practices or spiritual beliefs (UNESCO, 2017). This legal gap leaves Indigenous communities vulnerable to exploitation and disenfranchisement.

The Nagoya Protocol and Access and Benefit-Sharing

One promising international framework aimed at addressing these challenges is the Nagoya Protocol, which was adopted under the Convention on Biological Diversity in 2010. The protocol establishes principles of access and benefit-sharing (ABS), requiring that entities seeking to use Indigenous knowledge or genetic resources obtain prior informed consent (PIC) from the knowledge holders and share the benefits derived from their use fairly and equitably (Nagoya Protocol, 2010).

While the Nagoya Protocol represents a significant step forward, its implementation has been uneven, and many Indigenous communities remain unaware of their rights under this framework. Additionally, the protocol relies heavily on national governments to enforce its provisions, which can be problematic in countries where Indigenous rights are not adequately recognized or protected (UNESCO, 2017).

Ownership, Control, and Agency

Ensuring that Indigenous communities retain ownership and control over their knowledge is essential for addressing the legal challenges they face. Digitisation efforts and the creation of digital archives, while useful for preservation, can raise questions about who has the right to access and use this knowledge. Without clear legal agreements, there is a risk that digitized content may be misappropriated by external parties (Smith, 2012).

In some cases, external organizations assisting with Digitisation projects have laid claim to the resulting digital content, further marginalizing Indigenous communities. This underscores the need for robust legal frameworks that prioritize community sovereignty and ensure that Indigenous peoples retain agency over their knowledge (Meertens, 2018).

Toward Inclusive and Robust Legal Frameworks

Addressing the legal challenges of preserving Indigenous knowledge requires inclusive and robust legal frameworks that are specifically designed to meet the needs of Indigenous communities. Legal protections must ensure that Indigenous communities retain ownership and control over their knowledge, whether it is digitized or maintained in traditional forms. This includes the right to decide how their knowledge is used and who has access to it (Chiwanza, Musingafi, & Mupa, 2012).

International agreements like the Nagoya Protocol should be strengthened and implemented more effectively to ensure that benefits derived from the use of Indigenous knowledge are shared fairly and equitably (Nagoya Protocol, 2010). Legal frameworks should account for the cultural and spiritual dimensions of Indigenous knowledge, respecting traditional protocols and ensuring that knowledge is preserved in ways that align with community values (Smith, 2012; Meertens, 2018). Indigenous communities must be empowered with the resources and knowledge needed to navigate legal systems and assert their rights effectively. This includes training in IP laws, advocacy, and negotiation (UNESCO, 2017). Governments, NGOs, and international organizations should work together to develop global standards for the protection of Indigenous knowledge, ensuring consistency and fairness across borders (UNESCO, 2017).

The preservation and protection of Indigenous knowledge face numerous legal challenges rooted in the misalignment between conventional IP laws and the unique characteristics of Indigenous knowledge. By developing inclusive legal frameworks, strengthening international agreements, and empowering Indigenous communities, we can address these challenges and ensure that this invaluable knowledge is preserved for future generations.

The Lack of Formal Documentation

One of the most profound legal challenges lies in the absence of formal documentation of Indigenous knowledge. Unlike patents, copyrights, and trademarks that rely on written or otherwise tangible proof of ownership, Indigenous knowledge has traditionally been shared orally and collectively within communities (Chiwanza, Musingafi, & Mupa, 2012). The lack of documentation makes it difficult to prove ownership and protect this knowledge under existing IP laws, leaving it vulnerable to misappropriation and misuse.

For example, traditional medicinal knowledge passed down through generations can be exploited by external entities, such as pharmaceutical companies, which may use this knowledge to develop new products without acknowledging or compensating the Indigenous communities involved. This practice, commonly referred to as bio-piracy, highlights the inadequacies of existing legal systems in safeguarding Indigenous intellectual property (Chiwanza et al., 2012).

Additionally, the collective nature of Indigenous knowledge poses another challenge. Many IP laws require a single owner or entity to claim rights over an idea or creation. However, Indigenous knowledge often belongs to entire communities rather than individuals, further complicating efforts to protect it within the confines of conventional legal frameworks (UNESCO, 2017).

The Risk of Bio-Piracy and Exploitation

Bio-piracy is a pressing issue that exemplifies the exploitation of Indigenous knowledge. It occurs when external organizations or individuals, often from more technologically advanced societies, use Indigenous knowledge to develop commercial products without proper consent or benefit-sharing with the communities involved. For example, numerous pharmaceutical companies have relied on traditional healing practices and natural remedies to develop new drugs, reaping significant profits while excluding Indigenous peoples from any form of compensation or acknowledgment (Chiwanza et al., 2012).

The Nagoya Protocol (2010), an international treaty under the Convention on Biological Diversity, seeks to address this issue by establishing guidelines for access and benefit-sharing (ABS). It requires entities seeking to use Indigenous knowledge to obtain prior informed consent from communities and share the benefits derived from this knowledge fairly. While the protocol is a step in the right direction, its implementation has been uneven, and many Indigenous groups remain unaware of their rights under the treaty (Nagoya Protocol, 2010).

Ownership, Control, and Access

Ensuring that Indigenous communities retain ownership and control over their knowledge is a critical legal challenge. Digitisation efforts often raise questions about who has the right to access and use digitized knowledge. Without clear legal agreements, there is a risk that digitized content may be appropriated by external parties, leading to the potential for misuse or commercialization without community consent (Nagoya Protocol, 2010).

The lack of robust legal frameworks exacerbates these issues. Many existing laws fail to recognize communal ownership, which is a cornerstone of Indigenous knowledge systems. This leaves Indigenous communities without adequate legal recourse to prevent unauthorized use of their knowledge (UNESCO, 2017).

Additionally, the Digitisation of Indigenous knowledge raises concerns about intellectual property rights over digital formats. For example, if an external organization assists in the Digitisation process, there may be disputes over who owns the digitized content—the Indigenous community, the organization, or both. Resolving these disputes requires clear legal frameworks that prioritize community sovereignty and consent (Chiwanza et al., 2012).

Weaknesses in Legal Frameworks

Existing legal frameworks often fail to address the unique characteristics of Indigenous knowledge. Intellectual property laws such as patents and copyrights are designed for knowledge or creations that are new, original, and fixed in a tangible form. Indigenous knowledge, in contrast, is dynamic, evolving, and often exists in oral or non-material forms (UNESCO, 2017).

Moreover, these laws are inherently individualistic, designed to protect the rights of individual creators or corporations. This approach conflicts with the collective and intergenerational nature of Indigenous knowledge. For example, a specific traditional practice may have been developed over centuries by an entire community, making it difficult to assign ownership to a single individual or entity (Nagoya Protocol, 2010).

These shortcomings create significant loopholes that enable the misappropriation of Indigenous knowledge. For instance, multinational corporations can exploit traditional knowledge by patenting slightly modified versions of it, effectively excluding the original custodians from any benefits. The absence of international standards to protect Indigenous intellectual property further complicates efforts to prevent such exploitation (UNESCO, 2017).

The Role of International Agreements

International agreements such as the Nagoya Protocol play a vital role in addressing these legal challenges. The protocol provides a framework for ensuring that Indigenous communities are involved in decisions regarding the use of their knowledge and receive a fair share of any benefits. It emphasizes the principles of prior informed consent (PIC) and access and benefit-sharing (ABS), which are essential for protecting Indigenous intellectual property (Nagoya Protocol, 2010).

However, the effectiveness of these agreements depends on their implementation. Many countries lack the legal infrastructure or political will to enforce the provisions of the Nagoya Protocol. Additionally, there is often a lack of awareness among Indigenous communities about their rights under these agreements, limiting their ability to assert control over their knowledge (UNESCO, 2017).

Developing Inclusive Legal Frameworks

To address the legal challenges surrounding the preservation and protection of Indigenous knowledge, it is essential to develop inclusive legal frameworks that recognize the unique characteristics of this knowledge. These frameworks should prioritize community ownership, consent, and benefit-sharing. They should also provide mechanisms for resolving disputes and ensuring accountability for misuse or misappropriation (Chiwanza et al., 2012).

One approach is to integrate traditional governance structures into legal frameworks. For example, Indigenous councils or elders' committees can play a central role in deciding how knowledge is digitized, accessed, and shared. This ensures that legal protections align with the cultural values and priorities of the community (Meertens, 2018).

Education and awareness are also critical. Many Indigenous communities remain unaware of their legal rights and the mechanisms available to protect their knowledge. Providing training and resources to these communities can empower them to navigate legal systems effectively and assert control over their intellectual property (Smith, 2012).

Legal challenges in preserving Indigenous knowledge are deeply rooted in the misalignment between conventional intellectual property laws and the unique characteristics of Indigenous knowledge systems. The lack of formal documentation, the risk of bio-piracy, and the weaknesses in existing legal frameworks underscore the need for inclusive and community-centered approaches. International agreements such as the Nagoya Protocol provide a foundation for addressing these issues but require effective implementation and greater awareness among Indigenous communities. By developing robust legal frameworks that prioritize the rights and interests of Indigenous peoples, we can ensure the preservation and protection of their invaluable knowledge for future generations.

Cultural Sensitivities and Ethical Considerations

Legal challenges are often compounded by cultural sensitivities and ethical considerations. Indigenous knowledge is deeply tied to the cultural identities and spiritual practices of the communities that hold it. Misusing or misrepresenting this knowledge can have profound cultural repercussions, eroding the trust between Indigenous communities and external stakeholders (Smith, 2012).

For example, some knowledge is considered sacred and is restricted to specific individuals or groups within a community. Legal frameworks must account for these cultural protocols and ensure that knowledge is preserved and shared in ways that align with the values and priorities of the communities involved (Meertens, 2018). This requires close collaboration between Indigenous peoples, legal experts, and policymakers.

RESOURCE-RELATED CHALLENGES IN DIGITIZING INDIGENOUS KNOWLEDGE

Digitisation of Indigenous knowledge is a powerful tool for preservation and accessibility, yet it poses significant resource-related challenges. These include financial constraints, the shortage of trained personnel, and the need for long-term sustainability of preservation efforts. Addressing these challenges is crucial for ensuring that Indigenous knowledge is safeguarded for future generations. This expanded section explores these resource-related barriers in greater depth and highlights potential strategies to overcome them.

Financial Constraints

One of the most pressing challenges in digitizing Indigenous knowledge is the significant financial investment required. Digitisation projects often demand substantial funding to cover the costs of equipment, software, infrastructure, and personnel. For memory institutions in developing countries, these costs can be prohibitive. Biyela et al. (2016) note that many institutions face restricted budgets, limiting their ability to undertake even small-scale Digitisation efforts.

The financial burden is particularly acute for rural Indigenous communities, where external funding opportunities may be scarce or unavailable. Most communities lack the resources to purchase advanced technology, such as high-resolution scanners, cameras, and storage systems. Even when funding is secured for initial Digitisation efforts, financial sustainability remains an issue. For instance, funding is often project-based and short-term, leaving communities without the means to maintain or expand digitized archives once the funding period ends (Meertens, 2018).

In addition, Digitisation projects may require ongoing expenses, such as subscriptions to cloud storage services, software licenses, and cybersecurity systems. Without consistent funding streams, many projects are at risk of stalling or failing entirely. Governments, non-governmental organizations (NGOs), and international donors must play a pivotal role in addressing this financial gap through sustained funding models. Strategies such as public-private partnerships or grants tailored specifically for Indigenous Digitisation projects could help alleviate these barriers.

Shortage of Trained Personnel

The successful execution of Digitisation projects depends on skilled personnel capable of carrying out technical and preservation tasks. Unfortunately, a significant challenge lies in the shortage of adequately trained individuals, particularly in developing countries and remote Indigenous regions. According to the First Nations Information Governance Centre (2014), the lack of human resources with expertise in areas such as document scanning, data archiving, and metadata creation severely hampers progress.

The complexity of digital preservation systems requires specialized training in not just technological tools but also cultural sensitivity. For instance, Digitisation staff must understand how to handle culturally sensitive materials appropriately, follow protocols for restricted knowledge, and engage with Indigenous communities in a respectful manner. Without proper training, there is a risk of mishandling or misrepresenting critical cultural artifacts, which could lead to a loss of trust between communities and preservation stakeholders.

Capacity-building initiatives are essential to addressing this human resource gap. Training programs can empower Indigenous community members to take active roles in Digitisation efforts, ensuring that they retain control and agency over their knowledge systems. For example, community-driven workshops on Digitisation techniques and digital archiving could help build local expertise while fostering a sense of ownership. Additionally, partnerships with universities, technical schools, and cultural institutions can provide the necessary education and mentorship to Indigenous participants.

Sustainability and Maintenance

Another critical resource-related challenge is ensuring the long-term sustainability of digitized knowledge. Creating digital archives is not a one-time effort—it requires regular maintenance to prevent data degradation, equipment failure, or technological obsolescence. Over time, digital storage devices such as hard drives, servers, and cloud platforms can experience failures or require upgrades. For resource-constrained communities, maintaining and updating these systems can be a daunting task (Meertens, 2018).

The rapid pace of technological advancement further exacerbates sustainability challenges. New file formats, software versions, and storage technologies are continually emerging, often rendering older systems obsolete. Indigenous communities that lack resources for periodic upgrades may find themselves unable to access or use their digitized content within a few years. Additionally, the risk of data corruption or loss due to inadequate backup systems underscores the importance of investing in robust digital preservation infrastructure.

To promote sustainability, Digitisation projects must include long-term planning and funding mechanisms. Strategies such as establishing endowment funds for maintenance, using open-source software with minimal licensing costs, and implementing community-led oversight committees can enhance sustainability. Collaboration with national archives or regional memory institutions could also provide technical support and ensure that digital repositories remain secure and accessible over time.

Technical Infrastructure

Adequate technical infrastructure is a prerequisite for successful Digitisation, yet many Indigenous communities struggle with insufficient or outdated facilities. Reliable electricity and internet connectivity, essential for Digitisation and online storage, are often lacking in rural or remote areas (First Nations Information Governance Centre, 2014). Communities without access to these basic utilities face significant delays or limitations in their Digitisation efforts.

Moreover, the infrastructure required for secure data storage and access—such as servers, backup systems, and firewalls—demands ongoing maintenance and expertise. Without this infrastructure, digitized materials may be at risk of data breaches, unauthorized access, or irreversible loss. For example, a community that relies on a single external hard drive for storage may lose years of work if the device becomes damaged or stolen. Strengthening infrastructure through targeted investments and technology-sharing initiatives is therefore essential.

Governments and NGOs should prioritize bridging the digital divide by investing in essential infrastructure, such as rural broadband networks, solar-powered electricity systems, and shared Digitisation centers. These initiatives can enable even geographically isolated communities to participate in Digitisation projects effectively.

Equity and Resource Distribution

Another resource-related challenge concerns the equitable distribution of Digitisation resources and funding. Frequently, larger or more prominent Indigenous groups may have greater access to external funding and institutional partnerships, while smaller or marginalized groups are left out. This disparity undermines the inclusivity of Digitisation efforts and perpetuates existing inequalities. Addressing this issue requires deliberate efforts to ensure that resources are distributed fairly across diverse Indigenous communities.

One approach to promoting equity is to establish grant programs specifically for underrepresented groups. Additionally, creating regional Digitisation hubs that serve multiple communities can help level the playing field, providing access to shared resources and expertise. Collaboration among Indigenous groups, facilitated by national or international organizations, can also enhance resource-sharing and knowledge exchange.

Addressing Resource-Related Challenges

By recognizing and addressing resource-related challenges, we can develop strategies to effectively preserve Governments, NGOs, and private donors should provide sustained funding for Digitisation efforts, prioritizing long-term maintenance and upgrades. Training programs and workshops should empower Indigenous community members to lead Digitisation projects, fostering local expertise and ownership. Digitisation initiatives must incorporate strategies for long-term preservation, including regular backups, periodic upgrades, and secure storage systems. Investments in electricity, internet connectivity, and Digitisation facilities are essential for overcoming technical barriers. Ensuring fair access to resources and funding is critical to promoting inclusivity and representation in Digitisation projects. By implementing these strategies, stakeholders can ensure that Indigenous knowledge is preserved in ways that respect cultural values, empower communities, and maintain sustainability.

POSSIBLE MAJOR SOLUTIONS TO CHALLENGES

Ethical practices should be advocated for in the Digitisation and sharing of Indigenous knowledge, ensuring that the rights and interests of Indigenous communities are respected and protected.

Informed consent from Indigenous communities should be obtained before digitizing and sharing their knowledge is crucial. This ensures that they are fully aware of how their knowledge will be used and have given explicit permission for its Digitisation and dissemination (UNESCO, 2017). Obtaining informed consent is a fundamental principle that ensures Indigenous communities are fully aware of how their knowledge will be used and have given explicit permission for its Digitisation and dissemination. This process involves transparent communication and mutual understanding, respecting the autonomy and rights of the communities.

The informed consent process typically includes explaining the purpose, benefits, and potential risks of digitizing and sharing Indigenous knowledge in a manner that is easily understandable to community members. This ensures that Indigenous communities understand the purpose, benefits, and potential risks of digitizing and sharing their knowledge. Clearly outline the objectives of the Digitisation project should be clearly explained. For example, emphasizing that the primary goal is to preserve Indigenous knowledge for future generations and to make it accessible for educational and research purposes. The community should know why it is important to preserve Indigenous knowledge. This can include reasons such as safeguarding cultural heritage, supporting biodiversity conservation, and contributing to global scientific knowledge (UNESCO, 2017).

Describe how Digitisation can make Indigenous knowledge more accessible to community members, researchers, and policymakers. This can enhance the visibility and appreciation of Indigenous cultures and traditions. Highlight that digital preservation helps protect knowledge from being lost due to physical deterioration or environmental factors. Digital archives can provide a lasting and easily replicable record. Emphasize that Digitisation can empower Indigenous communities by giving them control over how their knowledge is shared and used. It can also provide opportunities for economic development through partnerships and collaborations (First Nations Information Governance Centre, 2014).

Explain the potential risks of knowledge being misused or misappropriated if not properly safeguarded. This includes the risk of bio-piracy and exploitation by external entities. Address the concern that digitized knowledge may be taken out of its cultural context, leading to misinterpretation or misrepresentation. Discuss the importance of data security measures to protect digitized knowledge from unauthorized access, hacking, or other cyber threats (Smith, 2012).

Effective communication involves using language and explanations that are easily understandable to community members, avoiding technical jargon, and providing opportunities for questions and feedback. Engaging with community leaders and elders can help ensure that the information is communicated in a culturally appropriate manner within cultural norms. They need to be consulted to seek their approval and address any concerns or questions they may have. Documenting the consent in written or verbal form, ensures that it is legally recognized and respected. Ongoing dialogue with the community should be maintained throughout the Digitisation process to ensure continued consent and address any emerging issues. By obtaining informed consent, we can foster trust and collaboration with Indigenous communities, ensuring that their knowledge is preserved and shared ethically and respectfully (UNESCO, 2017).

Indigenous communities should retain ownership and control over their digitized knowledge. This includes determining who can access and use the information and how it can be shared. Ensuring community control helps protect against misappropriation and misuse (First Nations Information Governance Centre, 2014). Any benefits arising from the use of Indigenous knowledge, whether monetary or otherwise, should be shared with the communities that provided the knowledge. This can include financial compensation, capacity-building initiatives, or other forms of support (Nagoya Protocol, 2010). Establishing transparent and accountable processes for digitizing and sharing Indigenous knowledge is essential. This includes clear documentation of consent, data management practices, and mechanisms for addressing grievances (Canadian Institute of Health Research, 2018). By adhering to these ethical practices, we can ensure that the Digitisation and sharing of Indigenous knowledge are conducted in a manner that respects and protects the rights and interests of Indigenous communities.

Fostering collaboration between Indigenous communities, researchers, policymakers, and other stakeholders is crucial to developing effective strategies for preserving Indigenous knowledge. Establishing trust between Indigenous communities and external stakeholders is essential. This can be achieved through transparent communication, respecting cultural protocols, and recognizing the value of Indigenous knowledge (Smith, 2012; Meertens, 2018). Transparent communication ensures that all parties are aligned in their goals and processes, while respecting cultural protocols acknowledges the sovereignty and traditions of Indigenous communities

(Meertens, 2018). Recognizing the value of Indigenous knowledge further strengthens collaboration by validating its significance in sustainable development and cultural preservation (Smith, 2012).

Joint projects that involve Indigenous communities in decision-making processes should be initiated. Collaborative projects can include workshops, conferences, and research initiatives that bring together different stakeholders. Encourage interdisciplinary collaboration by involving experts from various fields such as anthropology, linguistics, digital technology, and law to create comprehensive strategies for preservation (Biyela et al., 2016).

Integrating open science principles with Indigenous knowledge systems can enhance accessibility, transparency, and inclusivity in the preservation process. Digitized Indigenous knowledge is easily accessible to both Indigenous communities and the wider public. This can be achieved through open-access repositories and user-friendly platforms. Maintain transparency in the Digitisation and sharing processes by documenting methodologies, consent agreements, and data management practices. This fosters trust and accountability (UNESCO, 2017). Promote inclusivity by involving Indigenous communities in the design and implementation of open science initiatives. This ensures that their perspectives and needs are considered (First Nations Information Governance Centre, 2014).

Offering practical solutions and recommendations is essential for overcoming the challenges in digital preservation. Utilize digital tools such as databases, cloud storage, and digital recording devices to capture and store Indigenous knowledge. Ensure that these tools are accessible and user-friendly for Indigenous communities. Develop and implement legal frameworks that protect Indigenous knowledge from exploitation and misuse. This includes intellectual property rights, data ownership, and benefit-sharing agreements (Nagoya Protocol, 2010). Engage Indigenous communities through participatory approaches that involve them in the entire Digitisation process. This includes obtaining informed consent, providing training, and addressing their concerns (Smith, 2012).

Increasing awareness about the importance of preserving Indigenous knowledge and the need for concerted efforts is vital. Launch public awareness campaigns that highlight the value of Indigenous knowledge and the challenges faced in its preservation. Use various media platforms, including social media, to reach a broader audience. Incorporate Indigenous knowledge into educational curricula at all levels. This helps to foster appreciation and respect for Indigenous cultures among younger generations (Meertens, 2018). Advocate for policies that support the preservation of Indigenous knowledge. Engage with policymakers to create a supportive environment for Digitisation efforts and ensure that Indigenous knowledge is recognized and protected (UNESCO, 2017).

By focusing on these areas, we can develop effective strategies for preserving Indigenous knowledge and ensure that it is respected, protected, and utilized in a manner that benefits both Indigenous communities and the broader society.

The 7C Model in Context

The 7C model—*Community, Content, Connectivity, Capacity, Culture, Cooperation, and Continuity*—is applied as a holistic approach to overcome challenges in digitizing and preserving Indigenous Knowledge.

1. **Co-design**

Engage indigenous communities in the design process to ensure that their needs, values, and perspectives are incorporated. Work collaboratively with community members, researchers, and technologists to create culturally sensitive and relevant digital preservation solutions. Utilizing collaborative platforms within the Open Science Framework can facilitate the sharing of best practices, resources, and knowledge between indigenous communities and researchers. These platforms should be designed to respect and protect the rights and interests of indigenous communities (Karjalainen & Okere, 2018).

2. **Conceptualization**

Karjalainen & Okere, (2018) stated that a deep understanding of the indigenous knowledge to be preserved has to b developed, including its cultural, social, and environmental context. Create a clear plan for the Digitisation process, outlining goals, methodologies, and expected outcomes . Developing a deep understanding of the indigenous knowledge to be preserved requires considering its cultural, social, and environmental context. This involves engaging with community members to gain insights into their traditions, social dynamics, and ecological practices (Karjalainen & Okere, 2018). Creating a clear plan for the Digitisation process includes outlining the goals, methodologies, and expected outcomes, ensuring that the project is culturally sensitive and sustainable (Karjalainen & Okere, 2018).

3. **Collection**

Furthermore the same author mentioned that indigenous knowledge should be collected through various means, such as interviews, recordings, and documentation. Ensure that the collection process respects cultural sensitivities and obtains informed consent from community members. Developing and adhering to ethical frameworks and guidelines for the digital preservation of indigenous knowledge is essential. These frameworks should prioritize respect for cultural sensitivities, informed consent, and the protection of intellectual property (Chukwuma & Mpofu, 2019).

4. **Correction**

Encouraging and supporting community-led digital preservation initiatives ensures that indigenous communities are actively involved in the process. This empowers communities to take ownership of their knowledge and ensures that preservation efforts are aligned with their values and priorities (Karjalainen & Okere, 2018). Validate the collected knowledge with the community to ensure accuracy and authenticity. Incorporate feedback from community members to address any inaccuracies or misrepresentations.

5. **Curation**

To organize and categorize the digitized knowledge in a way that is meaningful and accessible to the community, it is essential to develop a taxonomy and classification system in collaboration with the community, include descriptive metadata, and implement tagging and indexing (Karjalainen & Okere, 2018). Preservation strategies should include metadata creation, reliable storage solutions, backup systems, and digital preservation policies to ensure the long-term accessibility and integrity of the knowledge (Karjalainen & Okere, 2018).

6. **Circulation**

Share the digitized knowledge with the community and other stakeholders through appropriate channels. Ensure that access to the knowledge is controlled and respects the community's wishes regarding who can view and use the information. Engaging in advocacy and policy efforts to promote the recognition and protection of indigenous knowledge within the digital realm is vital. This includes advocating for policies that support indigenous sovereignty, intellectual property rights, and equitable access to digital resources (Chukwuma & Mpofu, 2019).

7. **Creation**

Encourage the creation of new knowledge and cultural expressions based on the preserved indigenous knowledge. Promote sustainable practices to ensure the ongoing preservation and use of the digitized knowledge. Providing capacity-building and training opportunities for indigenous communities is crucial to address technological barriers. This includes training in digital tools, data management, and the principles of open science (Mirza, 2020).

The 7C Model emphasizes the importance of involving indigenous communities throughout the Digitisation process, ensuring that their knowledge is preserved in a way that respects their cultural heritage and promotes their agency (Karjalainen & Okere, 2018).

SUMMARY AND RECOMMENDATIONS

This chapter explores the challenges involved in preserving Indigenous Knowledge (IK) within the Open Science Framework, emphasizing the unique tensions between the communal, intergenerational nature of IK and the principles of Open Science, such as accessibility and transparency. The preservation of IK faces ethical, technological, legal, cultural, and resource-related challenges that necessitate the use of innovative frameworks like the 7C model. The chapter underscores the importance of balancing Open Science principles with the unique characteristics of Indigenous Knowledge. By adopting the 7C model, stakeholders can navigate the challenges of ethical inclusion, legal protection, resource mobilization, and cultural respect. This model provides a roadmap for fostering trust, ensuring inclusivity, and safeguarding Indigenous Knowledge for future generations.

REFERENCES

Balogun, T., & Kalusopa, T. (2021, July 16). A framework for digital preservation of Indigenous knowledge system (IKS) in repositories in South Africa. *Records Management Journal, 31*(2), 176–196.

Biyela, N., Oyelude, A., & Haumba, E. (2016, April). Digital preservation of indigenous knowledge (IK) by cultural heritage institutions: a comparative study of Nigeria, South Africa and Uganda. In *The proceedings of the digital transformation and the changing role of libraries and information centres in the sustainable development of Africa at the 22nd Standing Conference of Eastern, Central and Southern Africa Library and Information Associations*.

Boamah, E., & Liew, C. L. (2017). Conceptualising the digitisation and preservation of Indigenous knowledge: The importance of attitudes. In *Digital Libraries: Data, Information, and Knowledge for Digital Lives: 19th International Conference on Asia-Pacific Digital Libraries, ICADL 2017, Bangkok, Thailand, November 13-15, 2017, Proceedings* (pp. 65-80). Springer International Publishing.

Borrero, R. M. (2013). *Innovation and technology for Indigenous peoples. United Nations*. Retrieved from UN Website.

Chiwanza, K., Musingafi, M. C. C., & Mupa, P. (2012). Challenges in preserving Indigenous knowledge systems: Learning from past experiences. [Retrieved from Scientific Research Publishing.]. *Information and Knowledge Management, 3*(2), 19–25.

Chukwuma, A., & Mpofu, B. (2019). The relevance of Indigenous knowledge systems in social work education and practice in Zimbabwe. *Journal of Social Development in Africa*.

EcoLiterateLaw. (2024). The impact of digital technology on Indigenous peoples. Retrieved from EcoLiterateLaw.

First Nations Information Governance Centre. (2014). Ownership, Control, Access, and Possession (OCAP): The Path to First Nations Information Governance. Retrieved from https://fnigc.ca/

Global Indigenous Data Alliance. (2019). CARE principles for Indigenous Data Governance.

Karjalainen, J., & Okere, I. (2018). Local institutions and Indigenous knowledge in climate-smart agricultural innovations. *International Journal of Climate Change Strategies and Management*.

le Roux, A. (2017). *Indigenous knowledge in a virtual context: Sustainable digital preservation. Unisa Library*. Retrieved from WiredSpace.

Maasz, D., Winschiers-Theophilus, H., Stanley, C., Rodil, K., & Mbinge, U. (2018). *A digital Indigenous knowledge preservation framework: The 7C model—Repositioning IK holders in the Digitisation of IK. Digitisation of Culture: Namibian and International Perspectives*. Retrieved from SpringerLink.

Meertens, D. (2018). Conceptualising the digitisation and preservation of Indigenous knowledge. In *The Digital Humanities and Indigenous Knowledge* (pp. 83–95). Springer., https://link.springer.com/chapter/10.1007/978-3-319-70232-2_6

Meertens, D. (2018). Respecting protocols: Cultural preservation and digitisation in Indigenous communities. *Journal of Cultural Heritage Management*.

Mirza, N. (2020). Nursing, Indigenous health, water, and climate change. *Witness: The Canadian Journal of Critical Nursing Discourse*.

MSN. (2025). *Indigenous knowledge and climate adaptation in Vanuatu*. MSN Reports.

Nagoya Protocol. (2010). The Nagoya Protocol on Access and Benefit-sharing. Retrieved from https://www.cbd.int/abs/

Nagoya Protocol. (2010). The Nagoya Protocol on Access to Genetic Resources and the Fair and Equitable Sharing of Benefits Arising from Their Utilization. *Convention on Biological Diversity. (Open Science Framework (OSF)*. Available from: https://www.researchgate.net/publication/315794339_Open_Science_Framework_OSF [accessed Mar 02 2025].

Smith, L. T. (2012). *Decolonizing methodologies: Research and Indigenous peoples* (2nd ed.). Zed Books.

UNESCO. (2016). *The role of Indigenous knowledge in sustainable development*. United Nations Educational, Scientific and Cultural Organization.

UNESCO. (2017). *Safeguarding Indigenous knowledge through digitisation: Challenges and opportunities*. United Nations Educational, Scientific and Cultural Organization.

KEY TERMS AND DEFINITIONS

Bio-Piracy: The exploitation of Indigenous Knowledge and genetic resources by external entities, such as corporations or researchers, without proper consent or benefit-sharing (UNESCO (2017) and the Nagoya Protocol (2010)).

CARE Principles: A framework for data governance emphasizing Collective Benefit, Authority to Control, Responsibility, and Ethics, specifically developed to address the needs of Indigenous data sovereignty (Global Indigenous Data Alliance, (2019)).

Cultural Protocols: Traditional rules and practices that govern the use, access, and dissemination of Indigenous Knowledge within communities Meertens (2018).

Digital Preservation: The process of storing, maintaining, and protecting digital information to ensure its long-term accessibility and usability (Biyela et al., 2016).

Ethical Challenges: Issues related to consent, ownership, and the risk of misrepresentation or exploitation when digitizing Indigenous Knowledge (Smith (2012) and Meertens (2018)),

Indigenous Knowledge (IK): A collective body of knowledge, practices, and beliefs developed by Indigenous communities over generations, often passed down orally and deeply tied to cultural, spiritual, and environmental contexts UNESCO (2017).

Intellectual Property (IP) Laws: Legal frameworks designed to protect creations of the mind, such as inventions or literary works, which often conflict with the communal and intergenerational nature of Indigenous Knowledge (Chiwanza, K., Musingafi, M., & Mupa, P. (2012).

Open Science Framework: An approach aimed at making scientific research and data openly accessible to everyone while promoting transparency, collaboration, and reproducibility.

Sustainability: The ability to ensure the long-term preservation, accessibility, and usability of digitized Indigenous Knowledge, accounting for resource and technological constraints (Meertens, 2018),

Traditional Knowledge Systems: Knowledge systems embedded in the cultural and ecological practices of Indigenous communities, often transmitted orally and adapted over generations (UNESCO, (2017) and Chiwanza et al., (2012).

Chapter 11
Ethical Priorities for Libraries Supporting Indigenous Knowledge in Selected Public Libraries in Zimbabwe

Notice Pasipamire
https://orcid.org/0000-0001-5148-9975
National University of Science and Technology, Bulawayo, Zimbabwe

Artwell Moyo
https://orcid.org/0000-0002-0722-7879
Judicial Service Commission of Zimbabwe, Bulawayo, Zimbabwe

ABSTRACT

This chapter examines ethical priorities in Zimbabwe's public libraries regarding indigenous knowledge. Public libraries safeguard indigenous knowledge through community collaboration. Key concerns include acknowledging sources (intellectual property), project impact, and considering indigenous languages in classification. Public libraries recognize a moral duty to integrate indigenous and foreign knowledge. Collaboration with Indigenous elders is essential for language revitalization. The study highlights the importance of ethical standards in ensuring respectful engagement with Indigenous communities. By prioritizing ethics, public libraries can preserve and promote indigenous knowledge while respecting Indigenous rights.

DOI: 10.4018/979-8-3373-0204-1.ch011

INTRODUCTION

Public libraries serve as crucial institutions for the preservation and dissemination of knowledge within communities. Their role transcends mere storage of information to encompass active engagement in fostering cultural and intellectual growth. Globally, public libraries have implemented initiatives to document and integrate Indigenous Knowledge (IK) into modern information systems, as highlighted by Ranasinghe (2008) and IFLA (2002). In Zimbabwe, public libraries have a responsibility in safeguarding indigenous knowledge (IK), which includes the traditions, practices, skills, and languages unique to various communities. The United Nations Educational, Scientific and Cultural Organization (UNESCO) (2025) defines Indigenous knowledge as the "understandings, skills and philosophies developed by societies with long histories of interaction with their natural surroundings." This knowledge is integral to the identity, resilience, and development of indigenous communities.

Indigenous Knowledge (IK) and practices are typically unwritten, relying on oral transmission and human memory. Libraries have focused on documenting and disseminating IK (Adeyemo & Adebayo, 2017), covering topics such as agriculture, traditional medicine, and cultural practices. Documentation methods include recordings and visual documentation, undertaken by library personnel. However, librarians often face ethical dilemmas in their daily work, lacking structured institutional support or training to address them (Hoffman, 2005; Luo, 2016). This challenge is exacerbated by limited familiarity with professional codes of ethics, as seen in Amoah and Yeboah's (2018) research at the Sam Jonah Library in Ghana.

The CARE Principles for Indigenous Data Governance (Carroll et al., 2020) emphasize the need for ethical and culturally sensitive approaches when managing Indigenous data. Given that most Indigenous data is controlled by non-Indigenous institutions, there is a responsibility to engage respectfully with communities, ensuring data use supports capacity development, community data capabilities, and the preservation of Indigenous languages and cultures. By aligning practices with principles of equity and respect, institutions can protect the integrity of Indigenous knowledge and empower these communities.

Public libraries in Zimbabwe have emerged as anchor institutions in preserving these invaluable resources, collaborating with indigenous communities to document oral histories, folklore, and cultural practices, often archiving them in physical and digital formats. The International Federation of Library Associations and Institutions (IFLA, 2002) underscores the importance of Indigenous Knowledge through its statement on IK, advocating for its protection and integration into modern information systems. However, public libraries face multifaceted challenges in preserving this heritage. Librarians, archivists and curators are often not adequately trained on how to handle the indigenous knowledge components of their collection. It is very important to be sensitive to this aspect (Oyelude, 2023). Studies have shown that libraries have not yet devised mechanisms to enhance the availability and access to IK (Mhlongo & Ngulube, 2018). Moreover, interactions with indigenous peoples necessitate a sensitive, respectful approach, recognizing their ownership, autonomy, and cultural protocols (Penna, 2023). Public libraries must navigate issues such as consent, intellectual property rights, and the potential commercialization of indigenous resources. An institution's indigenous knowledge preservation team may actually need to employ local research aids or members of the local community who are well-versed in the local history, traditions and ways of handling the indigenous knowledge on a permanent or temporary basis but Cultural heritage institutions often lack the funding (or enough funding) to carry out indigenous knowledge preservation (Oyelude, 2023).

Indigenous or traditional knowledge (IK) holders face a lack of respect and appreciation for such knowledge (Oyelude, 2023). As custodians of indigenous knowledge, public libraries must prioritize ethical engagement to foster trust and collaboration with indigenous communities. Ethical considerations are crucial in ensuring that the voices of indigenous communities remain central in the decision-making process surrounding the documentation, archiving, and dissemination of their knowledge. Previous studies on IK have focused on the role of libraries in promoting indigenous innovations and preserving cultural heritage. Chigwada and Chiparausha (2016) emphasized the need for coordinated approaches to IK management, including the establishment of databases and community engagement programs. Mahwasane (2017) explores the role of contemporary library services in the utilization of indigenous knowledge systems. However, existing studies often lack a focused examination of ethical priorities in this process. Therefore, this study addresses the gap in knowledge by exploring the ethical considerations that libraries must navigate when acquiring, preserving, and disseminating IK.

Purpose of the Study

The study was carried out to assess the ethical priorities that public libraries in Zimbabwe deal with when it comes to acquisition, preservation and dissemination of indigenous knowledge.

Research Objectives

In order to understand ethical priorities for libraries supporting indigenous knowledge, the study was premised on the following research objectives:

1. To identify the forms in which public libraries in Zimbabwe preserve indigenous knowledge;
2. To explore the ethical priorities for public libraries when dealing with indigenous knowledge;
3. To determine the key ethical considerations prioritized by public libraries when handling indigenous knowledge.

LITERATURE REVIEW

Concept of Indigenous Knowledge

The concept of IK has been subject to critique, with some arguing that it involves an incomplete or questionable understanding of knowledge (Horsthemke, 2008). Horsthemke (2008) contends that IK may not be an effective tool in anti-discrimination and anti-repression discourse, and that there may be greater common ground in the development of knowledge than admitted by theorists. The concept of Indigenous Knowledge (IK) refers to the unique, traditional, and local knowledge existing within and developed around specific conditions of women and men indigenous to a particular geographic area (Grenier, 1998 cited by Kamboya, 2018). IK encompasses beliefs, rituals, and perceptions, ways of learning, local technology, stocks of knowledge, and practices of acquiring and transmitting it (Chambers, 2014). This knowledge is deeply rooted in personal experiences, traditions, and cultural expressions, making it inherently difficult to document and transfer through conventional means (Nakata & Langton, 2009). Indigenous knowledge is rooted in cultural experience and guides relations among human, nonhuman, and other-than-human beings in specific ecosystems (Bruchac, 2014). Archaeologists can gain valuable insights by consulting Indigenous traditions and localized knowledges, which contain crucial information that can explain and contextualize scientific data (Bruchac, 2014). However, researchers must strive to avoid interference with esoteric knowledges, sacred sites, and ritual landscapes.

The Role of Public Libraries in Indigenous Knowledge Preservation

Public libraries play a crucial role in preserving indigenous knowledge by documenting and archiving local traditions, stories, and languages (Gundakanal & Kaddipujar, 2024; Chigwada & Ngulube, 2023). They facilitate the collection and dissemination of indigenous knowledge, ensuring that intellectual property rights are respected and protected (Osagie et al., 2023). Libraries also contribute to the revitalization of endangered indigenous languages by providing educational resources, language learning programs, and digital archives (Hamwaalwa, 2021). Digital preservation has emerged as a popular method for revitalizing and safeguarding IK (Masenya, 2023). However, research has shown that indigenous knowledge owners and their communities often lack mechanisms and policies for effective digital preservation (Masenya, 2023). This highlights the need for anticolonial strategies and policies to protect IK systems, as well as training and awareness-raising programs for community members (Masenya, 2023).

Research has shown that rural libraries can actively participate in preserving IK and contribute to its documentation (Forutnani et al., 2018). However, librarians need motivation and skill development to effectively preserve IK. The literature highlights the importance of community engagement and participation in IK preservation. Rural libraries can serve as focal points for IK preservation and promotion, and their inherent nature allows them to aid and lead the process (Forutnani et al., 2018). Effective IK preservation requires a collaborative approach, involving librarians, community members, and IK specialists. A study on preserving and using indigenous knowledge (IK) in public libraries in Southwest Nigeria Owolabi et al. (2021) found that Educational purposes were the primary reason for using IK in libraries, Video and audio recordings were significant sources for preserving IK, Users accessed IK resources through ICT tools like computers and compact discs, Inadequate funding was the major challenge facing IK development.

Role of Public Libraries in Safeguarding Indigenous Knowledge and Revitalizing Endangered Indigenous Languages

Public libraries play a vital role in preserving indigenous knowledge and revitalizing endangered indigenous languages. Through collaboration with indigenous communities, libraries serve as repositories of cultural heritage, ensuring that traditional knowledge is documented, protected, and made accessible for future generations (Osagie et al., 2023). Libraries engage with indigenous communities to collect, record, and organize traditional knowledge, ensuring that it remains an integral part of cultural heritage (Sarkhel, 2017). The preservation and promotion of Indigenous knowledge and languages are critical issues that have garnered significant attention in recent years. Libraries have emerged as key institutions in this endeavor, playing a vital role in the documentation, archiving, and dissemination of Indigenous cultural materials. According to Thorpe and Galassi (2014), libraries like the State Library of New South Wales are instrumental in cultural revitalization through initiatives such as the Rediscovering Indigenous Languages project. Research has highlighted the importance of libraries establishing proper protocols and respecting Indigenous notions of ownership in the management of Indigenous cultural materials (Roy, 2013). This involves ensuring that Indigenous communities are actively involved in decision-making processes related to the handling, access, and sharing of cultural materials. By adopting a collaborative approach, libraries can foster trust and build strong relationships with Indigenous communities (Roy, 2013). Libraries are also playing a crucial role in preserving endangered languages through documentation and archiving (Gundakanal & Kaddipujar, 2024). The creation of digital language archives has enabled the preservation of linguistic materials, including audio recordings and written texts, for future generations. Furthermore, libraries are engaging in community initiatives to promote the use of endangered languages and raise awareness about their cultural significance (Gundakanal & Kaddipujar, 2024). The literature emphasizes the need for libraries to adopt reflexive approaches and evaluate the impact of their projects on Indigenous communities (Thorpe & Galassi, 2014). This involves developing frameworks to measure the effectiveness of library initiatives and ensuring that they are responsive to the needs of Indigenous communities.

Ethical Considerations in Indigenous Knowledge Systems

Ethical considerations are paramount in indigenous knowledge systems. Buthelezi, Ocholla, and Dlamini (2024) emphasize the need for just recognition of indigenous knowledge within library repositories and databases. Oyelude (2023) highlights the importance of respectful practices that honour the cultural significance of the knowledge being preserved. The CARE Principles for Indigenous Data Governance (Carroll et al., 2020) underscore the critical need for ethical and culturally sensitive approaches when managing indigenous data. Kovach (2005) argues for indigenous methodologies which include consulting indigenous peoples, not as objects but rather as participants, to predict possible negative outcomes, to share and protect knowledge, to use appropriate language and form in order to communicate results back to the people.

Gaps in Indigenous Knowledge Ethics

There is a notable literature gap on ethical priorities for libraries supporting IK, particularly from the perspective of librarians' lived experiences. Indigenous knowledge (IK), often regarded as undocumented tacit knowledge (Ebijuwa, 2015), has gained significant recognition in the library field. Research has focused on preserving and disseminating IK (Chigwada & Ngulube, 2023), innovative applications (Mahwasane, 2017), documentation (Adeyemo & Adebayo, 2017), and the role of information technologies in IK management (Hunter, 2005). However, public libraries face challenges in supporting IK, including inadequate infrastructure, funding constraints, and reluctance to disclose cultural information (Osagie et al., 2023). These challenges raise ethical concerns, highlighting the need to explore ethical priorities in IK through the lived experiences of librarians. Ethics are fundamental to librarianship, grounded in principles of trust, intellectual freedom, and equitable access to information (Ray, 2022). Ethical dilemmas in librarianship vary across regions, influenced by cultural and institutional contexts (Pivec, 2003). As libraries play a central role in supporting IK, defining and identifying ethical dilemmas through librarians' lived experiences is crucial.

METHODS

The study employed a qualitative approach. This approach was suitable as it allowed for a deeper understanding of the subject matter, which dealt with the subjective experiences and perspectives of the respondents, expressed in the form of language rather than numerical data (Silverman, 2013), utilizing interviews as the primary data collection method. Managers from four selected public libraries located in Gwanda, Bulawayo, Gweru and Harare in Zimbabwe were interviewed to gain insights into the ethical challenges they face when dealing with indigenous knowledge systems. The interviews were semi-structured, allowing for in-depth discussions while also addressing specific ethical concerns. Furthermore, the qualitative research enabled the investigation of the day-to-day experiences and behaviours of the research participants, rather than relying on laboratory settings (Creswell, 2009). The qualitative approach was appropriate for this study because it facilitated an in-depth exploration of the social world, as stated by Creswell (2009). This was particularly relevant as there was limited prior research on Zimbabwe public libraries ethical priorities when engaging with indigenous knowledge systems. Indigenous Knowledge (IK) is deeply rooted in personal experiences, traditions, and cultural expressions, hence ethical priorities in this regard are social phenomenon that cannot be fully captured through statistical means and requires an understanding of the perspectives and experiences of librarians as they deal with indigenous knowledge systems.

The target population comprised of librarians working in public / community libraries in Bulawayo, Gwanda, Gweru and Harare, serving different communities in Zimbabwe. This study focused on public libraries run by boards of trustees in specific locations, namely Bulawayo, Gwanda, Harare, and Gweru. The study did not include public libraries managed by municipal authorities. The scope was limited to exploring the ethical priorities for libraries supporting indigenous knowledge, with a specific emphasis on understanding the challenges, opportunities, and best practices in this area. The study aimed to identify key ethical considerations and priorities for public libraries in supporting indigenous knowledge, and the findings may inform policy and practice in similar contexts. Purposive sampling was used to select one public library in each city mentioned as well as two librarians per each institution for semi-structured interviews. Before data collection, the researcher sought permission from the library authorities and explained the purpose of the study. To ensure the credibility and trustworthiness of the study, the researcher employed member checking to verify the accuracy of the data. Key informants were asked to review the transcripts of the dialogues in which they had participated. Additionally, the researcher sought peer scrutiny before going to the field, which helped determine the appropriateness of the data collection instruments and ensure that the intended information would be collected.

The data analysis process involved a thorough and systematic examination of the qualitative data collected, guided by the main research question. Through a qualitative data analysis approach, the vast volumes of data were coded, structured, and organized to extract meaningful insights, identify patterns and themes, and draw connections between the data and the research question. The coding process involved assigning labels and categories to the data, allowing the researcher to identify and analyse recurring themes and concepts related to ethical priorities for libraries supporting Indigenous knowledge. This analysis produced key findings that highlighted the forms of preservation, ethical priorities, and key considerations that public libraries should prioritize when working with Indigenous knowledge and communities, including respect for Indigenous protocols, cultural sensitivity, and community-led decision-making.

RESULTS

A total of eight interviews were conducted with eight managers in four public libraries, providing valuable insights ethical priorities for libraries supporting indigenous knowledge. The findings of the study are organized according to the themes that emerged during the analysis of the results.

Preservation Methods

In Zimbabwe, libraries employ diverse methods to preserve indigenous knowledge. Experiences from selected public libraries revealed that main IK preservation methods used by public libraries, including poetry and folklore, drama and performance, as well exhibitions and artefacts:

Poetry and Folklore: Libraries archive traditional stories and poems that reflect the cultural heritage of local communities. One librarian in Bulawayo noted that

"One of the methods we use as a library to preserve IK is through archiving poetry and folklore. These traditional stories and poems are deeply tied to our community's heritage, especially among Ndebele people, we have different authors who we work with to write these stories and we order books from them. By having these documented, we as a public library we ensure that these narratives remain alive and accessible for future generations."

Librarian in Gweru also affirmed the use of poetry and folklore as a preservation method by stating that *"we have been working with various elders in the community to document their oral histories. We established a story telling session where we invite parents (mainly those whose children who are library members) to come and have a library time, were they take the role of librarian in educating children. This is a way to ensure that parents and children are engaged in library activities, from these sessions we noted that folklore and poetry were the common methods used by parents in teaching children, we then moved to document some of these stories as a library and from time to time we use them to conduct our own session to share indigenous knowledge to children, therefore I can safely conclude that we use poetry and folklore in preserving indigenous knowledge"*

Drama and Performance: Live performances of traditional dances and plays are documented as a means of preserving indigenous artistic expressions. Librarians interviewed highlighted that they have different sessions where they invite different people in the community to share their artistic performances especially with children. The lived experience of one respondent in Gwanda was captured as *"Our library supports different communities and cultures who all have different IK systems and information, hence in celebrating and preserving these communities IK, we often collaborate with local artists offering them library as a venue to hold their performances of traditional dances and plays. As they carry out their work, library members mostly our children patronage are allowed to watch and not only do they get entertainment; but they learn about their culture, thus getting cultural wisdom. For those who document and record their session (one group once did a recording) we ask them to share with library a copy of their performance, that we have kept and from time to time we play it to children instead of only using western movies".*

In Harare, one librarian noted, *"we have been working with various stakeholders in different communities to organise various events such as poetry competitions, focusing on cultural heritage of local communities. These sessions are recorded and we keep the recording within our collection which allows us to preserve community knowledge".*

Exhibitions and Artefacts: Libraries curate exhibitions showcasing indigenous artefacts, providing a platform for community members to engage with their cultural history. According to one librarian in Gwanda, *"When I started working in this community, I realized that there are important days like Africa day which were* *celebrated in schools, copying that model as a library we came up with the idea to organize exhibitions showcasing indigenous artefacts. We serve one of the minority communities in Zimbabwe which are Sotho, and Venda. During the exhibition, as a library we collect exhibitions and artifacts with the authority from the owners who came to exhibit, we keep these a way for the community to reconnect with their history. These exhibitions act as bridges between the past and present, helping young people learn about the cultural significance of these items."*

From the experiences of Bulawayo librarian, it was stated that, *"we have formed collaborative relationships with the community and other stakeholders in facilitating exhibitions"*.

Safeguarding Indigenous Knowledge

The study sought to evaluate the role of public libraries in safeguarding indigenous knowledge through collaboration with Indigenous communities and their contribution to revitalizing endangered Indigenous languages, the findings revealed that public libraries work closely with Indigenous communities to preserve and showcase their knowledge and heritage. Librarians from Gwanda and Harare highlighted that as libraries they have developed strong relationships with local elders who share their traditions during different events which are hosted by the library.

A remark from one of them noted that *"As a library we fully understand that we are part of a bigger community and we don't work in isolation, we have collaborated with different stakeholders in the community. This is a way for us to preserve indigenous knowledge while fostering respect for our cultural heritage within communities we serve"*.

In Bulawayo, collaboration with indigenous communities was revealed in this way *"We invite community members to lead workshops on indigenous practices. These partnerships are essential for safeguarding our unique traditions and passing them on."* While in Gweru, collaboration is captured by one librarian stating that *"By working directly with Indigenous authors and elders, we've been able to document oral histories and create resources that honor their wisdom"*

The findings of the study on the lived experiences of public libraries, revealed that libraries play a crucial role in supporting the revival and use of endangered Indigenous languages;

A librarian in Gwanda stated, *"We've started curating books and audio resources in Indigenous languages to give people the tools to learn and reconnect with their roots. We also fund and run local language writing competitions"*

In Harare, another librarian explained, *"We host storytelling sessions in Indigenous languages. Seeing children captivated by these stories shows how vital this work is for keeping our languages alive."*

From Bulawayo and Gweru, respondents shared the same sentiment where they have promoted language use by children in order to spark interest in Indigenous languages. One librarian noted that *"With the coming in of the Heritage based curriculum in Zimbabwe education, we have taken measures to support this curriculum by promoting use of local languages by children, partnering with linguists to create various resources such as dictionaries and grammar guides in endangered languages, which have been invaluable for preservation efforts."*

Ethical Priorities

The findings highlighted several ethical priorities that public libraries in Zimbabwe deal with, including:

Acknowledging Sources and Origins: There is a critical need to recognize the intellectual property of indigenous knowledge holders, ensuring that proper credit is given to the sources of knowledge. The lived experience of Harare librarian was captured as follows *"we actively collaborate with indigenous communities to ensure their knowledge is documented respectfully and credited appropriately. This we achieve through implementation of protocols for intellectual property rights to safeguard indigenous knowledge holders' contributions, we try by all means to acknowledge members of the community contributing their knowledge towards IK".*

In Gwanda and Gweru similar sentiments were echoed by the librarians, which can be summed up as follows *"as a library we have developed a tendency of sharing on our social media pages what we do in the library. Therefore, we tag people associated with the programs we do, for example if we highlight that we hosted a story telling session we acknowledge those who participated or were telling the stories. This is how indigenous knowledge holders are acknowledged of their knowledge, encouraging knowledge sharing"*

Impact Assessment: From the lived experiences of librarians, they highlighted that their libraries evaluate the potential impact of their projects on Indigenous communities, considering both positive and negative implications. One librarian in Harare community noted that *"before initiating any project to do with IK, we conduct community consultations to understand potential impacts and address concerns.* Another one from Bulawayo added that *"I strive to use ethical guidelines to evaluate how our projects might affect indigenous communities, aiming for positive outcomes."*

Language Considerations: The use of indigenous languages and local terminology in classification and indexing practices is essential to ensure that materials are accessible and relevant to the communities they serve. According to one Gweru respondent *"this is where we face challenges as a library, I have noted that as a library we use cataloguing standards which don't accommodate cataloguing of IK to* *make it accessible, hence we need to prioritize the inclusion of indigenous languages in our cataloging systems to make resources more accessible.* In addition, another respondent added that *"there is need to have librarians undergo training to understand and use local terminology effectively in classification practices, this allows for library professionals to be well equipped when it comes to dealing with IK."*

From one Librarian from Gwanda it was gathered that *"as we work with local communities in IK preservation, one challenge that arise is how to classify that knowledge. I am more competent in classification using the Dewey decimal classification scheme which doesn't consider IK, hence it becomes difficult to classify IK. What we have resorted to is keeping a separate section of IK collection, then on our reference services to clients we try to listen to what they want and if possible direct them first to IK section especially those looking for information on history and heritage. In actual fact we have been lucky with the introduction of Heritage based curriculum, we have managed to point more school going kids to IK, and we hope to develop a richer IK collection."*

DISCUSSION OF RESULTS

One pattern that emerges from the findings is the importance of community engagement in preserving indigenous knowledge. The study highlights the role of public libraries in collaborating with Indigenous communities, inviting community members to lead workshops, and documenting oral histories. This suggests that community engagement is a crucial aspect of preserving indigenous knowledge, and that public libraries are well-positioned to facilitate this process. These findings collaborate with Oyelude (2023) study on Indigenous knowledge preservation as a sign of respect for culture: concerns of libraries, archives and museums. The study noted that cultural heritage institutions and their staff have a huge stake in the preservation of indigenous knowledge for future generations. What they do or leave undone, how they do what needs to be done and the maintenance of respect for culture is in their hands, while a lot depends on them, especially in this digital era. Therefore, public libraries in Zimbabwe need to priorities collaboration and respect indigenous communities in order to preserve IK.

Another theme that also emerges from the findings is the role of public libraries in cultural revitalization. The study highlights the importance of preserving indigenous languages, traditions, and cultural practices. The use of exhibitions, performances, and storytelling sessions suggests that public libraries are not only preserving indigenous knowledge but also promoting cultural revitalization and community pride. A study by Chigwada and Ngulube (2023) in Zimbabwe revealed that libraries preserved indigenous knowledge in the form of poetry, folklore, drama and artefacts; demonstrated traditional dances; and sometimes recorded live sessions and special collections. The integration of indigenous languages into library practices is critical for revitalizing languages that are at risk of extinction. According to a study by Sarkhel (2017), it was concluded that library professionals need to be proactive in devising strategies for the management and preservation of IK with the help of their professional knowledge and skills in order to ensure access to this valuable resource. Libraries can play a transformative role by supporting language preservation initiatives and providing resources that reflect the linguistic diversity of their communities.

A relationship that emerges from the findings is the importance of collaboration and trust between public libraries and Indigenous communities. The study highlights the need for public libraries to work closely with Indigenous communities to ensure that their knowledge is documented respectfully and credited appropriately. One of the guiding philosophies of libraries is to strengthen community ties and foster a sense of connection among the people they serve (Amanze-Unagha & Unagha, 2023). Community collaboration as a concrete way to bring community members together and combine the strength of people who might otherwise not have the chance to interact and help out. This suggests that building trust and collaborative relationships is essential for effective preservation and promotion of indigenous knowledge.

One nuance that emerges from the findings is the challenge of classifying and indexing indigenous knowledge using traditional cataloguing systems. The study highlights the limitations of the Dewey Decimal Classification system in accommodating indigenous knowledge, and the need for more flexible and culturally sensitive classification systems. Garwood-Houng (2005) lamented the outdated and inappropriate description and classification of indigenous materials evident in subject headings, indexing terminology and classification systems. Lee (2011) observed that there is lack of published information on modified classification systems and thesauri for describing and organizing. This suggests that public libraries need to adapt their cataloguing practices to better accommodate indigenous knowledge and ensure that it is accessible to users.

Another pattern that emerges from the findings is the importance of capacity building for public librarians in preserving indigenous knowledge. The study highlights the need for training and professional development to ensure that librarians are equipped to handle indigenous knowledge collections and provide culturally sensitive services. Cultural competence training helps individuals develop the skills to navigate and communicate effectively across diverse cultures, promoting inclusivity, understanding, and mutual respect (Cultureally, 2025). This suggests that public libraries need to invest in capacity building initiatives to ensure that their staff have the necessary skills and knowledge to effectively preserve and promote indigenous knowledge.

CONCLUSION

This study highlights the crucial role public libraries in Zimbabwe play in preserving indigenous knowledge and revitalizing endangered indigenous languages. The findings reveal that public libraries employ diverse methods to preserve indigenous knowledge, including poetry and folklore, drama and performance, and exhibitions and artefacts. These efforts not only safeguard cultural heritage but also promote community engagement and cultural revitalization. The study also underscores the importance of collaboration between public libraries and indigenous communities. By working together, libraries can ensure that indigenous knowledge is documented, protected, and made accessible in a respectful and culturally sensitive manner. This collaboration is essential for building trust and promoting the preservation of indigenous knowledge. Furthermore, the study identifies several ethical challenges that public libraries face in preserving indigenous knowledge, including acknowledging sources and origins, impact assessment, and language considerations. To address these challenges, public libraries must prioritize culturally sensitive practices, engage in ongoing training and capacity building, and foster strong partnerships with indigenous communities. Ultimately, this study demonstrates that public libraries in Zimbabwe have the potential to play a transformative role in preserving indigenous knowledge and promoting cultural heritage. By embracing their role as custodians of cultural knowledge, public libraries can contribute to the revitalization of indigenous languages, the preservation of cultural traditions, and the promotion of community engagement and cultural identity.

RECOMMENDATIONS

Increased support for indigenous knowledge preservation and cultural revitalization initiatives

Public libraries can play a vital role in preserving and promoting Zimbabwe's rich cultural heritage by providing dedicated resources and funding for indigenous knowledge initiatives.

Enhanced Collaboration Between Public Libraries and Indigenous Communities

Building strong partnerships with indigenous communities is crucial for ensuring that library services are responsive to community needs and priorities. This collaboration can facilitate the co-curation of collections, programming, and services that reflect the community's cultural values and knowledge.

Culturally Sensitive Practices and Ongoing Training for Library Staff

Library staff should receive ongoing training and professional development opportunities to enhance their cultural competence and understanding of indigenous knowledge systems. This will enable them to provide more effective and respectful services to indigenous communities.

Development of Policies and Guidelines for the Preservation and Dissemination of Indigenous Knowledge

Public libraries in Zimbabwe should develop and implement policies and guidelines that prioritize the preservation and dissemination of indigenous knowledge in a way that respects the rights and interests of indigenous communities. These policies should be developed in consultation with indigenous communities and should address issues such as intellectual property rights, cultural protocols, and benefit-sharing.

Funding Declaration

The authors declare that no funding was received from any organization or individual to support the research, writing, or publication of this study. The study was conducted without any external financial support.

Conflict of Interest Declaration

The authors declare that they have no known conflicts of interest or competing interests that could have influenced the design, conduct, or reporting of this study. The authors have no financial, personal, or professional relationships that could be perceived as constituting a conflict of interest.

REFERENCES

Adeyemo, A.A. and Adebayo, J.O. (2017) Documentation and Dissemination of Indigenous Knowledge by Library Personnel in Selected Research Institutes in Nigeria. *Library Philosophy and Practice* (e-journal)

Amanze-Unagha, B. N., & Unagha, A. O. (2023). Building community partnerships and collaboration for effective. Library and Information Services Delivery. *International Journal of Applied Technologies in Library and Information Management, 9*(3), 31–39.

Amoah, J., & Yeboah, W. (2018). Ethical dilemmas among academic librarians: Evidence from Sam Jonah Library, University of Cape Coast, Ghana. *Ghana Library Journal, 30*(1), 14–27.

Bruchac, M. M. (2014). Indigenous Knowledge and Traditional Knowledge. In Smith, C. (Ed.), *Encyclopedia of Global Archaeology*. Springer.

Buthelezi, S. C., Ocholla, D., & Dlamini, P. (2024). Strategies for documenting and disseminating indigenous knowledge at a South African university'. *South African Journal of Information Management, 26*(1), a1648.

Carroll, S. R., Garba, I., Figueroa-Rodríguez, O. L., Holbrook, J., Lovett, R., Materechera, S., Parsons, M., Raseroka, K., Rodriguez-Lonebear, D., Rowe, R., Sara, R., Walker, J. D., Anderson, J., & Hudson, M. (2020). The CARE Principles for Indigenous Data Governance. *Data Science Journal*, 19.

Chambers, R. (2014). *Rural development: Putting the last first*. Routledge.

Chigwada, J., & Ngulube, P. (2023). Librarians' role in the preservation and dissemination of indigenous knowledge. *IFLA Journal, 50*(2), 242–256.

Creswell, J. W. (2009). *Research Design: Qualitative, quantitative, and mixed method approaches* (3rd ed.). Sage.

Culturally. (2025). What is cultural competence training? https://www.cultureally.com/blog/what-is-cultural-competence-training. (Accessed 30 April 2025).

Forutnani, S., Nowkarizi, M., & Kiani, M. R.. (2018). The role of rural libraries in preserving the indigenous knowledge of rural residents. *World Journal of Science Technology and Sustainable Development, 15*(3), 245–256.

Garwood-Houng, A. (2005). Protocols: Meeting the challenges of indigenous information needs. *Australian Academic and Research Libraries, 36*(2), 143–151.

Grenier, L. (1998). *Working with indigenous knowledge: a guide for researchers*. International Development Research Centre. https://idrc-crdi.ca/en/books/working-indigenous-knowledge-guide-researchers

Gundakanal, S. S., & Kaddipujar, M. (2024). Cultural custodians: the role of libraries in preserving and promoting language, literature, and heritage. http://dspace.rri.res.in/handle/2289/8297?mode=full

Hamwaalwa, N. (2021). Indigenous knowledge management: roles of public libraries and archives in preserving and providing access to indigenous knowledge (IK) in Namibia: Case study of "Preserve Namibia Indigenous Knowledge" project. In C. Mnubi-Mchombu, C. Nengomasha, & W. Uutoni (Eds.), *Inclusive libraries and information services towards achieving prosperity for sustainable development in Africa: Proceedings of the 24th Standing Conference of Eastern, Central and Southern African Library and Information Associations* (pp. 289-292). National Information Workers Association of Namibia. https://www.scecsal.org/publications/papers2021/scecsal_papers2021_hamwaalwa.pdf

Hoffman, A. (2005). Ethical challenges in librarianship. *Journal of Academic Librarianship, 31*(4), 271–274.

Horsthemke, K. (2008). The idea of indigenous knowledge. *Archaeologies, 4*, 129–143.

Hunter, J. (2005). The Role of Information Technologies in Indigenous Knowledge Management. *Australian Academic and Research Libraries, 36*(2), 109–124.

International Federation of Library Associations and Institutions (IFLA). (2002). *IFLA Statement on Indigenous Traditional Knowledge.* The Hague: IFLA. https://www.ifla.org/publications/ifla-statement-on-indigenous-traditional-knowledge

Kamboya, U. (2018). Indigenous knowledge and sustainable development: Role of libraries. *International Journal of Library and Information Studies, 8*(1), 495–499.

Kovach, M. (2005). Emerging from the margins: indigenous methodologies. In Brown, L., & Strega, S. (Eds.), *Research as resistance: Critical, Indigenous and anti-oppressive approaches* (pp. 19–36). Canadian Scholars' Press.

Lee, D. (2011). Indigenous knowledge organization: a study of concepts, terminology, structure and (Mostly) indigenous voices. *Partnership the Canadian Journal of Library and Information Practice and Research*, 6(1).

Luo, L. (2016). Professional ethics in librarianship: A case study of ethical dilemmas faced by reference librarians. *The Library Quarterly, 86*(2), 174–189.

Mahwasane, N. P. (2017). Library support and the innovative application of indigenous knowledge systems. *Journal of Sociology and Social Anthropology*, *8*(2), 77–81.

Masenya, T. M. (2023). Revitalization and Digital Preservation of Indigenous Knowledge Systems for Sustainable Development of Indigenous Communities in South Africa. *The Serials Librarian*, *84*(5–8), 86–102.

Mhlongo, M., & Ngulube, P. (2018). Public libraries as conduits for indigenous knowledge in South Africa. DigitalCommons@University of Nebraska - Lincoln. https://digitalcommons.unl.edu/libphilprac/1757/

Nakata, M., & Langton, M. (2009). Australian indigenous knowledge and libraries. In *UTS ePRESS eBooks*.

Osagie, F. O., Fagbemi, V. Y., Agubata, N. M., & Dogara, P. D. (2023). Public libraries as pivotal agents in preservation of indigenous knowledge for sustainable development. [RIST]. *Review of Information Science and Technology*, *9*, 27–34.

Owolabi, K. A., Ovwasa, D. E., Ajayi, T. B., & Odewale, M. O. (2021). Preservation and Use of Indigenous Knowledge Practices in Public Libraries in Nigeria. *Public Library Quarterly*, *41*(5), 485–502. DOI: 10.1080/01616846.2021.1938897

Oyelude, A. A. (2023). Indigenous knowledge preservation as a sign of respect for culture: concerns of libraries, archives and museums', *Insights: the UKSG journal*, 36(1), 1-21.

Penna, K. (2023). Culturally sensitive practices and intercultural communication competence for the protection of indigenous peoples' cultures. ENCATC Congress - Artificial Intelligence embraced: the future of the culture and creative sector. ISBN: 978-92-990088-5-0 At: Helsinki, Finland. https://www.encatc.org/media/7193-2023-book-of-proceedings.pdf

Pivec, F. (2003). *The Ethics of Librarianship: An International Survey*. Academia.edu.

Ranasinghe, P. (2008). Preservation and provision of access to indigenous knowledge in Sri Lanka. World Library and Information Congress: 74th IFLA General Conference and Council, 10-14 August 2008, Quebec, Canada. https://www.ifla.org/iu/ifla7u/index

Ray, D. (2022). *Ethics in Librarianship*. Quest Journals. *Journal of Research in Humanities and Social Science*, *10*(5), 53–56.

Roy, L. 2013. The Role of Tribal Libraries and Archives in the Preservation of Indigenous Cultural Identity Through Supporting Native Language Revitalization. *International Preservation News* No. 61.

Sarkhel, J. K. (2017). Strategies of indigenous knowledge management in libraries. *Qualitative and Quantitative Methods in Libraries, 5*(2), 427–439.

Silverman, D. (2013). *Interpreting Qualitative Data*. 5th Ed. Sage, London The United Nations Educational, Scientific and Cultural Organization (UNESCO) (2025). Local and Indigenous Knowledge Systems (LINKS). https://www.unesco.org/en/links

The United Nations Educational, Thorpe K and Galassi M. (2014). Rediscovering Indigenous Languages: The role and impact of libraries and Archives in cultural revitalisation. *Australian Academic and Research Libraries, 45*(2), 81–100.

Chapter 12

Recognition of African Indigenous Knowledge Systems (AIKS) Through Legislation:
Kenyan and South African Laws as a Case

Witness Roya
https://orcid.org/0000-0002-1394-4428
Graduate School of Business Leadership, University of South Africa, Midrand, South Africa

ABSTRACT

The Protection, Promotion, Development and Management of Indigenous Knowledge Act (South Africa) and the Protection of Traditional Knowledge and Cultural Expressions Act No 33 of 2016 (Kenya) were thematically analysed guided by the following research questions: What are the laws' main aims? (ii) How do the acts seek to promote preservation of African indigenous knowledge systems (AIKS)? and (iii) How can technology be leveraged to promote AIKS? The laws were found to make provisions for state regulation and demonstrate a commitment to reach out to diverse groups. Kenyan stakeholders are urged to avail copies of the law in at least a single indigenous language like Swahili which is understood across Eastern Africa. Governments, cultural organisations and businesses are urged to capacitate citizens with devices, gadgets and training on how to access and preserve AIKS electronically.

DOI: 10.4018/979-8-3373-0204-1.ch012

INTRODUCTION

Knowledge is culture specific, so AIKS are perfectly poised to deal with challenges facing African people. However, during colonization, white settlers inculcated, among locals, the perception that their indigenous knowledges were inferior to Western forms and technology is a western (modern) concept. This was fallacious because, at any given time, ways that societies improvise to deal with their issues include what are effectively forms of technology (Sithole, 2016). In his 1972 novel, *Mumbo Jumbo*, Reed satirizes Western misconceptions that Africans do not have meaningful value systems. Laws are key instruments in regulating and promoting cultures and knowledges and African countries have made a raft of legislative changes to reflect democratic societies.

Disregard for African cultures created distrust for Western aspects among African people (Ali, & Rose, 2022) which partly explains African countries' initial reluctance to adopt new forms of media. However, human beings have generally shown a tendency to distrust new technology. Writing from a Western context, Cohen popularized the term, "moral panic" to describe "an episode, often triggered by alarming media stories and reinforced by reactive laws and public policy, of exaggerated or misdirected public concern, anxiety, fear, or anger over a perceived threat to social order" (1972, p.1). Similarly African cultural purists were initially cynical of cellphones, AI and other technological affordances which have since proved to be effective vehicles for communication and preserving African cultures (Walsh, 2020).

Although technological affordances have revolutionized communication, oral traditions which encompass myths, songs, proverbs, folktales and taboos among other African aspects have largely remained undocumented (Osman, 2009; Ayeni & Aborisade, 2022, p.156). Deaths (literally and figuratively) of storytellers from the old generation, a reluctance to use digital media to project African cultures because of the perception that western cultures are superior, are the main contributory factors. Although there is an old African adage, "when an elder dies, a library burns to the ground", technological affordances such as YouTube, social media networks like Facebook and X, websites, blogs, to name a few, can be leveraged to preserve such wisdom from senior citizens before their demise. However, high illiteracy levels and poverty have delayed initiatives to promote the growth of technology in Africa. Increasing inequalities and "economic malaise" led to "the shift from industrialization in the early post-independence period, to the current focus on poverty reduction" across the continent (Duarte, 2021).

THEORETICAL FRAMEWORK

Afrocentricity, the theory which undergirds this study, was popularized by Molefi Kete Asante in his 1980 groundbreaking publication, Afrocentricity The Theory of Social Change (Walker & Burbanks, 2010, p.1). Afrocentricity places at the centre, African culture and history (Monteiro-Ferreira, 2009, Mazama, 2021). Afrocentricity's key tenets include "agency, location, centeredness, and orientation" because they argue that African history was distorted by slavery and colonialism. The idea of space which is embedded in these aspects is literal and figurative because black people were physically dislocated from their environments and assimilated into foreign cultures respectively (Asante, 2017). Therefore, having African ancestry does not guarantee one's Afrocentricity just as one can be in the diaspora but maintain one's Afrocentricity (Mazama, 2003; Asante, 2007). Afrocentricity scholars seek to "relocate" or restore African cultures from the "pervasiveness of European evolutionary thinking" (Mazama, 1998: 1).

Afrocentricity "is a theory of agency, that is, it is predicated on the idea that African people must be viewed and view themselves as agents rather than spectators to historical revolution and change" (Asante, 2007, p.15). Asante doubles down with the assertion that "Afrocentricity refers to African agency and the centrality of African interests, ideas and perspectives" (2020, p.48), Afrocentricity argues that civilization originated in Egypt but Europeans fraudulently appropriated it for themselves (Diop, 1974; Asante, 2020). From the outset, in the first chapter of *The African Origin of Civilization: Myth or Reality* (1974), Diop contends that "the Egyptians were Black"[1] (1974, p.3).Karenga's "Kawaida project" is premised on "recovering and exploring historical and current African texts as a way of dialoging with African culture, asking it questions, and seeking answers from it to the fundamental issues of humankind (2008, p.1). Mazama (2001) reckons that "the Afrocentric idea rests on the assertion of the primacy of the African experience for African people. Its aim is to give us our African, victorious consciousness back" (p.388).

In calling for the decolonization of the mind, Ngugi (1986, p.19) lamented the recruitment of exclusively British staff at an academy named after then Malawian president Kamuzu Banda ostensibly because "a Malawian might lower the standards, or rather, the purity of the English language". Mazrui gives Japan as an example of a country whose success story is a result of anchoring its industrialization policies on Japanese language and culture and urges African countries to follow suit (Akoda, 2015, p.66). Several scholars (Wiredu, 1998; Achebe, 2009; Ngugi, 2009) acknowledge the inevitability of contact between African cultures with western ones on condition that the former retains primacy. Mazrui's contribution to different disciplines such as political science, Islamic and African studies and cultural studies earned him the moniker "multiple Mazrui" (Akoda, 2015 p.66).

Proposed Framework

Much as Afrocentricity provides historical and cultural grounding, its proponents tend to expend too much energy towards dispelling the notion that it is not a reaction to Eurocentrism which it ironically seems to prove. Walker (2001) dismisses Afrocentricity, which he confuses with Afrocentrism, as "a mythology that is racist, reactionary, and essentially therapeutic" because Afrocentricity emulates colonial narratives although the only difference is that they privilege black people. Walker (2001) takes great exception to Afrocentrism scholars who dismiss as "insane or somehow not black" (p.xxiv) anyone who differs with them which are all fair points. With the premise that the water should not be thrown away with the bathwater, the proposed theoretical framework, "Afrotechnicity" is influenced by Afrocentricity in placing at the centre, African culture and history although it recognizes the need for adding technology as a vital cog in the drive for social inclusion. Afrotechnicity is premised on the notion of "the horse and the cart" where the horse represents African culture and history and the cart is technology. Although the prefix "Afri" somewhat creates an impression of narrowly focusing on a single group, like Afrocentricity, the theory recognizes positive aspects from other canons on condition that African values retain their status as the locus of enunciation.

LITERATURE REVIEW

Indigenous knowledge systems (IKS) are a people's distinguishing features. Odora Hoppers (2021, p.310) notes that IKS "are referred to differently in different contexts and circumstances" where they are variously referred to as "Traditional Knowledge Systems" (TKS), "Endogenous Knowledge Systems" (EKS) and "Classical Knowledge Systems" (CKS). Crossman and Devisch are quoted by Breidlid (2009, p.141) acknowledging that indigenous knowledge "has been passed on from generation to generation within an indigenous community". Wilson (1987, p.89) defines indigenous knowledge as "a continuous process of information dissemination, entertainment and education used in societies which have not been seriously dislocated by western culture or any other external influences as in the case of many parts of the world". This not only speaks to the originality of values cherished by indigenous people but it also shows the link between Afrocentricity and AIKS. Indigenous languages are "a priceless heritage and a vehicle of ancestral knowledge" (Espinosa cited in the United Nations Department of Economic and Social Affairs, 2019). Language is the driving force behind IKS (Langmia & Sani, 2023). Horsthemke (2008, 2017) strangely describes indigenous languages as a recent phenomenon.

Groping for Recognition

Indigenous knowledge remains lamentably underestimated. In 2019, during a "UN summit alongside other young campaigners furious at the West's failure to tackle the crisis", then teenage environmental activist, Greta Thunberg bemoaned the marginalization of indigenous people in climate discourse (The Strait Times, 2019). This was despite the fact that cultures of indigenous people secure "ecosystem sustainability" (World Bank, 2025, p.3) because their lifestyles are literally and figuratively organic as they live in harmony with nature. Indigenous people who are also "known as the "first" or "original" people were recognized as far back as the 1960s when UNESCO's office in Jakarta, Indonesia, released a report in Malayo-Oceania to raise awareness of the rich botanic knowledge embedded in traditional societies. From the 1970s to the 1980s, UNESCO initiated several programmes to address "traditional agro-piscicultural systems in Mexico through the Man and the Biosphere programme, and traditional knowledge and customary marine resource management in the Asia-Pacific region through the Coastal Marine Programme" (Nakashima, 2010, p.2). In 1987 "the World Commission on Environment and Development (WCED) in their report on sustainable development "recognized indigenous knowledges (Malapane, Musakwa, Chanza & Radinger-Peer, 2022, p.1).

Nonetheless, indigenous people have practically remained marginalized in both developed and developing countries. They are probably the most marginalized group in the world. Even in research, which is a driver for social inclusion, indigenous people remain underrepresented. In South Africa, research contributes just 0.82% of GDP compared to the Democratic Republic of Korea's 2.79% and USA and 4.55% respectively (World Bank, 2019; Nwonwu, 2020 Slide 3). There are approximately 476 million indigenous people around the world. Although this is just 6.2 percent of the global population, indigenous people make up 18.2 percent of those living in extreme poverty worldwide, their life expectancy is "up to 20 years lower than the rest of the population, and they face severe barriers to accessing basic services, education, infrastructure, and job opportunities" (Chanza, 2022).

Credibility Issues

Kessi, Marks and Ramugondo (2020) argue that scholars from non-western countries do not need validation from anyone as long as they proffer ideas that subvert colonial ideas. Indeed, there seem to be a recognition of value in African indigenous languages scholars even from their most staunch critics. Despite describing Credo Mutwa as a "sangoma, a writer and artist, and a controversial 'expert' on UFO-related conspiracy theories", Podolecka (2018, p.1) asserts that Chidester's allegation that Mutwa invented myths about South Africa's indigenous cultures are "harsh and not entirely true". Podolecka adds that Mutwa projects some rituals especially those performed by sangomas so accurately that this was confirmed by most of his ardent Zulu sangomas. Mutwa's ideas align with New Age ideology[2] (Podolecka 2018, p.139), at the turn of the century, he was recognized "as a Zulu shaman, environmentalist, healer, prophet, teacher, and authority on aliens from outer space" (p.68). Teish commends Mutwa for presenting "real knowledge about South Africa that was not polluted by some anthropologist's opinion" (cited in Podolecka, 2018, p.79).

Biopiracy

Ironically, despite the cynical attitude of western countries towards indigenous people, multinational corporations are accused of stealing knowledge owned by indigenous people. Although enhancing accessibility and benefit sharing (ABS) is a way of addressing biopiracy by facilitating incentives and rights of indigenous communities, this often benefits large businesses. However, thanks to biopiracy, misappropriation of genetic resources and knowledge which has been in existence for over two millenia (Wynberg, 2022), indigenous groups have been deprived of profits after the patenting of information emanating from their knowledge.

In 1870, Wilhem Bleek, a library curator sought permission from Cape governor, Philip Woderhouse for some Bushman prisoners to work at his home and spent years, "recording thousands of pages of folklore, mythology, and other texts from a succession of Xam informants (Finlay & Barnabas, p.71). Islam (2023) marvels at the way San people coexist with poisonous snakes which they eat and use their skins to make pouches. In 2022, the Kenyan government banned the "biopiracy export" of baobab trees to a Georgian company by Kenyan farmers until both parties regularised their terms (BBC News, 2022). Wynberg (2022) investigated four cases involving Aspalathus linearis (rooibos), in South Africa – Hoodia gordonii, Pelargonium sidoides and Sceletium tortuosum in South Africa. Instead, control is vested in land, markets and intellectual property. Table 1 below illustrates two of the cases. Khoi SASC accused Nestlé of patenting rooibos which belonged to TK of San. In the other lawsuit, SASC legal team represented the SASC against HG &H for using a San logo on some BSA-linked products BSA, creating the pretext that SASC had endorsed the marketing move. Sceletium tortuosum (kougoed, kanna) SASC benefits include 5% of net proceeds received by HG&H and an annual exclusivity payment of 1% on sales.

Table 1. Overview of key features characterizing the biopiracy charges, Source: Wynberg, 2022, p.6

Plant species	Biopiracy charge	Origin	Commercial use	Traditional use & knowledge	Benefit-sharing agreements (BSA) & governance	Governance
Aspalathus linearis (rooibos)	100-year old rooibos tea Industry based on TK of San and Khoi without sharing benefits. Challenge brought by SASC and NKC. Nestlé accused of biopiracy due to unauthorized use of rooibos in R&D program and the lodging of associated patents.	Western Cape, South Africa	Food and Beverage, Supplements, wide application for anti-oxidant and other properties	Used traditionally for various purposes. Many harvesting practices are traditionally based.	BSA between the Rooibos Council, SASC and NKC.	The Andries Steenkamp Trust administers benefits on behalf of the SASC. A "Khoikhoi Peoples Rooibos ABS Trust" administers benefits on behalf of the NKC.
Plant species	**Representation and support**	**Benefits**	**Patents**	**Other forms of intellectual property**	**Certifications**	**Custodianship and use**
Sceletium tortuosum (kougoed, kanna)	SASC Legal representative supporting the SASC. Nama communities not involved in negotiation process.	SASC benefits include 5% of net proceeds received by HG&H and an annual exclusivity payment of 1% on sales. R10 million received since 2010.	Multiple patents linked to treatments for depression and anxiety	A logo appears on certain products linked to the BSA, indicating endorsement by the SASC. An "exclusivity" payment is made in respect of product endorsement, marketing and branding assistance and the use of the San logo. Some farms are organically certified. There are no ethical trade	Some farms are organically certified. There are no ethical trade certificates	All cultivated on private farms.

Wynberg's findings revealed that ABS set a precedence for addressing biopiracy although the process seems to favour commercial organisations and unlike its neighbours, Botswana Namibia and Zimbabwe, community-based or owned approaches to biodiversity commercialization remain surprisingly absent from South Africa's landscapes. In 2019, studies by the Berne Declaration and Natural Justice found that Nestlé's use of Rooibos and Honeybush flouted South African Law and the Convention on Biological Diversity.

Growing Recognition

The aforegoing victories of indigenous groups are symptomatic of a growing recognition of their contribution to a sustainable lifestyle. The San's ethnobotanical knowledge is recognized by non-governmental organisations (NGO) and academics (UNESCO World Heritage Centre, 2016; UNESCO, 2017). UNESCO (2016 p, 1) has undertaken to support efforts to recognize that preserving indigenous knowledges "strengthens economic, environmental, social and cultural resilience within societies and forms the knowledge base for addressing critical sustainability problems of the 21st century". United Nations General Assembly (2015, p.4) noted that 70% of the world's population is fed by small-scale food systems which "maintain in situ genetic and seed diversity, biodiverse ecosystems and are mostly free of chemicals and genetically modified organism". Global fora have recently shown an appreciation of indigenous people and their knowledges. The Conference of the Parties (COP)'s last meeting in 2024 resolved to create the Cali Fund to support Indigenous biodiversity efforts and widened indigenous leadership roles in biodiversity policy (Hildebrand & Magnani, 2025). Universidad Indígena Intercultural (UII), a network of Indigenous Intercultural Universities was established In Latin America for indigenous students to conduct post-graduate courses which support promotion of indigenous cultures (Pedota, 2011).

African Indigenous Groups and the Environment

The environment is regarded as a source of life by indigenous groups in Africa. The Mau Forest (MF) is both the largest single block of closed-canopy forest ecosystem in East Africa and the largest indigenous forest block in Kenya and receives annual rainfall of approximately 2,000 millimetres (Muli, Nyambura & Onyango, 2023). Gitonga et al. (2020) observe that Prunus Africana (Hook f.) is used by the Meru of Kenya to treat inflammation which is associated with various diseases. Ndung'u et al. (2024, p.3) notes that the popularity of the deciduous tree extends to several African countries where its bark is used to treat illnesses like diabetes mellitus, benign prostatic hyperplasia, malaria, prostate cancer, snake bite and chest pain.

Participation of Indigenous Women's Knowledge Transmission

Within marginalized groups, women come out the worse for wear especially those who do not have strong financial and educational backgrounds. Many women across the African continent travel long distances looking for basic commodities such as firewood and clean water which stifles their participation in the "educational or economic development in their communities". Encouragingly, recognition that women are disproportionately affected by environmental destruction has found expression in "the emergence of ecological feminism, or ecofeminism" (Elverum, 2016, p.4). Adeola, Olaniyi and Ngare (2024) foreground the "intersection of gender equality, climate action, and technological innovation for sustainable development in Africa" (p.1). Kenyan environmental activist Wangari Maathai experienced patriarchy at several levels, from her former abusive husband to her political persecution from Kenyan president Daniel Arap Moi who was angered by Maathai's criticism of the government's move to convert a public park into an elite residential area. In a disheartening move, women's leadership of the ruling, KANU party supported Maathai's expulsion from the ruling party when Moi chided her for not behaving like a proper woman in "the African tradition" (Maathai, 2012, p. 215).

Adeola et al (2024) note that "interestingly, women are both victims of climate change and potential architects of climate action, a dyadic lens that supports the study of climate effects in Africa" (p.3). Against all odds, Maathai shrugged off her suffering to be the first African woman (in 2004) to receive the Nobel Peace Prize and the first laureate to be selected for activities focused primarily on "environmental sustainability" (Nobel Prize Outreach, 2025). Maathai leveraged her exposure to African culture, western religion and education to establish the Green Belt Movement (GBM) which revived deforested forests although she attributes her actions to "thinking literally and practically about how to solve problems on the ground... a desire to help rural populations, especially women, with the basic needs" (Maathai, 2009, p.1).

Maathai's steadfast pursuit for justice was influenced by other environmental activists like Chico Mendes, a Brazilian trade union leader who was assassinated in 1988 and Dorothy Stang, the American-born Brazilian nun who was killed in 2005. Maathai (2009, p.283) also acknowledges Lesotho women who make "keyhole gardens" in their homes, "growing large amounts of green vegetables, and protecting themselves from hunger".

Factoring in Technology

There is a growing recognition for the need to adapt technology whilst retaining aspects of performance (Merolla, 2014). To capture the interplay of "orature, auriture, and technology, in 2004 Kaschula, coined the term "techauriture" (Hautemo, Mushaandja, 2024, p.165) which Mostert, Lisney, Maroko and Kaschula, (2021) modified to "Cultauriture" to include a cultural component. Kaschula (2013, p.141) cites the design of the cover of the online copy of Finneghan's seminal book, *Oral Literature in Africa* which has a technologized etching of a performer/singer donning West African regalia and multiple award winner, Zolani Mkiva, "The President's Poet" or imbongi yesizwe ("The Poet of the Nation") who has a website (Kaschula (2017) as examples of technauriture. Opertti (2021) notes that the Scientific Council for National Education in the French Ministry of National Education and Youth capitalized on the COVID-19 outbreak by promoting technical skills competencies in areas such as indigenous and foreign languages and sciences.

However, digital inequalities remain a challenge for marginalized groups especially in Africa. The Kenya Economic Update released by the World Bank, access to the internet is disproportionately spread among urban and rural dwellers, with the latter constituting only 17% compared to the former's 44% (Kinyua, 2020). A study by Wyrzykowski (2023), revealed discrepancies between South Africa's urban and rural provinces…Western Cape is the only South African province where the proportion of mobile-only rural users is below 10%. Larrière and Maurice (2021) note that communication technology and globalisation have led to the emergence of 'neo-oralities', which give Africa's oral tradition a new lease of life. Melisa Achoko Allela co-founded LESO Stories, which seeks to leverage technology to "create new tellings and retellings of orally told stories while maintaining the performative, participatory, and communal elements of live oral storytelling" (melisallela.com; African Digital Heritage 2023).

Language

According to Jonnaert et al. (2021), inclusion of different languages in syllabi "creates a sense of identity at community, regional and national levels through shared customs and traditions, while also leading to an overall improvement in the quality of their school learning…without minimizing the importance of learning international languages to broaden career opportunities" (p.7). Cindi (2021, p.piv) notes that South Africa's National Curriculum Statement (NCS) and the Curriculum and Assessment Policy Statements (CAPS) seek to promote principles such as "human rights, inclusivity, environmental focus, social justice and valuing of IKS". The Department of Basic Education aligned the Revised National Curriculum Statement for Grades R-9, National Curriculum Statement (Grades 10-12), and subsequently, the Curriculum and Assessment Policy Statements, in line with the country's constitutional provision for "human rights, inclusivity, environmental and social justice, and valuing of Indigenous knowledge systems" (Khupe, 2014, p iii). Cindi

(2021, p.2) adds that this commitment was complemented by government policies such as the National Indigenous Knowledge Systems (IKS) Policy (2004) and the Indigenous Knowledge Act 6 of 2019. The Department of Basic Education stressed the importance of syllabi which blends science and indigenous knowledge systems from different times and cultures (Mutsvangwa & Zezekwa, 2021). According to Dalvit, Murray and Terzo (2008, p.287) indigenous knowledge and languages enhance learning for African students even for "Western" disciplines, such as Computer Science. Similarly, da Silva, Pereira and Amorim (2023) note "the need to teach vernacular languages to minority groups" (p. 1220).

Taylor and Cameron (2016, p.35) identify two main approaches to the relationship between science and IKS, with one viewing IKS as part of science whereas the other perspective regards the two as autonomous. The two scholars note a third perception in which science and IKS are perceived as interrelated concepts. Opertti, (2021, p.4) asserts that curricula must not only reflect society but contribute to social transformation.

South Africa's Research Initiatives

In the last few years, the South African government has made strides to promote indigenous knowledges. In 2019, the Department of Science, Technology and Innovation (DSTI), launched the Centre for Digital Language Resources (SADiLaR) to promote indigenous languages (DSTI, 2019). SADiLaR supports "the creation, management and distribution of digital language resources, as well as applicable software, which are freely available for research purposes through the Language Resource Catalogue" (sadilar.org).

The South African National Biodiversity Institute (SANBI)was established in September 2004 in terms of Section 10(1) of the National Biodiversity Management: Biodiversity Act (Act No. 10 of 2004) (SANBI, 2025). With funding to the tune of $90 million, SANBI has worked with marginalized groups to promote integration of science and indigenous knowledges to influence corporates and policymakers to implement policies that speak to "South Africa's just transition to a climate resilient society". SANBI successfully spearheaded a $7.5 million project involving the uMgungundlovu District Municipality and the university of KwaZulu-Natal in the Greater uMngeni Catchment project which ended in March 2024, significantly improving livelihoods of vulnerable communities and small-scale farmers in the area (p.74).

Notable attempts by the South African government to promote indigenous knowledges include:

The White paper on Arts, Culture and Heritage (1996)

National Policy of South Africa Living Heritage (2009)
Library and Information Service Transformation Charter (2009 and 2014) and
The National Policy on the Digitization of Heritage Resources (2010) (The Western Cape Government, Cultural Affairs and Sport cited in Tembe, 2019, p.41).

Tembe's study of the oral history records at Pietermaritzburg Archives found a need for documenting the "neglected and undocumented history and bridging the gaps" between information which is stored in archives and the information that is within the community about historical events (p.38). Tembe (2019, p, 38) and Tembe and Nsibirwa (2022 p151) found the Pietermaritzburg Archives Repository (PAR) Oral History Unit (OHU) "well-resourced with experienced and well-educated staff". However, only two of the seven employees who participated in the study "had qualifications related to archival science…most participants' qualifications were not related to archives, history or heritage and cultural studies as recommended qualifications

Glossary

Social inclusion refers to provision of services and other important aspects
Afrocentricity is the theory which is advanced by the likes of Ama Mazamsa, Molefi Kete Asante among others which places African culture and history at the centre of discourse.

Accessibility denotes the level of proximity and having something at one's disposal. Contrary to the popular but essentially flawed perception that the availability of something which marginalized groups do not have means to get is the panacea, training and connectivity are some of the factors that should be considered.

Afrotechnicity: The proposed conceptual framework for this study which avers that although cultural values are important, technology has also become a key component.

Techaurature is a portmanteau formed by a combination of "literature", "auriture", and "technology" to show the need for integrating these elements to the arts without compromising the performative aspect.

Methodology

I initially downloaded the Protection, Promotion, Development and Management of Indigenous Knowledge Act (South Africa) and The Protection of Traditional Knowledge and Cultural Expressions Act (Kenya) for use as the study's primary sources. I also variously typed "indigenous knowledge" (IK), "African indigenous knowledge systems", ("AIKS"), "indigenous knowledge systems" (IKS), "African cultures", "Western cultures/ values", "ICT" "leveraging" and "blending" as key words in different searches on Google, Google scholar and Web of Science. After noting a recurrence of results from various entries in Google search which indicate saturation, I examined the legislation act under study. The research questions, i) What are the laws' main aims? (ii) How do the acts seek to promote preservation of indigenous knowledge systems? and (iii) How can technology be leveraged to promote AIKS? guided the discussion of findings which appears in the next section. In the last segment, the conclusion sums the study by providing the main points, implications of findings, recommendations detail suggestions for improvement and consolidation of findings which stakeholders can consider.

DISCUSSION OF FINDINGS

The main themes were observed and placed into groups. However, these groups are interconnected so categorizes are only meant to provide readers with breaks. The following themes were observed: "Conceptualization of indigenous knowledge", "Regulation, registration and management of copyrights" and "Diversity."

Conceptualization of Indigenous Knowledge

The two laws under study define indigenous community as a distinct group of people with shared history, cultures, reside at the same location and regard themselves as a distinct group. Nonetheless, unlike the South African law which mentions "indigenous" 217 times, The term traditional features prominently in the Kenyan law, 200 times. In fact, the title, "Protection of Traditional Knowledge and Cultural Expressions Act" is instructive. Although "indigenous" appears once ("indigenous seeds", p.652), the South African law does not mention "traditional", apparently because of the latter term's reductionist connotations. Another synonym of traditional, "customary" appears once in "customary laws" in the SA law 8, 27, 2 but features five times on just one page, p.651 of the Kenyan law. Dalvit, Murray and Terzoli demonstrate that selection and omission of terms is a matter of preferred semantics as "traditional knowledge is what, in layman's terms, most people refer to as indigenous knowledge" (2008, p.288). Like the South African law, the Kenyan law is alive to prejudice against indigenous groups which is shown by its reference to "derogatory treatment" five times and "derogatory action" thrice. yet the SA law makes no mention of these terms or their cognates.

Section 6 and 11 of the Kenyan and SA law, respectively, describe indigenous knowledge as having been passed from "one generation to another" within an indigenous community, reflecting its historical and socio-economic experiences. In addition, the Kenyan law acknowledges that indigenous knowledge may be "individually or collectively generated".

Registration and Management of Copyrights

County governments register traditional artefacts in Kenya whilst the South African law mandates the National Indigenous Knowledge Systems Office (NIKSO, see Appendix). Section 8.3 of the Kenyan constitution empowers the government to partner county governments to maintain a digital repository which contains information on traditional knowledge and cultural expressions which have been documented and registered by county governments. On the other hand, Chapter 6 of PPDMIKA mandates the minister to appoint a curator of indigenous knowledge to control the Registration Office. However, one can conclude that both laws entrench government control in the recognition and preservation of indigenous knowledge which has its merits and demerits. Benefits of such an arrangement include enforceable regulation systems whilst the main pitfall is that the government may abuse its position in dealing with certain individuals and groups especially if they are critical of its administration.

Regulation, Registration and Management of Copyrights

For Kenya, the county or national government protects the rights to traditional knowledge by registering owners whilst in South Africa, this falls under the ambit of NIKSO which reports to a curator who is appointed by the minister. The two laws empower local authorities to control rights of knowledge which has an equal claim with other communities. However, whilst county and national governments retain their authority under the Kenyan legislation in situations in which one of the two communities is in a foreign country, the SA requires NIKSO to assist the two communities to find a way of equitably sharing proceeds. To use indigenous knowledge, both laws make it mandatory for one to apply whether one has sought consent from the owner or not. It is also important to point out that, in South Africa, copyright issues relating to traditional knowledge are covered by the Copyright Act 98 of 1978 and the SA law under study whereas, the Kenyan law under study incorporates copyright issues such as folklore which were "shifted from copyright law" (Spoor & Fisher, 2019). Part 1 5a of the Kenyan law under study mandates the national government to establish and maintain the Repository at the Kenya Copyright Board which further shows how the law empowers the government as explained in the previous segment. Assignment of different aspects into various laws enhances discreteness so citizens gain a better understanding of legal instruments. On the other hand, having various facets under the same laws enhances thorough handling of similar cases.

Diversity

Both laws endeavour to promote social inclusivity. In line with the "Marrakesh Treaty", the Kenyan law exempts people living with disabilities from copyright obligations. Amendments to the law also somewhat protect AI-generated knowledge (Spoor & Fisher, 2019). The South African law is available in different languages in line with the country's constitution which recognizes twelve languages which include sign language. This is commendable considering that, due to colonial African countries have high illiteracy rates. The Kenyan law, on the other hand, is only available in English, at least online which excludes not only Kenyan citizens but citizens of various Eastern countries who speak the same language and may be interested in understanding the Kenyan law.

CONCLUSION AND RECOMMENDATIONS

From the study's findings, it can be concluded that the two laws, like any other piece of legislation in any part of the world, have their benefits and limitations. State regulation enhances the law's enforceability which can be abused by ruling parties. Both laws strive to provide social inclusivity with the Kenyan law's recognition of the Marrekesh Treaty to encourage participation from people living with disabilities but the absence of a Swahili version of the law on the internet deprives many readers of an opportunity to access it in a language which they understand better than English. The researcher therefore urges responsible Kenyan authorities to ensure that the law is available in Kiswahili. The South African government is encouraged to align its laws with international conventions and developments like considering AI into issues to do with ownership rights. With the recent addition of sign language as the twelfth South African official language, there is need for availing a copy, through an accessible link, of the law in that language. Availing the two laws in more languages helps to promote interest in documentation among different groups.

REFERENCES

Achebe, C. (2009). Spelling our proper name. In *The education of a British-protected child* (pp. 79–94). Penguin Books.

Adeola, A., Olaniyi, E., & Ngare, I. (2024). *Gender equality, climate action, and technological innovation for sustainable development in Africa*. Palgrave Macmillan.

Ali, S. H., & Rose, J. R. (2022). The post-colonialist condition, suspicion and social resistance during the West African Ebola epidemic: The importance of Frantz Fanon for global health. *Social Science & Medicine*, *305*, 115066. DOI: 10.1016/j.socscimed.2022.115066 PMID: 35636052

Asante, M. K. (2007). *An Afrocentric manifesto: Toward an African renaissance*. Polity Press.

Asante, M. K. (2017). The philosophy of Afrocentricity. In Afolayan, A., & Falola, T. (Eds.), *The Palgrave handbook of African philosophy* (pp. 231–245). Palgrave Macmillan. DOI: 10.1057/978-1-137-59291-0_16

Asante, M. K. (2020). Africology, Afrocentricity, and what remains to be done. *The Black Scholar*, *50*(3), 48–57. DOI: 10.1080/00064246.2020.1780859

Ayeni, A. O., & Aborisade, A. G. (2022). African indigenous knowledge systems and the world. In Oloruntoba, S. O., & Falola, T. (Eds.), *The Palgrave handbook of Africa and the changing global order* (pp. 149–166). Palgrave Macmillan. DOI: 10.1007/978-3-030-77481-3_8

BBC News. (2022). Kenya bans 'biopiracy' export of notorious baobabs. https://www.bbc.com/news/world-africa-63716286

Breidlid, A. (2009). Culture, Indigenous Knowledge Systems and Sustainable Development: A Critical View of Education in An African Context. *International Journal of Educational Development*, *29*, 140–148.

Cindi, L. (2021). *Incorporating African Indigenous Knowledge Systems into the Basic Education curriculum: Experiences from two schools in the Gauteng and KwaZulu-Natal Provinces, South Africa. Master of Philosophy (MPHIL) in African Studies. Bloemfontein: University of the Free State Cohen, S. (1972). Folk Devils and Moral Panics: The Creation of the Mods and Rockers*. Routledge.

da Silva, C., Pereira, F., & Amorim, J. P. (2023). The integration of indigenous knowledge in school: A systematic review. *Compare: A Journal of Comparative Education*, *54*(7), 1210–1228. DOI: 10.1080/03057925.2023.2184200

Dalvit, L., Murray, S., & Terzoli, A. (2008). The role of indigenous knowledge in computer education in Africa. In IFIP International Federation for Information Processing. 281, pp. 287–294). DOI: 10.1007/978-0-387-09729-9_43

Department of Science. Technology and Innovation (DSTI). (2019). Government Establishes a New Digital Centre to Promote Indigenous Languages. https://www.dsti.gov.za/index.php/media-room/latest-news/2885-government-establishes-a-new-digital-centre-to-promote-indigenous-languages

Diop, C. A. (1974). *The African Origin of Civilization: Myth or Reality*. Lawrence Hill Books.

Duarte, C. (2021). Africa Goes Digital. Finance & Development. Washington DC. https://www.imf.org/en/Publications/fandd/issues/2021/03/africas-digital-future-after-COVID19-duarte

Finlay, K., & Barnabas, S. (2012). Shifting representations of the Bushmen. In *Cultural Tourism and Identity* (pp. 71–83). Brill.

Gitonga, G. M., Ngeranwa, J., Machocho, A. K., & Kiruki, S. (2020). Qualitative Phytochemical Screening and Antipyretic Activity of Methanolic Stem Bark Extract Of Mangifera Indica in Wistar Rats. *The Journal of Phytopharmacology*, *9*(4), 218–223.

Hautemo, A. M., & Mushaandja, T. (2024). Technauriture: The Intersection Between Orature and Technology. In Oojorah, V. A., & Udhin, W. (Eds.), *Imagining the Futures of Higher Education in Southern Africa. Future Education and Learning Spaces*. Springer. DOI: 10.1007/978-981-97-6473-0_10

Hildebrand, J., & Magnan, P. (2025). How Indigenous Communities Advance Climate Action at COP Conferences. https://www.weforum.org/stories/2025/01/indigenous-communities-advance-climate-action-cop-conferences/

Horsthemke, K. (2008). The Idea of Indigenous Knowledge. *Arch*, *4*, 129–143. DOI: 10.1007/s11759-008-9058-8

Horsthemke, K. (2017). Indigenous (African) Knowledge Systems, Science, and Technology. In Afolayan, A., & Falola, T. (Eds.), *The Palgrave Handbook of African Philosophy*., DOI: 10.1057/978-1-137-59291-0_38

Islam, R. (2023). The Gods Must Be Crazy (1980 & 1989): A Humorous Reflection on Modernity, Nature, and Contrasting Human Cruelty on Nature and Simplicity. https://www.probinism.com/the-gods-must-be-crazy-human-cruelty-on-nature/

Jonnaert, P., Ndinga, P., Ettayebi, M., Barry, A., Rabinovitch, L., & Malu, R. (2021). *Towards indigenous curricula. In-Progress Reflection. 41. On Current and Critical Issuesin Curriculum, Learningand Assessment.* UNESCO.

Karenga, M. (2008). *Kawaida and Questions of Life and Struggle: African American, Pan-African, and Global Issues.* University of Sankore Press.

Kaschula, R. H. (2004). Imbongi to slam: The emergence of a technologised Auriture. *Southern African Journal of Folklore Studies, 14*(2), 45–58.

Kaschula, R. H. (2017). Technauriture as a platform to create an inclusive environment for the sharing of research. *Searching for Sharing: Heritage and Multimedia in Africa, 7*(2), 41–60. DOI: 10.11647/OBP.0111

Kaschula (2013).Reviewed Work(s): Oral Literature in Africa by Ruth Finnegan. Review by: Russell H. Kaschula. Source: Journal of African Cultural Studies, March 2013, Vol. 25, No. 1, Special Issue: New Media Entrepreneurs and Changing Styles of Public Communication in Africa (March 2013), pp. 141-144 .Published by: Taylor & Francis, Ltd

Kessi, S., Marks, Z., & Ramugondo, E. (2020). Decolonizing African Studies. *Critical African Studies, 12*(3), 271–282. DOI: 10.1080/21681392.2020.1813413

Kinyua, J. (2020). Challenges in Implementing Digital Technologies in Rural Kenya. https://www.engineeringforchange.org/news/challenges-implementing-digital-technologies-rural-kenya/

Langmia, K., & Sani, M. H. (2023). Decolonizing the African Mind in the Digital Space. In Akpan, U. S. (Ed.), *African Media Space and Globalization.* Palgrave Macmillan. DOI: 10.1007/978-3-031-35060-3_15

Larrière, S., & Maurice, S. (2021). Orality in the Digital Age. https://leblob.fr/en/visions-africa/orality-digital-age

Malapane, O. L., Musakwa, W., Chanza, N., & Radinger-Peer, V. (2022). Bibliometric Analysis and Systematic Review of Indigenous Knowledge from a Comparative African Perspective: 1990–2020. *Land (Basel), 11*, 1167. DOI: 10.3390/land11081167

Mazama, A. (1998). The Eurocentric Discourse on Writing: An Exercise in Self-Glorification. *Journal of Black Studies, 29*(1), 3–16. https://www.jstor.org/stable/2668053

Mazama, A. (2001). The Afrocentric Paradigm: Contours and Definitions. *Journal of Black Studies, 31*(4), 387–405. DOI: 10.1177/002193470103100401

Mazama, A. (2003). The Afrocentric Paradigm. Africa World Press. melisallela.com. About Melisa Allela. https://melisallela.com/about

Merolla, D. (2014). Introduction. Orality and techauriture of African literatures. In. Dossier "Oraliteit en tegno-ouditUur van Afrikaliterature / Orality and Techauriture of African Literatures / Oralité et technoauriture de littératures africaines, D.Merolla (ed.), *Tydskrif vir Letterkunde – A Journal for African Literature,* 51, 1.

Mostert, A., Lisney, B.,MarokoG.M., & Kaschula., R.H. (2021). From Technauriture to Cultauriture: Developing a Coherent Digitisation Paradigm for Enhancing Cultural Impact. International Journal of Society, Culture and Language.

Muli, K. K., Nyambura, E., & Onyango, J. (2023). *Preservation of Knowledge Systems through Integration of Indigenous Knowledge System: A Case Study of Ogiek Community in Mau Forest.*

Nakashima, D. (Ed.). (2010). Indigenous Knowledge in Global Policies and Practice for Education, Science and Culture. https://unesdoc.unesco.org/ark:/48223/pf0000265855

National Indigenous Knowledge Systems Office. (2019). Porttfolio on Science and Technology. Cape Town: Department of Science and Technology Cape Town 29 August 2006. Decade of Traditional Medicine [2001 – 2010] Lusaka Summit of Heads of States and Government.https://www.slideserve.com/zalman/role-of-nikso-in-research-and-development-powerpoint-ppt-presentation

Ndung'u, J. K., Nguta, J. M., Mapenay, I. M., & Moriasi, G. A. (2024). A Comprehensive Review of Ethnomedicinal Uses, Phytochemistry, Pharmacology, and Toxicity of Prunus africana (Hook. F.) Kalkman from Africa. Scientifica (Cairo). DOI: 10.1155/2024/8862996

Ngugi, W. T. (1986). *Decolonising the Mind: The Politics of Language in African Literature. London: James Currey*. Heinemann Kenya.

Ngugi, W. T. (2009). *Something Torn and New: An African Renaissance.* Basic Civitas Books.

Nobel Prize Outreach. 2025. Wangari Maathai-Speed Read. nobelprize.org. https://www.nobelprize.org/prizes/peace/2004/maathai/speedread

Odora Hoppers, C. (2021). Research on Indigenous knowledge systems: The search for cognitive justice. *International Journal of Lifelong Education, 40*(4), 310–327. DOI: 10.1080/02601370.2021.1966109

Opertti, R. (2021). Ten clues for rethinking curriculum. In. On Current and Critical Issues in Curriculum, Learning and Assessment. In-Progress Reflection No. 42 BE/2021/WP/CD/42

Osman, A. (2009). Indigenous Knowledge in Africa: Challenges and Opportunities. An Inaugural Lecture. Centre for Africa Studies. https://www.ufs.ac.za/docs/librariesprovider20/centre-for-africa-studies-documents/all-documents/osman-lecture-1788-eng.pdf?sfvrsn=e436fb21_0

Pedota, L. (2011). *Master of Arts Program in Cultural and Educational Policy Studies*. Loyola University Chicago.

Reed, I. (1972). *Mumbo Jumbo*. Doubleday.

Sithole, M. (2016.) Opportunities of Incorporating African Indigenous Knowledge Systems (AIKS) in the Physics Curriculum. Alternation Special Edition 18. 255 - 294 255

South African Centre for Digital Language Resources (SADiLaR). About SADiLaR https://sadilar.org/en/about/

Spoor & Fisher. (2019). Kenya: Amendments to Copyright Law. https://spoor.com/kenya-amendments-to-copyright-law/

Taylor, D. L., & Cameron, A. (2016). Valuing IKS in successive South African Physical Sciences Curricula. *African Journal of Research in Mathematics. Science and Technology Education*, *20*(1), 35–44. DOI: 10.1080/10288457.2016.1147800

Tembe, M. A. (2019). *Preservation of, and access to oral history records at Pietermaritzburg Archives Repository. Master of Information Studies. Durban*. University of KwaZulu-Natal., https://researchspace.ukzn.ac.za/handle/10413/18140

Tembe, M.A.,& Nsibirwa, Z. (2022). Archiving the voices of the once voiceless: strategies for digital preservation of oral history at the KwaZulu-Natal Provincial Archives. *ESARBICA Journal: Journal of the Eastern and Southern Africa* Regional Branch of the International Council on Archives: African Journals Online (AJOL). DOI: 10.4314/ESARJO.V41I.10

The South African National Biodiversity Institute (SANBI). 2025. What SANBI does. https://www.sanbi.org/about/about-sanbi

The South African National Biodiversity Institute (SANBI). 2025. Annual Report 2023/24. https://www.sanbi.org/wp-content/uploads/2024/10/SANBI-Annual-Report-2023-24.pdf

The Strait Times. (2019). Activist Greta Thunberg turns spotlight on indigenous struggle at climate summit. https://www.straitstimes.com/world/europe/activist-thunberg-turns-spotlight-on-indigenous-struggle-at-climate-summit

United Nations Department of Economic and Social Affairs. (2019). 'Preserve, Revitalize and Promote' Indigenous Languages, or Lose A 'Wealth of Traditional Knowledge', UN Chief says. https://www.un.org/development/desa/en/news/social/indigenous-peoples-day-2019.html

United Nations General Assembly. (2015). *A/RES/70/1. Transforming Our World: The 2030 Agenda for Sustainable Development*. Resolution Adopted by the General Assembly. DOI: 10.1007/s13398-014-0173-7.2

Walker, C. D. B., & Burbanks, S. (2010). Afrocentricity Published in the Encyclopedia of Identity. In. Encyclopedia of Identity. Afrocentricity. Jackson, R.L II (Eds) Sage Publications

Walker, C. E. (2001). *We Can't Go Home Again: An Argument about Afrocentrism*. Oxford University Press.

Walsh, J. P. (2020). Social media and moral panics: Assessing the effects of technological change on societal reaction. *International Journal of Cultural Studies*, *23*(6), 840–859. DOI: 10.1177/1367877920912257 PMID: 40479194

Wilson, D. (1987). African Council on Communication Education Traditional Systems of Communication in Modern African Development an Analytical Viewpoint 89

World Bank. (2025). Executive Summary. 2024. https://thedocs.worldbank.org/en/doc/4cafc3a906669ba0d34d4a39dc903472-0090012024/original/IndigenousPeoplesResilience-ExecutiveSummary-FINAL-Aug-8-2024.pdf

Wynberg, R. (2022). Biopiracy: Crying wolf or a lever for equity and conservation? *Research Policy*, *52*(2), 104674. DOI: 10.1016/j.respol.2022.104674

Wyrzykowski, R. (2023). Rural-urban digital divide still poses a challenge in South Africa. https://www.opensignal.com/2023/12/13/rural-urban-digital-divide-still-poses-a-challenge-in-south-africa

ENDNOTES

1 This is, however, refuted by Walker (2001) who argues that the blackness which is denoted by Kemit refers to soil not people.

2 Steyn (2007, p.266) as a search for spirituality and meaning that contrasts with t7prevalent mainstream religious or scientific beliefs.

Chapter 13
Types of Symbolic Tacit Indigenous Knowledge Capture Processes of a Bull as Intangible Cultural Heritage Among Isukha Community, Kenya

Elphas Muruli Ngaira
https://orcid.org/0009-0004-5142-0168
Technical University of Kenya, Nairobi, Kenya

Ashah Owano
https://orcid.org/0000-0002-8370-1106
Technical University of Kenya, Nairobi, Kenya

Lilian Ingutia Oyieke
https://orcid.org/0009-0007-9559-214X
The Technical University of Kenya, Nairobi, Kenya

ABSTRACT

Tacit Indigenous Knowledge (TIK) forms part of people's history and cultural legacy, which is typically transmitted orally from elders in the community to younger generations. TIK is embedded in the Intangible Cultural Heritage (ICH) which requires interpretation of the process, rituals, experiences, and practices unique to indigenous communities. TIK of a bull in Isukha community is a critical aspect of ICH though it stands a chance of being eroded considering the present practices.

DOI: 10.4018/979-8-3373-0204-1.ch013

This chapter investigated the TIK capture processes of a bull as ICH among the Isukha Community in Kenya. It adopted the knowledge capture process model: identification, acquisition, refining, and storage. The study was guided by a descriptive qualitative research approach. The study was guided by a descriptive qualitative research approach. The data was collected through key in-depth interviews (KII) with community knowledge holders and focus discussion groups (FDG) focusing on observable interactions involving the bull

INTRODUCTION

The Tacit Indigenous Knowledge (TIK) capture is a process of extracting knowledge from people, communities, or organisations (Jia et al., 2022), methods used to collect, document, and store information for accessibility to same knowledge from both within and outside the organization (Manesh et al., 2020). Traditionally, qualitative methods are used to capture tacit indigenous knowledge from within as capturing TIK requires careful consideration and respect for the cultural context and traditions of the community involved (Hadjimichael & Tsoukas, 2019). According to Bacon et al. (2020), effective knowledge capture is crucial for preserving intangible community institutional knowledge such as kinship, marriage and inheritance, promoting collaboration, and facilitating informed decision-making and re-use. This study adopts the three stage process of Tacit indigenous knowledge capture process namely: Identification, acquisition, and refining.

IDENTIFICATION

Identification is a knowledge management capture process that distinguishes knowledge produced by a community (Mahdi et al., 2019). The process identifies the owners, users, uses and the key attributes of core knowledge assets (Wang et al., 2021). Mahdi et al. (2019) stated that identification provides a basis for creating a knowledge management plan to fulfil certain goals. One potential reason for the lack of attention to knowledge identification (KI) is difficulty to clearly define what KI entails. This is because knowledge is practically ingrained in all knowledge management (KM) operations, making it challenging to distinguish KI from other KM processes. KI is key process of KM whereby organisations take action to identify what knowledge is relevant and necessary inside their borders (Intezari et al., 2022). According to Kunduru & Kandepu (2023), knowledge identification helps in organizing vast amounts of information for archival into manageable categories, making it easier to navigate and comprehend. Examples of areas of KI include broader categories of oral traditions and rituals involving the bull among the Isukha Community of Kenya.

Figure 1. Knowledge Management Process

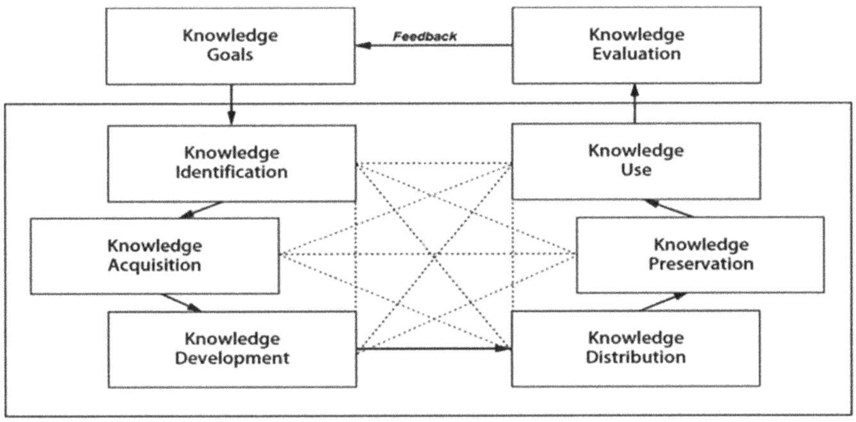

(adapted from Probst et al, 2000)

Some of the knowledge identification models include: Knowledge management assessment tool KMAT (Van Laar, 2023) that assist in determining KM maturity, knowledge transfer and barriers affecting decision making.

ORAL TRADITIONS

Oral traditions embody the collective wisdom, beliefs, values, and history of indigenous people thus a rich repository of knowledge that is passed down orally (Mahuika, 2019). Oral traditions allow communities to preserve, transmit, and perpetuate their cultural heritage through spoken stories, songs, chants, and rituals (Hipolito, 2024).

Bull Songs

In oral traditions, songs serve as repositories of cultural knowledge, transmitting values, beliefs, and traditions from one generation to the next (Viscogliosi et al, 2020). Songs play a significant role in oral tradition cultures serving various functions and classification (Miller, 2024). Narrative songs tell stories often recounting historical events (Tulius, 2020). Similarly, ceremonial songs performed during rituals, ceremonies or rites of passage (Walker & Berryman, 2023).

Dirges Sanka et al. (2020) are mournful songs or musical compositions performed in honour of the deceased, typically during funerals or memorial services. Symbolic tacit indigenous knowledge of a bull in dirges can encompass a range of meanings, including strength, sacrifice, resilience, and rite of passage.

War Songs Rickshaw Travel Group (2024), fighting bulls are well prepared to fight and subjected to a balance diet. *Mayo*, bull fighting is endowed with a series of pre and post-fight rituals performed by specially selected individuals based on set attributes.

Work Songs Qobilova (2024) are one of the ancient examples of folk art. These types of knowledge was created in various aspects of human activity such as ploughing and have gained relevance. These songs help carry out work process in an orderly manner, create an upbeat mood in the worker's heart and served as spiritual encouragement. Samarasinghe and Nethsinghe, (2023) noted singing by men while threshing paddy and ploughing paddy field.

Rituals

Sacrifice, a ritual among Isukha community done when an adult died. This is to show honour to him or her and the ancestors who have already passed on by slaughtering an animal from among the stock of cattle in the homestead. For a man, a bull would be slaughtered while a heifer would be slaughtered if a woman died. The slaughtering took place the night before burial. The blood from bull for sacrifice (*Ing'ombe yishisayi*) signified the appeasement of the deceased, ancestors and the mourners. Moreover, slaughtering of the animal and shading of the blood meant to ensure continuity of the generation of the dead (Ministry of State for National Heritage and Culture, n.d. p. 26)

Dance Landborn (2019) and bullfighting constitute a cultural complex internationally recognized. Further noted bullfight in the plaza and a dance that started in the afternoon and lasted into the night.

Sport Lopez et al. (2022) noted that since the founding of the nation that is now known as Spain, bulls in all of its forms, whether viewed as ritualistic or as a recreational sport, have been a part of Iberian culture

Dowry Taringa & Museka (2021), a father's bull was offered during *roora* (practice of paying bride price in Shona culture) ritual to symbolize reproductive capacity. *Eunwaebucha* (bull that bellows) was not given as bride price for it meant bad omen for the marriage and in most cases it was slaughtered (Wepukhulu et al., 2023). According to Cakata and Ramose (2023), the bile of the bull was hanged in the kraal to ask the ancestors to seal the bond between the bride and the groom.

ACQUISITION

The second TIK capture process is acquisition and transfer. It is the process of acquiring, assimilating, and integrating new knowledge. It includes both the collection of useful information from all relevant sources and to instil new acquired knowledge identified as necessary (Dias et al, 2023). Most TIK is embedded in local communities in form of indigenous knowledge. In order to be transmitted, this knowledge must be changed in the form of words, models or numbers so that they can be understood (Aksa, 2020). The traditional mechanism for transmitting ICH depends on oral knowledge rather than written texts (Lugmayr et al., 2017). The local way of protecting and preserving ICH, also known as traditional safeguarding practices, or local know-how ensures transfer and survival of the valued practices and materials from generation to generation (Bigambo, 2020). That is why, with this type of transfer, innovations might be very slow to spread in the population unless other modes of cultural transfer are employed. In this study some examples of TIK acquisition and transfer processes include apprenticeship, traditional practices, agricultural and healing practices involving the bull. During apprenticeship knowledge is transmitted from experienced elders or experts to younger generations through hands-on learning, observation, and direct participation (Sundararajan, 2023). Craftsmen in indigenous communities such as the Isukha recognize the demand for knowledge on production of various traditional tools associated with the bull. Examples of these tools include the yoke, flywhisk, bell, and drums are discussed.

APPRENTICESHIP

Apprenticeship serves as a traditional method for acquiring TIK. Knowledge is transmitted from experienced elders or experts to younger generations through hands-on learning, observation, and direct participation vide cultural practices and rituals. Apprenticeship emphasizes experiential learning, and mentorship. This process helps indigenous communities safeguard their cultural heritage, nurture future generations of knowledge holders, and ensure the continued vitality and relevance of their TIK (Sperveslage & Heagy, 2023).

TRADITIONAL CRAFTSMANSHIP

According to Sperveslage & Heagy (2023), the craftsman presumes that the bull's tail was purposefully added to the catfish as an iconographic element. The numerous iconographic references on the palette strongly associate the King with the bull, supporting him with strength and potency of a wild bull that allowed him to tame chaos and establish cosmic stability. For example, symbolic TIK of a bull props included traditional sticks, yokes, fly-whisk, bell, whip and plough to mention a few.

Traditional sticks Chiteyi et al. (2023) noted that African traditional sticks such as *lusui* which is used to guide the bull and not or hit or harm it. It is believed that touching a bull using *lusui* brings triumph and expels negative omens. On the other hand, *lusibwa* is used to control and shield adversaries should the bull become aggressive.

Yoke, both the yoke and the V-shaped wooden framework are included in the phrase "yoke structure." The payload carrier is connected to the V-shaped wooden frame that is typically utilised in twin bull drives and is attached to the yoke (Hombalmath & Mathad, 2021). According to SadanandaNayak (2021), *Juadi* (Yoke) worn around the animal neck, is one of the greatest inventions of a hand held tool for harnessing the locomotive energy of oxen.

Fly-whisk, Raven & Theuns (2021) are tools of healing work ranged from the fly-whisk and calabash, to bone-like forms. Further inscribed bull that once stood in front of a temple, tied with a bell and a fly-whisk. Similarly, Melka et al. (2022) argued that fly-whisk making and bamboo crafting were spaces for learning. Thornton (2020) observed lishoba or "fly-whisk" used throughout South Africa by craftsmen.

Bell Akpabio (2023) and gongs serve as disseminators of messages through their symbolic roles and meanings. They are reflective of the society and its knowledge system. Kumar et al. (2023) mentioned bell metal or Dhokra craft of Bastar is practiced by craftsmen. They cast Dhokra for their ceremonies and domestic use. Panopoulos and Chatzopoulou (2015) points to animal bells in every day ritual contexts.

Goad of Whip Conroy (2022) been designed for centuries after craftsmen realized there is a demand for ox yokes and related equipment outside of New England. Whips commonly used to compel and coerce horses and oxen (Gruen, 2024).

Plough Lawton (2023) craftsmen were the cornerstones on which a community's relatively more advanced civilisation depended, and their knowledge was carefully guarded. The craftsmen were required to create plough which required less strength to pull and was pulled by a pair of oxen (SadanandaNayak, 2021).

Transport and Decorative Art clay hearth featured a burial for two bull heads made of bronze. The design, which takes the shape of a bull or a bull horn, is similar to the presence of such items in the location (Avsharova, 2020). Carts yoked to bulls that resembles copper sculpture of a bull-cart (Parpola, 2020). Asadi (2023) encourages symbolic storage where wild bull skull on the wall of a building have been used as ritual objects.

Ritual Objects Miri and Arymanesh (2022) have always been in the study of symbols and objects. By examining these symbols and objects, researchers can structure the symbols and carry out cognitive analyses with the aim of discovering more about the customs, beliefs, and mental processes of prehistoric people.

Musical Tools Terzes (2020) enlighten missing links in the chains of tools technical history to understand the music of the past. Teffera (2021) argued that musical tools found in central and southern Asia, large parts of Africa are made of bronze and then covered with bull hide. Consequently, horns were used to make musical tools like trumpets and drinking vessels (Odulwa et al., 2021).

Household Utensils and Toys of Zoroastrian customs are reflected in the common representation of a bull as a pouring jug found in monuments throughout central Asia. This jug is characterised by a wide and high neck, four bean shaped handles and a bulls' pourer (Dvurechenskaya, 2022). Household utensils and toys can incorporate bull motifs or designs, adding a touch of symbolism and personality to everyday items.

Language and Gestures Associated with the Bull Grandin and Deesing (2022) are body language used to communicate with each other. Bulls produce a variety of vocalizations to convey different messages. The pitch, intensity, and frequency of vocalizations can carry symbolic meaning (Blank, 2021). Bulls may engage in marking behaviours to establish their territory that includes rubbing their scent glands on objects or spraying urine to leave their scent (Babber et al., 2022). These actions serve as symbolic markers of ownership and dominance. Chiteyi et al., (2023) noted athleticism in fighting bulls in the circular arena and testosterone-rich bulls display dominant behaviours; deep throbbing howls of soils with horns, scratching ground/ soil with front legs prior to charging at each other.

AGRICULTURAL PRACTICES

Indigenous knowledge of agricultural practices often encompasses a deep understanding of local ecosystems, including animal husbandry (Adade et al., (2020). In many traditional agricultural systems, bulls are utilized as draft animals for ploughing fields, hauling loads, and other tasks. Associated indigenous knowledge includes techniques for training bulls for such work (Woldeyohannes, 2020). Bulls often hold cultural and spiritual significance in indigenous societies, and agricultural practices related to bulls may be intertwined with traditional beliefs, rituals, and ceremonies such as to honour bulls, as well as taboos or restrictions related to their care and use (Osypinska et al., 2022).

HEALING PRACTICES

Indigenous healing practices portray a bull as a symbolic health and well-being using animals. It reflects a holistic approach that encompasses physical, emotional, spiritual, and ecological dimensions. The practices are deeply rooted in cultural traditions, ecological wisdom, and a profound respect for the interconnectedness of all life (Kalle & Kass, 2020). According to Babacan et al. (2022), rituals, ceremonies, and prayers may be performed to invoke spiritual aid in healing of bulls.

REFINING

The third process is refining knowledge. It involves removing unnecessary data and revealing contradicting or missing information (Guo et al., 2022). Once knowledge has been captured, it has to be classified, categorized, indexed and described. There is need to assess and verify stored knowledge to ensure that it is accurate, current, complete and consistent (Rafi et al., 2022). Refining includes Knowledge distillation which is one of the most widely used methods for in model acceleration and compression. This process distils knowledge into concise and relevant nuggets of key highlights, lessons learnt and practices of knowledge sources (Wiegers, 2021). The TIK will be distilled for transfer of generalization abilities Wang et al. (2021), performance improvement Mirzadeh et al. (2020), enhanced training stability Jin et al. (2019) and facilitates model understanding (Wang &Yoon, (2021).

Policy Integration

Masenya, (2022) advocated for protection of IK through policy and protocol framework using De colonial theory. Tacit indigenous knowledge preservation requires a holistic response that combines community-driven strategies, policy support, youth engagement, and the integration of indigenous epistemologies into education and cultural programming.

Contemporary Threats to IK Preservation

When a bull is used as a symbol, there are a number of potential threats that could prevent symbolic tacit indigenous knowledge of the bull from being passed down through the generations. These threats; modernization and Urbanization, generational disconnect, religious and educational influences, cultural shifts, and external influences are influenced by social, economic, environmental, and cultural dynamics (Rokka, 2021; Yeoh, 2020).

Gender and Youth in Sustaining STIKAB Heritage

Symbolic tacit indigenous knowledge of a bull holds deep cultural, social, and spiritual significance, embedded in rites of passage, marriage, clan identity, and ancestral connections (Mensah, 2021; Avrami, 2016). Youth Berio &Moore, (2023) play crucial yet distinct roles in preserving and transmitting intangible cultural heritage. Symbolic segregation by gender and age as seen in the exclusion of women from bullfighting grounds then or the ritual purity expected of bull raiser and caretakers reflects embodied epistemology, where physical behaviour and moral conduct are co-requisites for knowledge transmission. Assou-Dodji & Serwah (2025) observed that sustaining the IK depends on empowering women as active cultural transmitters and engaging youth through mentorship, cultural education, and technology. Supporting intergenerational dialogue and valuing the roles of both groups is essential for the continuity of this rich heritage (Yadav, 2025).

Indigenous Knowledge Transfer Model

Indigenous knowledge transfer model (IKTM) Parsons (2004) discussed hereunder is rooted in cultural traditions, oral histories, and experiential learning, and it plays a crucial role in preserving and transmitting essential knowledge (Moore & Nesterova, 2020). IKTM's prominent feature emphasis on oral tradition (Kovach, 2021) that ensures preservation of knowledge while fostering a sense of community and connection to the past heritage. Another key aspect of indigenous knowledge transfer is its holistic nature (Makgabo & Quitero, 2024). Turner et al. (2022) argued that indigenous cultures often view knowledge as interconnected and integrated with all aspects of life. Onwuatuegwu and Paul-Mgbeafulike (2023) pointed that indigenous knowledge transfer model often prioritize experiential learning and hands-on apprenticeship. Bihari (2023) recognised indigenous knowledge transfer model to face significant challenges of cultural assimilation, and the erosion of traditional practices. By honouring and preserving these traditions, practitioners and researchers can not only benefit from their practical wisdom but also contribute to the broader goal of culture 2030 indicators' diversity and sustainability (Dei, 2024).

METHODOLOGY

The research employed focused ethnography qualitative design. For instance, data was collected through interviews and focus discussion groups with community knowledge holders and content analysis focusing on observable interactions involving the bull such as: gestures, rituals, songs, traditional craftsmanship, agricultural and healing practices. Furthermore, the study was designed to use 15 key in-depth interviews, two focus group discussions and one participant observation. As a result, research questions formed the basis for thematic and narrative description analysis, interpretation and discussion for presentation.

Information of Participants

The context of this study was Isukha community in Kakamega East Sub County, Kenya with six locations. To achieve a manageable and informative sample size, the study employed purposive sampling technique, targeting a sample size of 63 bull raisers, elders, culture teachers, bull enthusiasts, and indigenous knowledge experts as respondents. Total response rate from the context of study was 31 (49.2%) of sample size. Inclusion criteria was individuals with at least five years' experience in bull raising and indigenous experts acknowledged for their knowledge within the Kakamega East Sub County. Key informants were identified through local agricultural Sub County and ward offices, county administrators and community elders. Using snowball sampling, these informants helped the researcher with referrals to other potential participants particularly those involved in bull enactments and fit the study inclusion criteria. The study aimed for maximum variation by including a diverse range of participants in terms of age, gender, geographical location, and the scale of their bull raising operations. This method assisted the researcher to gather comprehensive and representative insights from a group of 31 participants comprising of 15 Key in-depth interview IK experts and bull raisers, two FGDs of eight to ten participants one from location with high prevalence of bull raising and IK experts and lastly participant observation for bull symbolism in dowry negotiation, burial and bull fighting rituals.

Table 1. Sample framework

<table>
<tr><th>Location Name</th><th>Estimate no. of bull raisers&IK experts</th><th>Sample size = (35% total no. of bull raisers & IK experts)</th><th>Key In-depth interviews (KIIs)</th><th>Focus Group Discussions</th><th>Participant Observation (event)</th></tr>
<tr><td>Isukha West</td><td>30</td><td>11</td><td>2</td><td></td><td rowspan="6">Dowry negotiation, Burial, Bull fight</td></tr>
<tr><td>Isukha South</td><td>80</td><td>28</td><td>4</td><td>1</td></tr>
<tr><td>Murhanda</td><td>55</td><td>19</td><td>2</td><td></td></tr>
<tr><td>Isukha Central</td><td>70</td><td>25</td><td>3</td><td>1</td></tr>
<tr><td>Isukha North</td><td>40</td><td>14</td><td>2</td><td></td></tr>
<tr><td>Isukha East</td><td>40</td><td>14</td><td>2</td><td></td></tr>
<tr><td></td><td>315</td><td>111</td><td>15</td><td></td><td></td></tr>
</table>

FINDINGS

Data Presentation, Analysis and Interpretation

This section presents results of the study that examined processes used in capturing symbolic tacit indigenous knowledge of a bull as intangible cultural heritage in Isukha community, Kenya.

During data collection, the researcher reached data saturation with 23 of 15 KIIs and 1 FDG against projected 31 participants in the sample size. Data collected focused on a four stage process of tacit indigenous knowledge capture process namely: Identification, acquisition, transfer, and refining. Symbolism of the bull in dowry carries deep symbolic tacit indigenous knowledge across many cultures. A bull represents masculinity, power, and the ability to provide and protect a family. Moreover, symbolizes the groom's strength and capability to take care of bride. Dowry honourship is an entrenched symbolic custom that gives both men's and women's lives meaning, respect, and a sense of belonging (Musili, 2022). In many indigenous societies, cattle especially bulls are a form of wealth thus offering a bull in dowry demonstrates the groom's family's economic standing and respect for the bride's family (Chiteyi et al., 2023). Giving a bull in dowry is a sign of deep respect for the bride's family, reinforcing kinship ties and social harmony. Adam et al. (2021) observed that tacit indigenous knowledge embedded in the practice includes; spiritual beliefs, customary law and ancestral approval, social reciprocity and community cohesion. Dowry system, with part of prescribed bride-price of a pair of bulls (Pallathadka et al., 2022) as a key asset, ensures reciprocal relationships between families, strengthening community bonds.

Symbolic Tacit Indigenous Knowledge of a Bull in Rituals

IKTM has governance and management practices, ethics and values; observation of places, processes, and relationships as postulates that revealed IK as holistic, specific and situated knowledge and in the present study of STIKaB as ICH in Isukha Community, Kenya. IK is connected to continual resource use and maintenance heritage in bull fighting and shilembe rituals of Isukha community. Transfer of IK is largely through oral and participation in symbolic bull enactments.

Table 2. a) Dowry

Contextual information	
Date and time of observation	25/01/2025 (12:50 - to 3:30PM)
Location	Isukha South (Code S-PO1)
Name of event	Dowry negotiation (Event)
Observer name	Elphas Muruli
Aim was to gather information about symbolic tacit indigenous knowledge of a bull capture processes; identification, acquisition, and refining in (*bukhwi*) dowry ritual. In many cultures, a dowry negotiator (knowledge owner) is a respected individual who represents either the groom's or bride's family in dowry negotiations by facilitating discussions about the bride price and ensuring that both families reach a fair settlement. Roles of the dowry negotiators include; mediating between families as a bridge between the groom's and bride's families respect. A skilled negotiator is able to bargain dowry amount and terms, considering tradition, economic conditions, and the two families' expectations. Possess knowledge of customs of diverse cultural traditions, as different ethnic groups have specific dowry customs. Case in context was Luhya and Kikuyu cultures coming together. Negotiator Ensures respect and etiquette in negotiation sessions with humour, wisdom, and tact for the growth of newly found friendship. Negotiators help finalise agreement once both sides agreed on the dowry, including documenting and formalising the agreement. For example it was observed that a dowry settlement of eleven heads of cattle totalling Kenya shillings four hundred and forty thousand and Kenya shilling seven hundred thousand in cash. For example, Parents of the groom and bride too collect useful information regarding dowry from opposite family/culture to avoid embarrassing negotiation scenes on material day. It helps to organize information engrained in every dowry step/activities including songs and gifts symbolism. Parties from bride and groom's side had knowhow per their culture regarding dowry hence their presence. Inclusion or Young generation (three) in attendance (One chosen as secretary for the bride's side, the other two were groom and his cousin. This ensures apprenticeship, hands-on learning, observation and direct participation In a bull related narrative, "A man had been paid a cow as dowry and was in oestrus period, the man approached his brother that had a bullock for help. The brother allowed the request insultingly,"*yiyi shinukhwelwa khumukhana tawe, nakula*" translated as; "this bullock is not a daughter's dowry proceed, I bought". This provoked the borrower to begin cattle keeping that included bulls. Both brothers had raised huge herds of cattle that included bulls and one day when they met at a water point, two of their bulls engaged I a fierce fight while owners broke in song and dance to encourage the bulls. This is believed to be source of the present bull fighting ritual in Isukha to date concluded" S-PO1 dowry negotiator. In dowry negotiation, tacit indigenous knowledge is refined through repetitive experiential events negotiators have in life time. Bride and groom are automatic go-through process owners, parents are able to verify and remove (distil) unnecessary data from the captured information regarding dowry practice from either side.	

b) **Burial**

According to informant S-KII4, a bull in Isukha is for community's fame *lukhumu* in addition *to* meaning that one practised cattle keeping. E-KII1 observed that, in the ancient days, a bull was not for every adult who died in Isukha community but associated certain burial rituals depending on the old man's standing in society. Informant N-KII2, agrees that it is not compulsory that all the dead deserves bull honour. A Bull was for special symbolic meaning in Isukha tradition firstly, a man with a herd of cattle was a sign of wealth and when he died, immediately after burial by the church, family members organized for bulls, cows, oxen, bullocks, sheep, and goats to come over and mourn on late man's grave signifying the man was a cattle keeper and the herd continues to grow within the family generation. An 87 year-old S-KII4 advised that one should not rush into purchasing a bull based on factors such as fighting prowess or it's a champion bull. Using his example, S-KII4 visits and interacts with the bull at the source to test a few things before closing the deal. In the week starting 13th January, 2025, S-KII4 had just disposed of a seven year old black bull named J at a tune of two hundred and twenty thousands. Bulls are used for mourning specific heroes that had engaged in combat and in the process killed their enemies. Their cleansing *khwitisa* as a connecting factor was publicly known (C-KII1). According to N-KII1, the main relationship of a bull and burial is through *shilembe shia "Siyo" or "selebutswa.* S-KII4 concurs with N-KII1 on categorization of Isukha community *bilembe.* This was confirmed by S-FDG1 adding that "late raiser's bull do honour the raiser by leading funeral procession from the mortuary to the homestead" and corroborated shilembe shia *selebutswa* of 25/01/2025 in Irobo area of Isukha South, Kakamega East Sub County.

Qualifications for *Shilelembe shia"Siyo"* includes one to have been in war and killed enemies. This is supported by the averment of M-KII1 bull raiser informant noting that when his father a military veteran died, M-KII1 sought the official *Indumba yi shilembe* and bull raisers through an invitation letter. Majority of the qualifiers were former WWI or II veterans, early day's Chiefs M and I (*Bulukongo*) as popularly called to mention a few. Regalia for *shilembe shia sio* included spears, war swords worn by the late veterans' colleagues. Spears were thrust into the chest part of the late veteran's grave while singing war songs such as; *"ooooh simba ilalila shibala....simba ilalila....oooh simboa ilalila* ; *khusalanga! yieeeyieee! khusalanga nukhchelela khuli* (name of village) as they danced making pronouncements in the homestead such as; "so and so was a man, *bulolo bukhaluma bandu ta, bichinji bikhaluma bandu ta, tsisuli tsikhaluma bandu ta, khukhukholele mulukha kukhum-alilikha"* translated as (name of late) "mites should not itch people, fleas should inflict people, bedbugs should feed on people, we have performed your last tradition" N-KII2. Another significant element of "*shilembe shia siyo*" was a special drum "*Indumba*" still present in Isukha community. *Indumba yishilembe* has only one custodian at a time and never see the sun of the day, this drum is never transported openly but inside a sack at night, on routes without heavy human traffic and by the custodian (old man) who meets *shilembe shia siyo* and no longer siring children. When the drum arrive, a secluded round shaped place besides the late veteran's grave using fresh trees twigs and croton macrostachyus ex delile; *Musutsu* (luhya) pole with a hook erect therein from which the drum is hanged and beat. The drummer could be seen but not the drum because of its potent (M-KII2 bull raiser informant).

c) Shilembe ritual (Siyo or selebutswa)

With a coordinated signals between the drummer, a man calling the late uncle who breaks *yambobo* pot, veteran colleagues enter the late man's house, collecting the body and inter. At 5:00am the following day, *shilembe*i *shia siyo* is re-opened and the veteran's colleagues in war regalia, wild animal skins, natural creepers and climbers *malande* covering streamed into the home some holding goats, bulls and others ox singing as the drum sounded. And where homestead lacked a bull, an ox or a big he goat *imbuli ikhole* were and alternative symbols. Invited bull raiser from afar off would come the following day in the name that; *khutsitsa khuhuyitsa (name of late) yakhutsa* translated as; "we are going to see off (name) who passed on" (N-KII2 bull). We also note with observations in E-KII2 and S-KII4 that, Veteran colleagues of the late in their war regalia would join the fighting bulls in performing *Shilembe* ritual. They mourned dancing over the late man's grave as bulls fought, added E-KII2 indigenous knowledge informant. Each veteran carried in his hands a sword and the number of spears representing the enemies he had killed. In addition to the fighting bulls, veterans in a frenzy destroyed all grown plants, others chopping off bananas using war swords, a simulation of killing opponents in memory of the late veteran (S-KII4). Other veterans pick on one another in a wrestling match just as bull were doing. The idea was to create a real war scene where enemies are fought and killed. As future memorial elements, certain symbolic traditional plants ficus thonningii *mukumu*, dracaena fragrans *tsikhubu and* markhamia lutea *lusiola* were planted on the flattened grave site. It signified honour for *mundu yakhola mulukha* a hero, (E KII1&S-KII4). On the selection criteria of a bull for burial ritual in Isukha, 60 years-old N-KII1 bull raiser concurs with E-KII2 indigenous knowledge expert that such a bull "not *shihunwa*_bullock" mus t not have the following a malformations *livendi* (golden with dark spots) colour, hornless, castrated, blind and teeth less. N-KII2 adds by defining a bull "*ichilishi*" as uncastrated grown male cattle "*yakhulumana*" six and above years old. E-KII1 noted that, selection criteria for a burial bull begun with family meeting. Source of this animal was owner's homestead flock having been pointed out while the late lived. For an elder who owned no bull, any of the sons, or from where the first born daughter is married. E-KII2, indigenous knowledge expert informant opposed E-KII1 and S-KII4's assertion of purchasing a bull from the market noting that from time immemorial, a bull for an old man's burial was not bought on market. Living example, 82year-old E-KII2, has pointed from his herd, the bull "Nyati" name of bull, will be slaughtered in his funeral"to be slaughtered after goes the way of the ancestors. This knowledge is with the immediate family members and some of his age mates. In circumstances the late old man had not pointed a bull for lack, other forms of source were maximized; hiving off a portion of deceased's land, sale and buy a bull; from his first born daughters place of marriage and from living son(s). Another burial spectacular element was an old woman with in luhya (*tsimbande, tsinuni, makanda, matuma and mbele mmbohu*) translated as monkey nuts, sim sim, beans, maize and sour milk" spoke saying; (Late man's name) *uleshe bindu biame hanu, bitsilili khubaho khuli ulimienje. Ukhatsia navyo tawe* translated as "(late mans' name) allow plentiful of harvest to continue the way you have been cultivating, don't go with it".

d) ***Bull fighting***

Rope and men wrestling transformed into bull fighting as mwisukha form of entertainment extended to honour community heroes for their warrior achievements. Abisukhas' great grandfathers started the idea of bull fighting including those without cattle. This is in line with S-FDG1 indigenous knowledge expert respondent's statement that, bulls were raised by old men and not a common phenomenon compared to twenty first century. Bull raisers were sparsely distributed a cross Isukha. Many raisers in this century are driven by greed and money contrary to old day's objectives of pass time, friendship, entertainment and fame. N-KII2 argued that cattle keeping tradition forbid in breeding. E-KII1 noted that a bull was taken care by only one individual and children were not allowed near a bull. Bull fighting was born out of *shilembe* ritual which occurred once in five years. S-FDG1 concurs with E-KII1 and which they refers to as " long period for them to organize bull fighting". Present generation has made shilembe *selebutswa* as though football for watching every weekend.

Symbolic Significance of Bulls' Colour, Size, And Gestures in Rituals

According to E-KII1, S-KII4, and S-FDG1, symbolic significance of a bull's colour, size and gestures does not matter in burial only a bull be healthy and without defect. Colour has no significance for rituals such as burials asserted N-KII1 bull raiser informant and in contrast to E-KII1 indigenous knowledge expert and N-KII bull raiser informants. N-KII2 and S-FDG1 argued that colour, size have symbolic significance in bull rituals. For example there is a colour of bull breed "brown, hairy earlobes" that is fierce and a good fighter. Size really matters when it comes to power for pushing opponent during bull fighting and also avert conflicts when one bull is fatally injured. Next, size of a bull matters during bull fighting. Additionally, symbolic significance of bull's gestures is a preserve of bull fight other than burial. According to N-KII1 and 2, when a bull pulling soils backwards using its front legs and similarly dip horn into the ground and haul plus served hay respectively means high morale, is a sign of acceptance, ready for the opponent, created territory and any other bull is unwelcomed. While thrusting horn into the ground after raiser instruction denoted it's ready for immediate duty or asking for a fight encounter in due course, the raiser has to make haste. C-KII2 reported that the bull is well trained in *milukha* traditions which are demonstrated during enactment mourning of a raiser.

Bulls are very sensitive animals and understands whenever the raiser communicates hence call for monitoring as bulls know their owners and hired care taker. Bulls are able to detect raiser moods and respond accordingly (S-KII4). While in the homestead a bull may raise up its head and point the nose left and right and finally concentrated in one the direction after sensing danger from that direction argued S-FDG1. N-KII2 concurs with N-KII1 on gesture symbolism that when bull raiser receives an invitation for *shilembe shia siyo/selebutswa*, raiser relays the message to the bull saying "know that during the burial of so and so, you will attend". In response, a bull's ears erect or snorts as affirmation. Bull training starts at an early age from the day a raiser acquires as communication training of a bull is everything. Also the bull has mastery of all connected rituals artefacts such as Isukuti, Jingles, whistle, *lusui* stick, *lutari*, *iloboshi* and club *shikong'o* (S-KII4 bull raiser informant).

Accuracy In Tacit Indigenous Knowledge Transfer

According to E-KII2 and S-FDG1, TIK is important and worth preservation and hand over to the offsprings as it was handed over to holders by our ancestors. For example "circumcision" *bushebi* noted W-KII1indigenous knowledge informant. "Ancestor would not give us something to destructive" said N-KII2. Also, S-KII4 noted that, ancestral wisdom was and is still vital.

During the interview, 60 year-old, M-KII1 had this to say while in the pen of his bull Lumbe "bull fighting knowledge was not stolen but got through the grandfather M, who gave to M-KII1's father K who handed to M-KII1. For example noted S-KII4, the traditional herbs knowledge administered to the bull was an inheritance from his late father and grandfather and chief M. In event the only son of M-KII will be unwilling to receive this knowledge, any of M-KII1 sister's sons will be located by this *musambwa*". If this kind of knowledge is abandoned, that gap silently harms the generations thereafter. "*When it fails to have a successor, it will disturbs (inasumbua)*" M-KII1. To ensure its accuracy as a collective cultural memory for social identity, E-KII2 observed that those interested in acquiring in such knowledge, passionately pursues its creation. Such potential successors must experientially interact with the knowledge holder to see and practically do. Giving the biblical illustration of Jacobs red soup that consumed his brothers births right, E-KII2 noted that one thing stood out; "*Jacob was very close to the father and mothe*r" a key factor in the TIK transfer. M-KII1 added that transfer of TIK depends on a willing recipient a holder may spot from the first day.

N-KII1 affirmed believe in the importance of TIK and often uses this ancestral heritage lineage pronouncements before his bull leaves the homestead for a fight.

To ensure TIK's accuracy, knowledge holders must be cautious and watch out for willing son, NOT sons and begin experiential training through the daily routines of knowledge creations. In addition, N-KII2 cautioned that young generation should not be entrusted with TIK community mandate. Also, concoction of traditional energizer herbs made from barks of; *munyama, mutukuyio and shikhuma* by boiling well and serving their bulls when cold. Too, *lusamari* when roasted, mixed with ground half rotten maize, busaa residue *liseshe* and salt stimulate vigour/anger bulls. E-KII1 noted the importance of TIK transfer more in *shilembe* ritual in Isukha community for purposes of enhancing accuracy of TIK transfer.

Mentorship and Intergenerational Guidance

"Mentorship existed in the 70s and 80s that was well-structured. Fathers mentored sons from childhood in handling bulls, understanding their symbolism, and preserving rituals." Bull Raiser, Isukha North – KII

Role Of Elders, Ethical Data Stewardship, Gender and Generational Dynamics

Elders (bakhulu or IKEs) are pivotal in knowledge capture, acting as custodians of tradition. They guide younger generations through participation in sacred and social spaces, validating the symbolic meanings through ceremonies and communal performances.

"Two-fold transmission: through oral tradition and performances. Elders are central in ensuring that knowledge passed is culturally accurate." IKE, Isukha East

"It involves experts who are knowledgeable in bull rituals. They decide what is taught, when and how." FGD Participant, Isukha North

"Many of these IKs are attached to Isukha bullfighting traditions, and rituals are the main theatre of learning." FGD Participant, Isukha South

Discussions On Value of a Bull in Burial Practices

Isukha community does not bring together elders, spiritual leaders, and younger generation to discuss the value of a bull in burial practices. A fact attributed to diminished closeness between parents and their children (S-KII4). C-KII3 adds, "In the early days, there were meetings, today, people (elders, youth) and church suffer divisions. Some old men have abandoned the bull practices for church". Parents have this individualistic attitude "I have my children in this homestead, what other help can outsiders offer?" E-KII2. S-FDG1 differs with E-KII2 that there exist bull fighting group networks that champion the needs of the bull raisers and followers. Political and County Government leaders do make one off inroads to members or associations. N-KII2 agrees with E-KII2 on gatherings to discuss bull burial practices. Noting that people have commercialized the entire bull fighting ritual contrary to what it stood for in the early days in Isukha community; friendship and entertainment. S-FDG1 adds that one cannot have a bull fight match against his enemy as the underlying essence is friendship. Also unstructured community governance for control and proper management of this valuable cultural knowledge resource. Besides, N-KII1 added that, it was the desire of many then to come together and discuss the values of bull practices as illustrated by Muungano bull fighting self-help group that started in 2008. Membership included bull and non-bull raisers. Within Muungano of bull fighting association, a number of alternative dispute resolution (ADR) meetings been held for cohesive family living. N-KII1 further noted that Isukha community may bring together elders, spiritual leaders but the conflict of bull fighting fraternity and the church is yet to find a lasting solution. In observation with E-KII1, spiritual leaders, elders and youth in Isukha community have an obligation to come together and discuss the values of a bull in burial practices and related ICH. Unfortunately, modernity and foreign influence is causing youth to abandon their heritage.

Gender and Youth in Sustaining STIKaB Heritage

Mentorship, particularly from fathers, uncles, or experienced elders, is portrayed as a structured yet informal apprenticeship system. These relationships carry not only technical knowledge but also cultural, moral, and spiritual teachings tied to the bull symbolism. In old days, every village had a renowned old man/woman for TIK holder and their home a sleeping assembly point for boys and girls respectively. "*This structure is long dead due to broken down family ties, increased population and urbanization and education system*" E-KII1.

Taboos and Effect on Sharing Symbolic Tacit Indigenous Knowledge of a Bull

One of the bull taboos is abstinence from for the bull raiser or care taker. Herbal medicines administered to a bull connect the raiser and the bull through "herbal oath " that makes the bull recognise its raiser and his voice when in the surrounding W-KII indigenous knowledge expert informant and including mood (S-KII4 bull). Procedure for "bull's oath" agrees W-KII2& S-KII4, is performed at around 6:00 pm the evening to the bull fighting. Raiser asks the wife to hand him the ground solid and liquid herbal medicine served on *luchio* a piece of broken pot. The raiser sips the liquid herb concoction and sprays between bull's horns, with cupped hand takes the liquid concoction and smears the neck of the bull, over the back to the tail and winds by pulling the tail down hard with one hand and swearing by his ancestors and taps the bull's back once. Site for this rite is inside the bull's pen so that it's able to lick medicine from the raisers hands as he pronounces blessings (W-KII2). Regarding pronouncement, C-KII3 adds, "*Nyati (bull name) chende buleyi wutsia khunyole machibu maleyi*" translated as " Nyati wherever you go, let us get good results". On who to administer "bull's oath" procedure, S-KII4 disagreed with W-KII2 and preferred raiser administration to delegated care taker administration. For N-KII1 having been born of a bull raising father, knowledge on taboos for bull fighting was automatic and enabled to sustain the lineage legacy to date. According to N-KII2, some of the taboos surrounding bull fighting ritual includes; not to be used for breeding as a bull loose close to three kilograms after mating which may be an uphill task for bull to replenish, though E-KII2 asserted that, *mwisukha* inhabitant of Isukha kept bulls for breeding *khuruma tsitwasi* as the main reason. Secondly when any former World War I (WWI) veterans died bulls turned up at their burial and engage in a fight (*tsikhamarana*) in honour symbolism of veterans' bravery in killing their enemies in the battle field. On sexual abstinence highlighted by E-KII1, N-KII2pinpointed specific period; two weeks from the day a raiser received bull fighting invitation. No alcoholism is permitted for all bull handlers as bull and alcohol are enemies. Another feature is a specialised three to four lead squad of strong men whom the raiser entrust the bull to for protection and control every moment it leaves the homestead for fight a fact S-FDG1 & N-KII2 are in agreement with. This team is also subject to all taboos governing the bull care and management. S-KII4 observed that taboos firstly preserve TIK to safeguard against knowledge piracy that can affect accuracy of the original ancestral knowledge. Moreover, taboos are in fact enablers of TIK transfer lest Isukha community face a taunting indigenous court query and lack an answer.

IMPLICATIONS

This study demonstrates how ICH is a resource is created, preserved and capable of solving local problems through good practices and ideas. The study sheds light on extant typologies and value of TIK of a bull as ICH in life and why it must be safeguarded. The study is essential to IKM professionals' new replicable knowledge of a bull in addition to finding empirical gaps that can improve or add to existing theories. This study offers crucial symbolic data for policy makers in the Ministry of Gender, Culture, the Arts and Heritage, Ministry of Tourism and Wildlife and Department of Culture, the Arts and Heritage. Unique symbolic elements of a bull may find custody into the National Museums of Kenya as heritage resource that gives Abisukha a sense of continuity with prior generation.

CONCLUSION

Symbolic tacit indigenous knowledge among the Isukha is primarily captured through intergenerational mentorship, experiential immersion, and communal oral traditions. Elders function as key custodians, transmitting knowledge through rituals, mentorship, and symbolic enactments. Songs, rituals, artefacts, and ceremonies embed this knowledge in cultural memory, making it emotionally and morally resonant. Triangulation with field observations in Murhanda and Isukha South confirmed that these practices remain active, though fragile. The need for safeguarding is increasingly urgent as traditional pathways face disruption from shifting cultural, generational, and structural realities.

REFERENCES

Adade, W. P., Sikutshwa, L., & Shackleton, S. (2020). Acknowledging indigenous and local knowledge to facilitate collaboration in landscape approaches—Lessons from a systematic review. *Land (Basel)*, *9*(9), 331. https://www.mdpi.com/2073-445X/9/9/331

Adam, A. A., Othman, N., & Halim, A. A. (2021). Indigenous knowledge transfer among Dusuns and Bajaus in Kota Belud, Sabah, Malaysia: Approaches and challenges. *International Journal of Academic Research in Business & Social Sciences*, *11*(11), 1722–1734. https://www.researchgate.net/profile/Anis-Amalina Adam/publication/

Akpabio, E. (2023). Toolal Communication. In *Indigenous Communication: A Global Perspective* (pp. 31–55). Springer International Publishing., https://books.google.com/books?hl=en&lr=&id=ArrmEA

Aksa, F. I. (2020). Wisdom of indigenous and tacit knowledge for disaster risk reduction. *The Indonesian Journal of Geography*, *52*(3), 418–426. https://www.researchgate.net/profile/Furqan-Ishak-Aksa/publication/3483

Asadi Ojaei, S. K., & Abbasnejad Seresti, R. (2023). Symbols for the Neolithization Process: Ritual Animals of the Eastern Fertile Crescent. *International Journal of the Humanities*, *30*(2), 72–97. https://eijh.modares.ac.ir/article-27-58950-fa.html

Assou-Dodji, H., & Serwah Boateng, J. (2025). Voices of empowerment: women's role in safeguarding cultural heritage and advancing human rights in Africa. https://strathprints.strath.ac.uk/93364/

Avsharova, I. (2020). The reflection of" Bull" cult at the Bronze Age material culture of Azerbaijan. *International Journal of Psychosocial Rehabilitation*, *24*(01). https://www.researchgate.net/profile/Imcra

Babacan, E. Y., Polat, R., Güler, O., Moyan, A., Paksoy, M. Y., & Cakilcioglu, U. (2022). An ethno-veterinary study on plants used for the treatment of livestock diseases in Genç (Bingöl-Turkey). [IJTK]. *Indian Journal of Traditional Knowledge*, *21*(1), 81–88. http://op.niscpr.res.in/index.php/IJTK/article/view/41752

Bacon, E., Williams, M. D., & Davies, G. (2020). Coopetition in innovation ecosystems: A comparative analysis of knowledge transfer configurations. *Journal of Business Research*, *115*, 307–316. https://www.sciencedirect.com/science/article/pii/S0148296319306745

Berio, L., & Moore, R. (2023). Great ape enculturation studies: A neglected resource in cognitive development research. *Biology & Philosophy*, *38*(2), 17. https://link.springer.com/article/10.1007/s10539-023-09908-y

Bigambo, R. (2020). *Safeguarding practices for intangible cultural heritage in Tanzania: national vs local perspectives* (Doctoral dissertation, University of Birmingham). https://etheses.bham.ac.uk/id/eprint/

Bihari, S. (2023). Cultural Heritage and Indigenous Knowledge: Reviving Traditions for Future Generations. *Sustainable Development Goals in SAARC Countries: Key Issues, Opportunities and Challenges, 1*, 24-32. https://www.researchgate.net/profile/Santosh-Mane-2/publication/

Blank, D. A. (2021). Artiodactyl vocalization. *Neuroendocrine Regulation of Animal Vocalization*, 159-188. https://www.sciencedirect.com/science/article/pii/B9780128151600000062

Cakata, Z., & Ramose, M. B. (2023). When ukucelwa ukuzalwa becomes bride price: Spiritual meaning lost in translation. *African Identities*, *21*(3), 478–490. https://doi/abs/10.1080/14725843.2021.1940091

Chiteyi, L. I., Lusambili, K. M., & Ochieng, A. (2023). Assessing The Origins and Development of Bullfighting Culture among the Abakakamega of Western Kenya. *East African Journal of Arts and Social Sciences*, *6*(2), 80–100.

Conroy, A. B. (2022). Oxen: Status, Uses and Practices in the USA, Encouraging a Historic Tradition to Thrive. https://scholars.unh.edu/faculty_pubs/1523/

Dei, D. G. J. (2024). Strategies for capturing, managing, and sharing indigenous knowledge. *Information Development*, *02666669241248832*, •••. https://doi/abs/10.1177/02666669241248832

Dias, Á., Silva, G. M., Patuleia, M., & González-Rodríguez, M. R. (2023). Developing sustainable business models: Local knowledge acquisition and tourism lifestyle entrepreneurship. In *Knowledge Management, Organisational Learning and Sustainability in Tourism* (pp. 37-56). Routledge. https://www.taylorfrancis.com/chapters/edit/10.4324/9781032626512

Dvurechenskaya, N. D. (2022). Ceramic collections from the citadel of the Uzundara fortress have been dated as the first third of the 3rd century BCE. *Archaeological Research in Asia, 31*, 100394. https://www.sciencedirect.com/science/article/pii/S2352226722000459

Grandin, T., & Deesing, M. J. (2022). Genetics and behavior during handling, restraint, and herding. In *Genetics and the behavior of domestic animals* (pp. 131–181). Academic press., https://www.sciencedirect.com/science/article/pii/B9780323857529000032

Gruen, L. (2024). Tangled Manes, Tangled Chains. *The Georgic Mode in Twentieth-century American Literature: The Satisfactions of Soil and Sweat*, 137. https://books.google.com/books?hl=en&lr=&id=

Guo, L., Yan, F., Li, T., Yang, T., & Lu, Y. (2022). An automatic method for constructing machining process knowledge base from knowledge graph. *Robotics and Computer-integrated Manufacturing, 73*, 102222. https://www.sciencedirect.com/science/article/pii/S0736584521001058

Hadjimichael, D., & Tsoukas, H. (2019). Toward a better understanding of tacit knowledge in organizations: Taking stock and moving forward. *The Academy of Management Annals, 13*(2), 672–703. https://doi/abs/10.5465/annals.2017.0084

Hipolito, M. F. G. (2024). A Systematic Literature Review of the Significance of the School of Living Tradition in the Preservation of Cultural Heritage of the Indigenous Peoples. *International Journal of Multidisciplinary: Applied Business and Education Research, 5*(2), 605–614. https://www.ijmaberjournal.org/index.php/ijmaber/article/view/1401

Hombalmath, M. M., & Mathad, M. A. (2021, November). Design Optimization of Bullock Cart Yoke. [). IOP Publishing. https://iopscience.iop.org/article/10.1088/1742-6596/2070/1/012154/meta]. *Journal of Physics: Conference Series, 2070*(1), 012154.

Intezari, A., Namvar, M., & Taghinejad, R. (2022). Knowledge identity (KI): A determining factor in the effective use of analytics. *Knowledge Management Research and Practice, 20*(1), 14–33. https://doi/abs/10.1080/14778238.2021.1967213

Jia, J., Zhang, Y., & Saad, M. (2022). An approach to capturing and reusing tacit design knowledge using relational learning for knowledge graphs. *Advanced Engineering Informatics, 51*, 101505. https://www.sciencedirect.com/science/article/pii/S1474034621002548

Jin, X., Peng, B., Wu, Y., Liu, Y., Liu, J., Liang, D., . . . Hu, X. (2019). Knowledge distillation via route constrained optimization. In *Proceedings of the IEEE/CVF International Conference on Computer Vision* (pp. 1345-1354). https://openaccess.thecvf.com/content_ICCV_2019/html/Jin_Knowledge_

Kalle, R., & Kass, M. (2020). Local practice of cattle farming and ethnoveterinary medicine in Estonia: Case study of Saaremaa and Muhumaa. *Ethnoveterinary medicine: Present and future concepts*, 345-374. https://link.springer.com/chapter/10.1007/978-3-030-32270-0_15

Kovach, M. (2021). *Indigenous methodologies: Characteristics, conversations, and contexts*. University of Toronto press. https://books.google.com/books?hl=en&lr=&id=0V47EAA

Kumar, P., Madaan, S., & Bhargava, G. (2023, January). An Insight About Bell Metal Craft—Scopes of Improvement and Promotion. In *International Conference on Research into Design* (pp. 965–975). Springer Nature Singapore., https://link.springer.com/chapter/10.1007/978-981-99-0264-4_79

Kunduru, A. R., & Kandepu, R. (2023). Data archival methodology in enterprise resource planning applications (Oracle ERP, People soft). *Journal of Advances in Mathematics and Computer Science, 38*(9), 115–127. http://public.paper4promo.com/id/eprint/1106/

Landborn, A. (2019). Chapter Sixteen La Malagueña Y El Torero: Flamenco Dance and Bullfighting. *Transatlantic Malagueñas and Zapateados in Music, Song and Dance: Spaniards, Natives, Africans, Roma*, 294. https://books.google.com/books?hl=en&lr=&id=i0OeDwAAQBAJ&oi=

Lawton, B. (2023). *Various and Ingenious Machines. Volume Two: Manufacturing and Weapons Technology* (Vol. 6). Brill. https://books.google.com/books?hl=en&lr=&id=

Lopez, J. M. C., Canalejo, A. M. C., & Márquez, C. M. (2022). Perception of and motivation for attending bullfights: A comparative analysis of tourists and residents. *Cuadernos de Turismo*, (50), 333–353. https://revistas.um.es/turismo/article/view/542001

Lugmayr, A., Sutinen, E., Suhonen, J., Sedano, C. I., Hlavacs, H., & Montero, C. S. (2017). Serious storytelling–a first definition and review. *Multimedia Tools and Applications, 76*, 15707–15733. https://link.springer.com/article/

Mahdi, O. R., Nassar, I. A., & Almsafir, M. K. (2019). Knowledge management processes and sustainable competitive advantage: An empirical examination in private universities. *Journal of Business Research, 94*, 320–334. https://www.sciencedirect.com/science/article/pii/S0148296318300882

Mahuika, N. (2019). *Rethinking oral history and tradition: An Indigenous perspective.* Oxford University Press., https://books.google.com/books?hl=en&lr=&id=z5KwDwAAQBAJ&oi=

Makgabo, C., & Quintero, G. J. (2024). Philippine and South African Experiences on Folk Literature Research: Relevance, Gains, and Challenges. *Journal of Languages and Language Teaching*, *12*(2), 985–995. https://e-journal.undikma.ac.id/index.php/jollt/article/view/10667

Manesh, M. F., Pellegrini, M. M., Marzi, G., & Dabic, M. (2020). Knowledge management in the fourth industrial revolution: Mapping the literature and scoping future avenues. *IEEE Transactions on Engineering Management*, *68*(1), 289–300. https://ieeexplore.ieee.org/abstract/document/8964410/

Melka, T. L., Warkineh, T. Z., Gizaw, A. M., Bedada, Y. M., & Birhanu, E. T. (2022). How Can We Achieve Lifelong Learning Opportunities for All in Ethiopia? Making a Case for Intergenerational Approaches to Literacy and Learning. Policy Brief. *Online Submission.* https://eric.ed.gov/?id=ED647423

Miller, B. G. (2024). *Oral history on trial: Recognizing Aboriginal narratives in the courts.* UBC Press. https://books.google.com/books?hl=en&lr=&id=SjOvU2iJ6JwC&oi=fnd&pg=PP2

Miri, S. J., & Aryamanesh, S. (2022). Ritual as a Social Phenomenon in Ancient Near Eastern Societies. *International Journal of the Humanities*, *29*(4), 60–74. https://eijh.modares.ac.ir/article-27-59542-en.html

Mirzadeh, S. I., Farajtabar, M., Li, A., Levine, N., Matsukawa, A., & Ghasemzadeh, H. (2020, April). Improved knowledge distillation via teacher assistant.). https://aaai.org/ojs/index.php/AAAI/article/view/5963. *Proceedings of the AAAI Conference on Artificial Intelligence*, *34*(04), 5191–5198.

Musili, T. K. (2022). Ndwae ngone mwaitu: A Postmodern Cultural Phenomenon of Dowry among the Akamba and its influence on Spousal Violence. *Journal of International Women's Studies*, *24*(4), 11. https://vc.bridgew.edu/jiws/vol24/iss4/11/

Odulwa, K. K. K., Wafula, P., & Choti, C. (2021). The Pre-Colonial Socio-Economic Practices of the Khayo Community of Busia County in Western Kenya. https://www.academia.edu/download/90892691/ijsrp-p11725.pdf

Onwuatuegwu, I. N., & Paul-Mgbeafulike, V. S. (2023). African traditional educational framework: Unveiling the wisdom beyond western education. *Nnadiebube Journal ofPhilosophy, 6*(2). https://www.acjol.org/index.php/NJP/article/view/3675

Osypinska, M., Żurawski, B., Belka, Z., Osypinski, P., & Łopaciuk, R. (2022). "Oh, My Beloved, Great Bull!" An Interdisciplinary Approach to Cattle as a Marker of Economic and Political Status in Medieval Sub-Saharan Societies. *Journal of African Archaeology*, *21*(1), 46–62. https://brill.com/view/journals/jaa/21/1/article-p46_4.xml

Pallathadka, L. K., Pallathadka, H., & Devi, M. S. (2022). A Review of marriage rituals in different cultures. *Integrated Journal for Research in Arts and Humanities,* *2*(5), 152-160. https://www.academia.edu/download/102036846/234.pdf

Panopoulos, P., & Chatzopoulou, V. (2015). Homeland as Sound and Sound as Homeland: Cultural and Personal Soundscapes in Christos Christovasilis's Short Stories. By Panayotis Panopoulos. Translated by Vasiliki Chatzopoulou. *Ethnomusicology Translations*, (1), 1-19. https://scholarworks.iu.edu/journals/index.php/emt/article/view/20266

Parsons, W. (2004). Not just steering but weaving: Relevant knowledge and the craft of building policy capacity and coherence. *Australian Journal of Public Administration*, *63*(1), 43–57. https://onlinelibrary.wiley.com/doi/abs/10.1111/j.

Qobilova, E. (2024). The Place of Folk Songs in Isajan Sultan's Stories. *American Journal of Social Sciences and Humanity Research*, *4*(02), 44–49. https://theusajournals.com/index.php/ajsshr/article/view/2450

Rafi, N., Ahmed, A., Shafique, I., & Kalyar, M. N. (2022). Knowledge management capabilities and organizational agility as liaisons of business performance. *South Asian Journal of Business Studies*, *11*(4), 397–417. https://doi/10.1108/SAJBS-05-2020-0145/full/html

Raven, E., & Theuns-de Boer, G. (Eds.). (2021). *Reading Śiva: An Illustrated Selection from the ABIA Online Bibliography on the Arts and Material Culture of South and Southeast Asia* (Vol. 38). Brill., https://books.google.com/books?hl=en&lr=&id=CAxWEAAAQBAJ&oi=

RickShaw Travel Group. (2024). Bull fighting in western Kenya. https://rickshawtravels.com/about/articledetails/Bull-Fighting-in-Western-

Rokka, J. (2021). Consumer Culture Theory's future in marketing. *Journal of Marketing Theory and Practice*, *29*(1), 114–124.

SadanandaNayak. G. (2021). Historical Analysis on traditional Tools and Technology for Farming and Food Processing: A Study of South Odisha. *PalArch's Journal of Archaeology of Egypt/Egyptology, 18*(4), 1627-1635. https://archives.palarch.nl/index.php/jae/article/view/6540

Samarasinghe, K., & Nethsinghe, R. (2023). Social processes that shaped Sri Lankan Sinhala folk music and its characteristics based on the purposes of usage. http://repository.lib.vpa.ac.lk/handle/123456789/1836

Sanka, C. G., Adu-Gyamfi, S., & Marfo, C. O. (2020). Tales in the Paasaali Dirge: Structure and Moral Lessons from the Past. *Manusya: Journal of Humanities, 23*(1), 19-39. https://brill.com/view/journals/mnya/23/1/article-p19_19.xml

Sperveslage, G., & Heagy, T. C. (2023). A Tail's Tale: Narmer, the Catfish, and Bovine Symbolism. *The Journal of Egyptian Archaeology*, *109*(1-2), 65–75. https://doi/abs/10.1177/03075133231188144

Sundararajan, S. A. (2023). CONCEPTUAL STUDY ON INDIGENOUS KNOWLEDGE MANAGEMENT.https://isarpublisher.com/backend/public/assets/articles/

Taringa, N. T., & Museka, G. (2021). No to bride price/bride wealth, yes to Roora: Investigating the meaning, function and purpose of Roora as a ritual. *Lobola (Bride-wealth) in Contemporary Southern Africa: Implications for Gender Equality*, 13-27. https://link.springer.com/chapter/10.1007/978-3-030-59523-4_2

Teffera, T. (2021). The Nägarit in Past and Present Day Ethiopia. Drums and Drum Ensembles of the Silk Road. Edited by Xiao Mai & Gisa Jähnichen. 28–30. December 2020, Shanghai Conservatory of Music. Logos, Berlin, pp. 39-56. https://www.academia.edu/45644565/Timkehet_Teffera_2021_The_

Terzes, C. (2020). Musical tools of Greek and Roman antiquity. *A Companion to Ancient Greek and Roman Music*, 213-227. https://doi/abs/10.1002/9781119275510.ch16

Thornton, R. J. (2020). Therianthropes as human-animal relations: Contextualising rock art in anthropological perspective. *Anthropology Southern Africa*, *43*(3), 181–196. https://www.tandfonline.com/doi/abs/10.1080/23323256.2020.1773290?

Tulius, J. (2020). Lesson from the past, knowledge for the future: roles of human memories in earthquake and tsunami narratives in Mentawai, Indonesia. *Paradigma, Jurnal Kajian Budaya*. https://dr.ntu.edu.sg/handle/10356/165267

Turner, N. J., Cuerrier, A., & Joseph, L. (2022). Well grounded: Indigenous Peoples' knowledge, ethnobiology and sustainability. *People and Nature*, *4*(3), 627–651. https://doi/abs/10.1002/PAN3.10321

Van Laar, D. M. (2023). *Knowledge Management Implementation in US Army Headquarters: A Case Study* (Doctoral dissertation, Walden University). https://search.proquest.com/openview/1d5dddccf01d7e287b8a9913c

Viscogliosi, C., Asselin, H., Basile, S., Borwick, K., Couturier, Y., Drolet, M. J., & Levasseur, M. (2020). Importance of Indigenous elders' contributions to individual and community wellness: Results from a scoping review on social participation and intergenerational solidarity. *Canadian Journal of Public Health = Revue Canadienne de Santé Publique, 111*, 667–681. https://link.springer.com/article/10.17269/s41997-019- PMID: 32109314

Walker, W. H., & Berryman, J. (2023). Ritual closure: Rites de passage and apotropaic magic in an animate world. *Journal of Archaeological Method and Theory, 30*(2), 449–494. https://link.springer.com/article/10.1007/s10816-022-09565-7 PMID: 35757009

Wang, L., & Yoon, K. J. (2021). Knowledge distillation and student-teacher learning for visual intelligence: A review and new outlooks. *IEEE Transactions on Pattern Analysis and Machine Intelligence, 44*(6), 3048–3068. https://ieeexplore.ieee.org/abstract/document/9340578/ PMID: 33513099

Wang, Y., Li, H., Chau, L. P., & Kot, A. C. (2021, October). Embracing the dark knowledge: Domain generalization using regularized knowledge distillation. In *Proceedings of the 29th ACM International Conference on Multimedia* (pp. 2595-2604). https://doi/abs/10.1145/3474085.3475434

Wepukhulu, A. N., Matisi, M. N., & Ngome, S. (2023). Evolution of Beliefs and Practices of Bride Wealth and Their Role on Marriage among the Bukusu. https://www.topflightvalet.com/formatedpaper/1704025.pdf

Wiegers, K. (2021). *Software Development Pearls: Lessons from Fifty Years of Software Experience*. Addison-Wesley Professional. https://books.google.com/books?hl=en&lr=&id=

Woldeyohannes, T. (2020). Assessment of husbandry practices, production and reproductive performance of indigenous cattle in Hadiya zone, southern Ethiopia. *International Research Journal of Science and Technology, 1*(3), 177-198. https://www.researchgate.net/profile/Tariku-Woldeyohannes/

Yadav, R. (2025). The Intergenerational Transfer of Environmental Knowledge and Leadership. In *Generation Alpha and Next Generation Leadership for Greener Futures* (pp. 237–254). IGI Global Scientific Publishing., https://www.igi-global.com/chapter/the-intergenerational-

Yeoh, B. S. (2020). The global cultural city? Spatial imagineering and politics in the (multi) cultural marketplaces of South-east Asia. *Culture-Led Urban Regeneration*, 102-115. https://www.taylorfrancis.com/chapters/edit/10.4324/9781315878768-6/

Chapter 14

The Erosion of Traditional Marital Beliefs: A Consequence of Modern Influences

Dadirayi Mudzengerere
https://orcid.org/0009-0005-0810-7288
University of Zimbabwe, Harare, Zimbabwe

ABSTRACT

In many African cultures traditional marital beliefs have passed through generations, providing a sense of continuity and identity. Modern influences continue to erode these beliefs, leaving young people navigating the complexities of marriages without guidance of traditional wisdom. This chapter examines the erosion of traditional marital beliefs as a consequence of modern influences, focusing on how modern technologies amplify gender-based violence in marital relationships. Through desktop research, the study explores the shifting dynamics of respect and power within marriages, highlighting how digital platforms and contemporary societal norms contribute to the normalization of violence. It reveals that the intersection of traditional expectations and modern realities creates a complex environment where gender-based violence can thrive, often exacerbated by the anonymity and reach of online interactions. The findings underscore the urgent need for interventions that address these modern influences while respecting cultural contexts, aiming to foster healthier marital relationships.

DOI: 10.4018/979-8-3373-0204-1.ch014

1.0 INTRODUCTION

In an era defined by rapid technological growth, the landscape of romantic relationships has undergone a profound transformation, particularly among young generation. This transformation and the widespread adoption of technology has enabled new forms of connections, fundamentally changing the way people meet, interact, and form romantic relationships (Kakembo, 2024). According to Stewart and Linden (2022), social media such as online networking sites have been successfully integrated into everyday lives of almost every individual on the planet (Baloyi, 2022; Kakembo, 2024). The shift towards social media interactions has fostered a new era of relation-building, as it facilitates communication and connections mostly among young people (Kakembo, 2024; Stewart & Linden, 2022). Online platforms are re-shaping the way individuals initiate and develop romantic relationships, by passing conventional methods of courtship. Stewart and Linden (2022) say that, a few of the usual and extensively made use of social media sites platforms include Facebook, Twitter, WhatsApp, Snapchat, and several among others. These platforms resulted in a transformation of personal life and connection (Kakembo, 2024). Although these platforms have opened up new opportunities for social connection, they have also contributed to the erosion of traditional marital beliefs.

With the shifting cultural paradigms, and advanced globalization, the traditional beliefs surrounding marriages are increasingly being eroded by modern influences (Ainamon et al., 2022). The fabric of traditional beliefs that once guided societal norms and values in marriages is increasingly frayed by forces of modernity. Societal norms are what people believe "should" happen. These beliefs, often steeped in historical context and societal norms, have long served as the bedrock of marital relationships, guiding behaviors, expectations, and roles within the union (Ainamon et al., 2022; Ogoma, 2014). The profound shifts in communication, lifestyle, and cultural exchange have led to an unprecedented erosion of traditional beliefs that once served as cornerstones of identity, community, and moral guidance (Ogoma, 2014). As societies become increasingly intertwined with technology, the foundational beliefs that once guided interpersonal relationships are being changed, and in many cases, eroded. This erosion of foundational cultural tenets, particularly within indigenous and marginalized communities, is not merely a matter of cultural loss; it carries profound implications for interpersonal relationships and social dynamics.

In the Zimbabwean Shona culture, marriage has traditionally been a communal affair, involving not just the couple but also their families, and communities. According to Mawere and Mawere (2010), African traditional marriages are considered sacred and highly respected institutions, governed by traditional customs and rituals that emphasis the importance of family, community, and social harmony. Ainamon et al. (2022) assert that marriage in traditional African society is performed with great importance and dignity. Despite their importance, these traditional Shona marriages are facing increasing pressure from modernizing forces that threaten to erode their significance. The rises of social media, online interactions, and the pervasive influence of western culture have created new dynamics that often clash with age-old practices and values (Ainamon et al., 2022; Kakembo, 2024). As modern influences permeate daily life, the longstanding reverence for ancestral wisdom and communal practices are confronted with the stark realities of contemporary existence. Young people are now forming relationships and getting married without involving their families, which has created challenges in marriages (Ogoma, 2024). With the absence of communal involvement and support, young couples may find navigating the complexities of relationships without the wisdom that traditional structures provide. Lack of mediation from families can lead to a rise in unhealthy dynamics including divorce, violence and even more severe outcomes, such as murder (Ainamon et al., 2022). The most alarming consequences of this shift are the rise of gender-based violence within marital relationships, exacerbated by the pervasive reach of modern technologies. The very tools that promise connection and empowerment can also become instruments of harm, facilitating the normalization of violence and undermining the sanctity of traditional roles and beliefs (Stewart and Linden, 2022). By delving into these dynamics, this chapter seeks to explore the intricate relationship between modern influences and the decline of traditional beliefs, particularly in Shona indigenous societies, where cultural practices and spiritual values have been handed down through generations.

This chapter aims to shed light on the potential consequences of a society where traditional marital beliefs are increasingly sidelined in favour of modern, digital interactions. As the chapter embarks on this exploration, it will uncover the layers of vulnerability that modern technologies expose, particularly for women who may find themselves in precarious situations where traditional support systems are weakened. The chapter examines the ways in which the erosion of traditional marital beliefs is contributing to the breakdown of social cohesion and the rise of violence in marital relationships. By exploring the nuances of this phenomenon, the chapter aims to contribute to a deeper understanding of the challenges facing the institution of marriage in contemporary Africa.

1.2 CONCEPTUAL FRAMEWORK

In many African cultures, including the Shona people of Zimbabwe, traditional beliefs and cultural practices play a significant role in shaping marital relationships. These beliefs and cultural practices often serve as guiding frameworks that influence the dynamics of marital relationships, shaping expectations, communication styles, and conflict resolution strategies among couples (Baloyi, 2022; Stewart and Linden, 2022). The arrangement of marriages by elders has long been integral to shaping and strengthening marital relationships. Ainamon et al (2022) define marriage as a socially recognized institution that acknowledges both universally and socially the union between a man and a woman. Similarly, Ogoma (2014) define marriage as a socially or ritually recognized union or legal contract between spouses that establishes rights and obligations between them and their children, and between them and their in-laws. The commitment inherent in marriage can contribute to a stable environment for raising children, offering them emotional and financial support (Ainamon et al., 2022; Baloyi, 2022). This stable environment not only fosters healthy emotional development in children but also equips them with the resources they need to thrive academically and socially. However, the breakdown of this traditional family structure, largely due to modern influences has led to a shift on how children are raised and supported.

Modern influences including technological advancement have contributed to the erosion of these traditional marital beliefs. The adoption of social media platforms, online dating, and other digital technologies has led to significant shift away from traditional courtship and marriage practices (Kakembo, 2024; Stewart & Linden, 2022). Stewart and Linden (2022) say social media platforms encompasses a range of online platforms that facilitate interaction, content sharing, and collaboration, allowing users to connect and engage with others through networking. They further stated that, the popularity and growth of Social Networking Sites (SNS) has actually reinvented the way of interaction especially among young people. Within social media platforms, members connect by creating digital profiles that showcase information they are comfortable disclosing, enabling recognition and interaction within these online communities. As a result, these online interactions and connections are influencing young people's attitudes towards relationships, identity, and community, leading them to explore alternative forms of social connection and intimacy (Kakembo, 2024). They are increasingly forming romantic relationships and make life choices outside the traditional family and community structures. Montemayor (2019) in Stewart and Linden (2022) define a romantic relationship as a relationship that is based on mutual, intimacy, commitment and love. Akram and Stewart and Linden (2022) state that, romantic relationship encompasses feelings of closeness, sexual attraction, connectedness and decision to remain together. As youth and young adults seek deeper connections and prioritise convenience in their interactions, their approach to romantic relationships increasingly reflects a shift towards valuing emotional intimacy and sexual attraction over traditional cultural expectations. They are prioritising convenience and global connectivity over established cultural norms. This disconnection from traditional cultural values and practices, has contributed to an increase in gender-based violence. This paper raises questions on the validity of the assumption. The vital question is how do modern influences, particularly technological advancement, contribute to the erosion of traditional marital beliefs and the amplification of gender-based violence in marital relationships? What role does technology play in shaping perceptions of marriage and gender roles? Are traditional marital beliefs relevant today? Who wants to learn them and for what use? What is the general attitude of young people towards cultural beliefs? What are psychological impacts of modern marital dynamics on individuals, particularly women? What strategies can be implemented to prevent and support victims of gender-based violence (GBV)?

By examining the intersection of modern influence and traditional marital beliefs, it is important to explore nuanced dynamics that emerge in this digital age. There is a need to assess how the tools designed to enhance communication can facilitate aggression, undermining the very essence of partnerships and mutual respect. Through this exploration, stakeholders who include researchers and educators can raise awareness of the urgent need for a re-evaluation of both personal and societal approach to marriages. It is important to advocate for a framework that not only honours the sanctity of relationships but also promotes safety, equality, and mutual support.

1.3 METHODOLOGY

To navigate this study, document analysis or desktop research was be adopted. Cardno (2018) says documentary analysis is a qualitative systematic procedure for reviewing or evaluating documents both printed and electronic material. According to Cardno (2018) the advantage of document analysis to qualitative researchers is that it is straightforward, efficient, cost-effective and manageable. However, the researcher needs to be aware of some of the shortfalls of relying on documents as sources of evidence. Therefore to ensure a comprehensive understanding of online relationships, the researcher considered multiple sources of data, acknowledging the limitations of each, in order to triangulate findings and increase the validity of the study. The study analysed social media platforms, articles, academic journals and research papers, on online relationships. A comprehensive analysis of social media platforms and online forums was conducted, examining their structures, user interactions, and the types of content that are prevalent, in order to understand how these digital environments influence public discourse, shape social behaviours, and impact user relationships within various communities. The chapter is meant to examine the complexities of marital relationships today and identify pathways for fostering healthier, more equitable partnerships in the face of evolving societal norms.

1.4 UNDERSTANDING SHONA TRADITIONAL MARITAL BELIEFS

Traditional marital beliefs in the Shona culture encompass various practices and norms that historically define marriage (Mawere & Mawere, 2010). Among these traditional marital beliefs are the issue of the education of boys and girls, gender roles, the involvement of families and communities, and the paying of bride price. According to the Shona people, these are significant elements that illustrate the cultural framework surrounding marriages. The significance of these elements in Shona marriage culture is mirrored in other African societies, where marriage is often seen as a key aspect of cultural identity. Emelone (2020) in Baloyi (2022) says within the African context, marriage has always been viewed as an inseparable institution from cultural and traditional value systems. This is evident in the numerous rituals, ceremonies, and taboos surrounding marriages in the African cultures, which serve to reinforce its importance and sanctity. These cultural practices and traditions not only celebrate the union but also establish a framework that fosters mutual respect and commitment. The African traditional marriage system contained the proper ingredient to a healthy marital relationship. The emphasis was on communal and familial bonds, mutual respect, and shared responsibilities, which fostered a sense of unity and cooperation between spouses.

Marriage was not just a social contract but a profound institution that intertwines procreation, love and companionship, reflecting the values and beliefs of the Shona-speaking people. Mawere and Mawere (2010) state that, in the Shona culture as in some other African traditional cultures, marriage is underpinned by procreation, love and companionship. There are many African proverbs and sayings that celebrate the institution of marriage as "marriage is a garden that must be watered" or "good marriages are like good harvests". This further highlight the importance and value placed on marriages in African cultures. The proverbs and sayings highlighted the notion that a successful marriage union requires attention, care, and dedication from both partners. Baloyi (2022) says that, marriage and family remains the most valuable institution in the African context. The traditional values that were associated with marriages attract a lot of attention for they were meant to protect and preserve the institution. According to Ainamon et al (2022), the custom of marriage is a significant component in the African culture and is also one of the most thought about issues in African literary works. The works of different African writers often portray marriage as a symbol of cultural heritage, highlighting its significance in maintaining social order and continuity. For example many African authors examine the role of marriage in shaping identity, community, and social norms. This literally exploitation of marriage reflects the broader African cultural context, where socialization and education plays a crucial role in preparing individuals for their future roles. African cultures place a strong emphasis on socializing and educating young people to take on their future roles as spouses, parents and respected community leaders. The socialization process is essential for building strong and harmonious communities. Baloyi (2022) says that, through cultural dialogue, both men and women can be friends and enjoy the positive aspects afforded by African cultural values. This supports collaboration and understanding in marriages, allowing both genders to appreciate and embrace the strengths of African cultural values that promote friendship and mutual respect.

1.4.1 Education for future roles and community involvement

The concept of education for future roles is deeply ingrained in African societies, where young people are groomed from a tender age to take on specific roles and responsibilities. The education system is reflected in various initiation rites, training programs, and socialization processes. The education of boys and girls in the Shona culture has historically played a crucial role in shaping traditional marital beliefs. The education system is a holistic approach that encompasses social, emotional, and cultural development. The education was provided by the community in a collective manner as it took a village to raise a child. A study carried in Nigeria by Ogoma (2014) stated that, in the Yoruba culture efforts are made to train the child from birth on how to be responsible and relevant, not only to himself, family but also to the entire community. The study further highlighted that, a child was taught to be hardworking, to be moral, and above all, to manage his/her family in future. This education was not solely the responsibility of individual families but was provided by the community, emphasizing the important of social cohesion and shared values. Community members who include elders played a vital role in imparting these teachings, reinforcing cultural norms and expectations. The involvement of the whole community helped to instill a sense of responsibility and accountability among young people, as they were not only preparing for their personal futures but also contributing to the continuity of cultural practices. The collective nature of this education reflects the Shona belief that marriage is not just a union between two individuals but a significant alliance between families and communities. This approach not only reinforced individual roles within marriage, but also strengthened the cultural fabric of the community, ensuring the preservation of Shona values and practices.

The Shona culture encompasses various practices and norms that have historically defined marriages. In line with traditional gender roles, the education and socialization of Shona boys and girls were designed to equip them with relevant knowledge and skills to fulfill their respective roles in societies. According to Fafunwa and Aisiku (2022) the education received was not the same for girls and boys as they were socialized to assume roles appropriate for their sexes. The education shaped the dynamics of marital relationships, influencing how individuals approach their roles as couples. Roles within the domestic sphere of the home and in the community in African traditional societies are divided along gender lines. The division of roles influences the dynamics of family life and community interactions, reinforcing societal norms. While these traditional divisions of labour may shape expectations within marriages, there is a need for a more equitable approach that allows both partners to thrive. Baloyi (2022) states that, the institution of marriage ought to provide space for partners to have total freedom to act to the maximum of their abilities. This freedom was often cultivated through the socialization of boys and girls into their respective roles. Boys were taught to be good husbands, focusing on responsibilities such as providing for the family, managing livestock and making decisions that would benefit the household. In the Shona culture, the concept of manhood was deeply intertwined with the institution of marriage, influencing how men perceive and experience their roles and responsibilities. Baloyi (2022) says men's roles include disciplinary, home technology expert, and ultimate decision maker in the household. Therefore, boys undergo initiation stages where education about manhood is imparted by elderly men, some of whom are regarded as custodians of culture (Baloyi, 2022). Boys were taught to control their emotions as a man who shows emotions is usually regarded as weak and not part of the larger group of real African men. Baloyi (2022) states that, the interpretations of what it means to be a married woman were explained within the confines of sociological gender construct. Girls were educated to become good wives, by learning skills that would prepare them for their future roles in the household. This included lessons in cooking, child-rearing, and maintaining family harmony. They were taught to be loving, nurturing, self-sacrifice, home-making and availability. Women were also expected to engage in subsistence farming and food processing. The education of boys and girls fostered a harmonious and mutually supportive partnership that is essential to the success and stability of the tradition Shona family.

1.4.2 Courtship and paying of bride price (lobola)

In many African cultures, the journey towards marriage is marked by intricate traditions and customs that reflect the values and beliefs of the community. In the Shona culture, the issue of courtship and paying of lobola played a vital role in shaping the social, cultural, and economic fabric of the society. Courtship and marriage are intricately woven processes that involve not only the union of two individuals, but also the bonding of two families. Ainamon et al (2022) state that, courtships served as vital period of acquaintance allowing the prospective bride and groom to foster a deeper understanding of each other and their respective families. The courtship process involved not only the bride and the groom but also their parents, who sought to ensure that their children were marrying into families with good morals and free from harmful spiritual influences. The assessment was crucial in determining the suitability of the partner and the potential success of the marriage. A study by Ogoma (2014) revealed that, in the olden days, two people meet two times and got married on the third time and these marriages lasted a lifetime. In those few days, partners would scrutinize their partner's conducts, asking friends and neighbours about their morals, while also investigating their family history to ensure that they were not from a lineage believed to be cursed by bad spirits. Courtship was viewed as a critical phase as choosing the right partner was the key to a successful and harmonious marriage. An important aspect of the courtship process was the discussion and agreement on the bride price, a critical custom that established the groom's commitment to his new family.

The paying of bride price is a significant factor that illustrates the cultural framework surrounding marriages in Africa. Traditional African societies have consistently practiced some form of payment or exchange during marriage ceremonies. Ainamon et al, (2022) say the paying of bride price are interlinked and form an important part in the lives of African men and women from whatever ethnic groups they come from. The payment of bride price is deeply rooted in traditional marriage customs transcending ethnic boundaries. Similarly, among the Shona-speaking people, just in other ethnic groups, the payment of bride price was a crucial aspect of the marriage process. Mawere and Mawere (2010) postulate that, the normative marriage customs of the Shona-speaking people are characterized by the negotiation and payment of lobola, or bride wealth (kukumbira). It was a symbol of acknowledgement of the groom to the bride's parents to show appreciation of allowing him and his family to have their daughter. It showed respect and gratitude to the bride's family for raising their daughter. In the Shona culture, the amount of bride price was negotiable and dependent on the groom's standing as adjustments were made based on individual circumstances. Mawere and Mawere (2010) posit that, a boy and a girl were and are supposed to have sex after the payment of lobola. Sex was considered a sacred act that should only occur within the context of marriage, ensuring spiritual purity and blessings. Sharing the same view, Ainamon et al (2022) postulate that, the culture of paying bride price before marriage was of great importance in the traditional African set up. In many African societies, the payment of bride price is considered an essential precursor to marriage as it carries significant social, cultural, and spiritual implications. This emphasis on bride price as a cornerstone of marriages is equally reflected in the customs of the Shona people, who have long viewed it as an important factor that reinforces marital stability and responsibility. Bere-Chikara in Mawere and Mawere (2010) says the Shona people considered lobola or bride price as a noble custom that functioned as a safeguard against marital dissolution because it generally needs to be repaid upon divorce. The payment of lobola and its return after divorce served as a strong incentive for both families, particularly the bride's family who knew they would have to refund the bride price if their daughter's behavior led to divorce. While these beliefs hold cultural significance, they are increasingly being scrutinized and re-evaluated by younger generations seeking more equitable and modern approaches to marriage.

1.5 YOUTHS PERSPECTIVES ON TRADITIONAL MARITAL BELIEFS

The perspectives of youth regarding traditional marital beliefs in the Shona culture are increasingly complex and nuanced. Most young people see it as an outdated and oppressive custom that reinforces patriarchal values. They are navigating the tension between cultural heritage and modern values, often leading to a redefinition of what marriage means to them. According to Mawere and Mawere (2010), African traditional forms of marriage (tsika dzekuroorana, in Shona), is one part of the traditional culture that has been dramatically affected under the guise of modernity. Many young people view certain aspects of these beliefs as outdated or "backward," in the context of modern societal changes, globalization, and shifting gender norms. For example, they advocate for relationships that prioritize mutual respect and support, challenging traditional roles that have historically defined marital dynamics. Ogoma (2014) says westernization, modernization and Christianity have all affected the marriage system in Africa. These influences have led to the blending of traditional practices, causing evolving of marriage customs that reflect both local heritage and global trends. Many young people are drawn to more egalitarian relationships models that contrast sharply with traditional Shona customs. The influence of modernization has brought new marriage concepts such as homosexual relationships which are not accepted in the African cultures. A homosexual relationship is a romantic or sexual relationship between people of the same sex. Younger generations embracing in homosexuality remain closeted, fearing rejection by family and community (Kaylo, 2012). This shows that, modernity and globalization are creating tension with African traditional marriage practices. Therefore, it is difficult to preserve African traditional marriage practices in the face of modernization and globalization.

In the pursuit of equal rights, young people are challenging the traditional gender roles that dictate specific responsibilities for men and women within marriages. In today's cultural context, women are encouraged to view themselves as equals to men, with a focus of mutual support and shared participation of gender roles. Emboldened by the growing momentum of the equality movement, young Africans especially women are increasingly challenging traditional gender roles and expectations. Women are demanding a more nuanced understanding of masculinity and femininity that prioritizes mutual respect and equal rights. They are now advocating for a more inclusive understanding of identity and partnership that transcends outdated societal norms and foster a more equitable and diverse framework for relationships. As both men and women are now sharing financial responsibilities within families, the notion of equal rights is becoming more pressing. Women are also highlighting the need to share domestic duties, such as cooking and washing clothes, in the same way that they are sharing financial responsibilities. Despite the growing emphasis on equal rights, many men still resist the notion of sharing domestic roles in marriage. They are clinging to traditional patriarchal attitudes and expectations, which can create tension, conflict, and significant challenges in modern marriages. This resistance to equal rights in marriages is particularly pronounced in situations where men can feel threatened by the shifting dynamics of power and responsibility. According to Ogoma (2014), nowadays, social media is encouraging women to expect and receive loving treatment from their husbands, who are reminded to be nurturing and supportive, rather than being harsh or domineering. Modern societal values emphasize the importance of equal partnership in marriage, with men being encouraged to adopt more nurturing and empathetic roles, prioritizing their wives' emotional wellbeing. Women are increasingly asserting their independence, autonomy, and expectations for more empathetic and supportive partnerships. Ogoma (2014) states that, technology never taught women that they should submit to their own husbands. It has equipped women with knowledge that empower them to assert their independence and equality, rather than promoting the idea of submission to their spouses. Ultimately, the quest for equal rights reflects a broader shift towards a more inclusive and equitable understanding of marriage, that values mutual respect, and partnership.

The paying of bride price (lobola) is another issue being often viewed critically by the youth. Most young people believe that, the paying of bride price is a dehumanizing and paternalistic practice. Mawere and Mawere (2010) say most young people believe that the paying of bride price is now valueless and a thing of the past. Some perceive the paying of bride price as a transactional aspect of marriage that devalues women and reinforces patriarchal structures. They believe that the paying of bride price reduces women to mere commodities, contradicting modern values of gender equality, mutual respect, and individual freedom. However, older generation believes that the failure of young people to pay the bride price is a sign of laziness. To them, paying bride price is a significant responsibility and crucial step in demonstrating a man's commitment to marriage and family. By failing to pay bride price, young people are seen as shirking this responsibility, which is perceived as lack of initiative. In this view, paying bride price is a way of demonstrating one's willingness to work hard, take responsibility, and make sacrifices for the sake of family and relationships. Apart from paying bride prices, young people are also valuing individual autonomy in choosing their life partners. The traditional practice of arranged marriages is often viewed as limiting personal freedom and choice. They are now opting for romantic relationships based on personal choice, compatibility, and emotional connection. The rise of social media, education, and urbanization has exposed young people to diverse perspectives and values, eroding the appeal of arranged marriages and fuelling a desire for more personalized and fulfilling relationships.

1.6 THE SHIFT TO MODERN MARRIAGES

In recent years, dating has transformed dramatically with the rise of technology and the internet which have changed on how people meet and form relationships. The proliferation of dating apps and social media platforms has revolutionized the way people meet, interact, and form romantic connections, fundamentally changing the landscape of modern dating. Stewart and Linden (2022) say social media use and western influence has eroded our African morals and beliefs towards marriages. They have contributed to the erosion of African morals surrounding marriage leading to a shift in societal norms and expectations. Sharing the same view, Ainamon et al (2022) say African marriage system has continuously undergone changes for generations. The advent of the digital age has revolutionized how individuals connect, interact and form romantic relationships. They have shifted away from the traditional African values, such as respect for elders, communal involvement, and spiritual guidance in marriage, to a more individualistic attitude towards relationships and matrimony. The proliferation of online dating platforms is offering unparalleled opportunities for individuals to meet and interact with potential partners. Exposure to different cultures through media, education, and travel has influenced individuals' perceptions of marriage. Dating apps such as Facebook, Bumble, and several others are helping people connect across geographical boundaries, increasing their chances of finding a compatible match. Individuals are meeting potential partners from diverse backgrounds, engage in conversations, and build connections without the traditional barriers that often accompany in-person interactions. The exposure to different cultures fosters a desire for relationships based on personal choice and emotional connection rather than societal expectations or familial arrangements. Social media platforms showcase alternative forms of relationships, marriages, and family structures, potentially inspiring individuals to question and challenge traditional norms and expectations. These influences are contributing to a more nuanced and inclusive understanding of what it means to build a life with someone.

The generational shift indicates a broader trend towards rethinking cultural traditions in light of modern societal changes. Young people are no longer meeting potential partners through traditional avenues like mutual friends, or social gatherings; instead they are relying on social media and mobile apps. The digital landscape fosters a sense of autonomy and choice, enabling individuals to explore relationships on their own terms. In recent years, there has been a significant transformation on how young people approach relationships and marriages. This shift is highly influenced by the rise of digital communication platforms, which have fundamentally changed the landscape of romantic interactions. Social media allows a level of connectivity and interaction that was previously unimaginable. Young people are meeting potential partners from diverse backgrounds, engage in conversations, and build connections without the traditional barriers that often accompany in-person interactions. This digital landscape fosters a sense of autonomy and choice, enabling individuals to explore relationships on their own terms. As a result of this shift, the involvement of elders in the matchmaking process has significantly decreased. Ogoma (2014) postulates that, the traditional place of investigation or courtship, which played a significant role in the success and stability of marriages in the past, has been lost. The traditional space for courtship has been replaced by social media, leading to a loss of meaningful, in-person interactions and relationships that were once essential to build strong, lasting marriages.

1.7 DANGERS OF ONLINE DATING AND SOCIAL MEDIA PLATFORMS

With the increase of digital spaces to find love, companionship, and connection, it is essential to critically examine the complexities and challenges of dating in the digital era. It is important to understand the nuances of digital dating, in order to develop effective strategies to mitigate its risks and harness its benefits. The anonymity and lack of familiarity inherent in online dating can lead to misunderstanding and conflicts, especially when individuals come from diverse cultural backgrounds and have not established a solid foundation of trust and communication. In this evolving landscape, understanding the impact of technology on human relationships, including issues of authenticity, communication barriers, and emotional well-being is crucial. While online dating may offer conveniences and increased opportunities for meeting potential partners, research indicated that its challenges may ultimately outweigh its benefits. Online dating platforms can pose significant risks, including human trafficking, cat fishing, romance scams, identity theft, online harassment, revenge porn, stalking, and financial scams. Scammers may create fake profiles, manipulate emotions, or propose fake relationships to gain financial benefits or personal information. Akram and Albalawi (2016) cited in Stewart and Linden (2022) state that, social media and online dating sites have facilitated fake lives as most youth just copy what they see in social media without accessing the challenges. Through tailored profiles, edited photos, and fabricated stories, social media platforms can present unrealistic and unattainable standards of relationships, love, and happiness. Individuals may misrepresent themselves which can lead to physical and emotional harm when the truth is revealed. This is leading many young people to emulate these fake representations without considering the real-life complexities and challenges that accompany genuine relationship. Additionally, young people are fabricating false narratives that deceive and misled others, putting them at risks of harm. In most cases, there are false stories shared on social media platforms which can lead others into danger, highlighting the need for critical thinking and media literacy. There are a lot of fake stories shared on social media that are negatively influencing other people's relationships; there is the:

- Sharing of overly romanticized portrayals of marriage, creating unrealistic expectations and potentially leading to disappointment or dissatisfaction.
- Presentation of divorce or separation as an easy or desirable solution, potentially encouraging others to give up on their own relationships.
- Sharing of content that normalizes abusive, controlling, or toxic behavior in marriages.
- Offering of misleading advice on marriage-related issues, such as conflict resolution, or intimacy
- Comparison and competition between relationships, potentially fostering feelings of insecurity.
- Over sharing of unrealistic marriage goals, such as perfect relationships, or an ideal partner.
- Presentation of infidelity or cheating as acceptable justifiable, potentially normalizing harmful behavior.
- Sharing of negative views on marriages, potentially discouraging others.

This highlights the importance of responsible social media use, and the need to promote healthy, realistic, and respectful portrayals of marriage and relationships. Therefore, young people need to be mindful to the content they share and consume on social media. By so doing, they can create a more positive and supportive online community that celebrates the diversity of human experience.

When venturing into online dating, young people are potentially exposing themselves to a multitude of dangers which can have long-lasting and devastating consequences. While online platforms offer opportunities for connection, they also introduce significant risks, particularly the issue of human trafficking which causes emotional stress to families. Social media platforms can facilitate international connections, which can increase the risk of human trafficking, particularly in cases where victims are lured to other countries with false promises of love or employment. According to United Nations (2014), human trafficking is a process of people being recruited in their community and country of origin and transported to the destination where they are being exploited for purposes of forced labour, prostitution, domestic servitude, and other forms of exploitation. Human trafficking involves the use of force, fraud, or coercion to exploit people to labour or commercial sex. Online dating increases the likelihood of interacting with strangers, which can heighten the risk of human trafficking, particularly for women. Human traffickers can create fake profiles, using stolen photos and fabricated information, to lure victims into romantic relationships or other situations that may lead to exploitation. Traffickers may use social media influencers or popular accounts to promote their fake opportunities or to recruit victims. They can employ different tactics to groom victims, gradually building trust and establishing a seemingly genuine relationship through online conversations, phone calls, and other forms of digital communication platforms. This will help traffickers gain the victim's confidence, making it more likely that the victim will share personal information, send money, or agree to meet in person, ultimately paving the way for exploitation and potential human trafficking. Some online dating platforms may lack robust moderation and reporting mechanisms, making it difficult to detect and prevent human trafficking activities. The myriad risks and negative consequences associated with online dating include cat fishing, harassment and violence. These dating platforms often lack robust mechanisms for reporting and addressing abusive behavior and violence. The dating apps can also lead to cyber stalking and online harassment, which can serious consequences for mental health and physical safety. Online dating can create unrealistic expectations which can lead to disappointment and dissatisfaction in marriage.

In many communities, there is a growing sense of unease and suspicion. There is an increase concern that some online dating are not for love or companionship, but for sinister and ritualistic purposes. The platforms have perfected the art of fostering superficial connections, leading to an overwhelming number of fake friends. According to Stewart and Linden (2022), social media has perfected the arts of having too many fake friends, cyber bullying, and an elevated risk of suicidal thoughts. There is growing concern that social media platforms may be contributing to an increase in suicidal ideation, attempts, and completions, particularly among vulnerable populations such as youth and young adults. Social media is facilitating and contributing to an elevated risk of suicidal thought, as users often grapple with the pressure of online validation and the negative impact of toxic interactions.

1.8 INCREASES IN GENDER-BASED VIOLENCE

The alarming rates of gender-based violence (GBV) within marriages are emerging as a pressing social issue. Gender-based violence encompasses a range of harmful behaviours, including physical violence, sexual violence, emotional or psychological abuse, economic abuse, and cultural or social violence. These forms of violence can manifest in various forms and impact individuals differently. The pervasive influence of social media on modern relationships is, in many cases, proving to be a corrosive force, eroding trust, fueling comparisons, and dissatisfaction. This is undermining the very bonds that couples are striving to strengthen, eroding the trust, intimacy, and communication that are essential for building healthy relationships. Stewart and Linden (2022) highlight that, the use of social media platforms has led to misunderstanding and destructed relationships among youth in societies. It has been linked to increased relationship conflicts and breakdowns among individuals, potentially undermining social cohesion and community bonds. A research study by Stewart and Linden (2022) showed that the use of social media has sometimes led to misunderstanding and destructed relationships among individuals in the society. Such feelings can escalate into violence, as individuals react to perceived threats to their relationships. When trust in relationships is broken, the emotional fallout can lead to aggressive behaviours, particularly among those who feel wronged or humiliated. In extreme cases, this can result in violent confrontations, and tragically, even, murder. Young individuals are engaging in relationships without fully understanding their partners' background, behaviour or intentions. When people enter into relationships without a solid foundation of trust and knowledge about each other, the potential for conflict and violence escalates. As individuals form relationships online, the lack of a solid foundation of trust and knowledge about each other can be exacerbated by the culture of social media platforms. The culture fostered by many social media and dating platforms can inadvertently encourage and influence infidelity. The ease of connecting with others online can facilitate cheating in marriages and relationships, leading to trust issues and conflicts. Infidelity has also emerged as a significant trigger for violence, with some case of marital murder being linked to suspicious or discoveries of cheating. Social media is influencing Individuals to resort to violence, including murder, as a way of resolving conflicts with their spouses. Even in marriages that began with traditional ceremonies, the discovery of a partner's infidelity – often facilitated by online connections can spark violence. The behaviour not only undermines the trust within existing relationships but can also provoke intense emotional responses, including jealousy and betrayal. This highlights the evolving nature of relationships and the challenges of adapting traditional values to modern realities. Stewart and Linden (2022) state that, social media has some websites which have pornographic videos that influence young people's mind and make them immoral. Watching phonographic videos can distort expectations and foster unrealistic comparisons, leading some spouses to measure their partner's sexual performance. When reality fails to live up to these expectations, some individuals may seek fulfillment elsewhere through online connections that promise excitement and novelty.

Apart from violence and murder, infidelity is increasing the likelihood of divorce due to factors such as unrealistic expectations, lack of compatibility, and insufficient relationship skills developed during the dating phase. The ease of online communication is increasing the risk of infidelity, which is a common predictor of divorce. Research has shown that most couples who meet online may have lower relationship quality and lower levels of commitment compared to couples who meet through traditional means. According to www.africanmarriage.info cited in Ogoma (2014), today's couples date for three years and get divorced six months after finally getting married. Mambure (2024) define divorce as an official separation of spouses, initiated by one partner or affected by the judgment or decree of a court, or by agreement of both spouses. Mambure (2024) says divorce rates in Zimbabwe have been steadily increasing over the past few decades sparking concerns about the stability of marriages and the welfare of families. Lack or poor communication was regarded as a major contributing factor that causes divorce. Many couples are increasingly finding themselves emotionally disengaged from each other as their attention and affection are on social media platforms. A growing trend is emerging where couples are becoming more enamored with their social media lives than with each other, leading to a decline in emotional and physical intimacy. This is also contributing to the rising rates of divorce. The evolution signifies a move towards egalitarian relationships, where both partners are seen as equal and love is prioritized over economic considerations. Stewart and Linden (2022) posit that, lack of physical commitment and laziness has increasingly become the norm among youth. They further stated that, young people spend most of their time on social media platforms wasting special time. As individuals become increasingly absorbed in their online lives, many are neglecting the fundamental responsibilities, including providing for their families, threatening the very foundations of their relationships and household stability. Excessive social media use is increasingly linked to decreased productivity and motivation among young people, as the significant amount of time spent on platforms can distract from academic, professional, and personal goals. A study by Mambure (2024) revealed that, another contributing factor of divorces specifically in Zimbabwe is economic hardships. The strain of economic hardship, exacerbated by the distractions and decreased productivity associated with excessive social media use, is contributing to increased marital stress and tension, ultimately leading to a rise in divorce rates. The study further highlighted that as a survival strategy, both young men and women end up looking for partners on different online platforms who are financially stable. Couples may lack social support and community ties that can help strengthen marriage. However, some couples who meet online may have happy, healthy, and long-lasting marriages because the success of a marriage depends on communication, commitment, and compatibility (Mambure, 2024).

1.9 THE ISSUE OF AVENGING SPIRITS "NGOZI"

In the Shona culture like other cultures, avenging spirits (ngozi) reflects deep-seated beliefs about justice and morality. Ngozi is the spirit of a deceased person, particularly one who has been wronged or died under tragic circumstances (Musanga, 2017). The spirit seeks justice or retribution for its death, often leading to disturbances in the lives of the living, especially those connected to the deceased. Traditionally, the communal emphasis on harmony, respect, and ancestral reverence instilled deep fear of engaging in violent acts, including suicide and murder. Suicide and murder was often viewed as a taboo that could provoke ancestral wrath and communal ostracism. African marriages placed a strong emphasis on community involvement and conflict resolution. Spouses sought counseling and intervention from respected elders, who would mediate disputes and work towards restoring balance and harmony within relationships. Young people were often cautioned about the grave consequences of taking another's life. Elders shared tales of the vengeful spirits known as 'Ngozi,' in Shona. Avenging spirits were believed to seek retribution for wrongs committed, particularly in cases of murder or betrayal. They were believed to haunt and punish those responsible for unjust deaths, serving as a powerful deterrent against violence and murder. These spirits are commonly believed to be manifestations of individuals who have suffered unjustly or have been wronged in life. According to Goredema (2016) in Musanga (2017), if a person's life is deliberately taken away through cold-blooded murder, (kupondwa, in Shona), and then the person's spirit will come back fighting for justice by haunting the offender or his/her family. The spirit of the person who has been brutally murdered is thought to linger on, seeking retribution and justice by haunting the perpetrator or their families. The avenging spirits bring suffering, misfortune, and even death to those responsible for the unjust killings. This belief acted as a powerful deterrent, instilling fear in individuals who might consider committing acts of violence. The Shona's belief in 'Ngozi' underscores a community's understanding of retribution and the moral order of the universe. They suggested that unresolved injustices can lead to spiritual unrest and societal imbalance. The belief in avenging spirits serves as a poignant reminder of the importance of addressing historical injustices, promoting cultural sensitivity, and fostering a culture of reconciliation and healing.

These beliefs shape community responses to violence, influence societal norms, and reflect the ongoing struggle for justice in the face of tragedy.

High rates of murder cases can amplify the presence and impact of avenging spirits (Ngozi) within communities, as the souls of the deceased may seek vengeance for their untimely demise. The Shona people believe that the phenomenon manifests in various ways, from cultural rituals intended to appease these spirits to social behaviours aimed at avoiding their wrath. Musanga (2017) says in Shona cosmology, once "ngozi" is committed, it has to be recompensed in order to restore justice to occur between individuals, families, and the whole community. The Shona believe that legal systems alone may not suffice in addressing the complexities of violence and retribution. Goredema (2016) cited in Musanga (2017) states that, appeasing "ngozi" is a way to recompense the victim and his/her family and ensure that restoration justice, peace, and reconciliation take place between individuals, families and the society in general. Appeasing avenging spirits facilitate a process of restorative justice, providing a measure of reparation and closure to the victims and their loved ones, while also promoting healing, peace and reconciliation. The Shona people used several ways to appease the spirits such as, rituals and offerings, seeking guidance from traditional healers, and restitution which was highly accepted by the deceased family. Restitution involves compensating the family of the deceased or performing ritual acts as goodwill to restore balance. Restitution in Shona culture often involved compensating the decease's family by paying a certain number of cattle or offering a girl from the perpetrator's family as a replacement for the lost life. The Shona people believed that engaging in those practices brought peace to both the spirit of the deceased and the living community. However, the practice of offering a girl from the perpetrator's family to the deceased's family is no longer common, as it now contravenes the law. Nowadays, people who commit murder face legal consequences, typically resulting in imprisonment for a specific number of years under hard labour.

Many young people today seem less influenced by these cultural beliefs. The allure of modern lifestyles, combined with a growing disconnection from traditional values, has led to a troubling trend where individuals no longer hesitate to commit acts of brutality, including murder. In today's society, there is a troubling phenomenon where some young individuals resort to violence, particularly in cases of infidelity or cheating. When a man discovers that his wife or spouses is cheating, the extreme reaction after harming her and subsequently sharing it on social media is deeply disturbing. They show a disturbing lack of remorse after the killings as they can even share the pictures of the brutal murder on social media platforms. Instead of facing condemnation, these actions can sometimes be applauded and glorified by certain online communities, especially when the victim is labeled negatively such as a prostitute. This not only perpetuates a cycle of violence but also raises alarming questions about societal values and the impact of social media in normalizing such behavior. This is raising serious concerns about societal values and accountability. After committing acts of violence, many young offenders do not face their actions. Instead, they flee seeking refuge in other countries. Musanga (2017) states that, engaging in transnational migration as a way of escaping "ngozi" is futile, as the only solution is to recompense the victim and his or her family and ensure that restorative justice takes place between individuals, families, and society in general. Young people need to be informed that transnational migration is a strategic means of escaping imprisonment not avenging spirits, as appeasing 'Ngozi' through traditional rituals and practices is the only effective way to address historical injustices and promotes healing.

1.10 STRATEGIES FOR PREVENTION AND SUPPORT

The prevention of gender-based (GBV) violence requires a multifaceted approach that respects traditional marital beliefs while embracing modern values of equality and mutual respect. This involves addressing societal norms, improving legal systems, and empowering individuals. There are several strategies to present GBV and support the victims. There is a need to implement education and awareness programs that teach communities, especially young people about healthy relationships, consent, boundaries and respect to break the cycles of violence. The programs should cover topics like gender equality, sexual orientation, and emotional intelligence. Public awareness campaigns are also needed to challenge harmful gender stereotypes and raise awareness about the impacts of GBV. Campaigns should involve young people and also encourage bystanders to intervene, support survivors, and educate communities on legal rights and services available.

Governments need to enforce laws that criminalize all forms of gender-based violence, sexual harassment, trafficking, among several other forms of violence. Survivors of violence need to be protected and perpetrators should be accountable. Law enforcements and judicial systems need to be trained to handle cases of gender-based violence with sensitivity, ensuring that survivors feel safe to report any form of violence and those cases are processed swiftly and fairly. Women and girls need to be empowered through education, economic opportunities, and leadership roles. This will help reduce their vulnerability to GBV. There is a need to challenge gender roles and expectations that perpetuate inequalities. This can involve promoting gender-neutral parenting, encouraging men and boys to take active roles in care giving. Men and boys need to be involved in programs that promote gender equality and ending violence. They should be encouraged to reject violence, practice non-violent communication, and intervene in situations of potential abuse.

Communities need to be supported and strengthened to play a role in preventing and responding to GBV. Different community programs that raise awareness, foster healthy relationships skills, and create the safety of vulnerable individuals. Religious and community leaders can be influential in shifting societal attitudes towards GBV. Communities need to support each other and intervene in potential cases of GBV. Different stakeholders such as governments and Non-Governmental Organisations (NGO) need to establish hotlines, shelters, and counseling services to support survivors as they need immediate access to medical, legal, and psychological support. Professionals who work with survivors should be trained to provide trauma-informed care. This will respect the survivors' experiences and promotes their healing and recovery.

Young people can be trained as peer educators who can challenge harmful behaviours within their own social circles. Peer groups can offer a safe space to discuss issues of gender violence and offer support to those affected. Data need to be collected on gender-based violence to understand its prevalence, impact, and patterns. The collected data can guide interventions, track progress, and ensure accountability. There is also a need to evaluate violence prevention programs and interventions in order to assess their effectiveness and make necessary improvements. Education about digital safety is critical as many forms of GBV now take place online. Cyber bulling and revenge porn are examples of forms of online GBV. Consent in the digital space, and the legal repercussions of online abuse can empower people to protect themselves and others. Governments need to provide mobile applications designed to help survivors of GBV to quickly access support and report incidents.

International organizations need to raise global awareness about gender-based violence (GBV), advocate for legal reforms, and push for an end to harmful cultural practices like child marriages. International cooperation is necessary to combat human trafficking, support refugee women, and ensure the protection of marginalized groups. Governments and organizations should collaborate to end human trafficking by improving boarder security, providing education on the signs of trafficking, and supporting survivors of trafficking with rehabilitation services.

Preventing gender-based violence requires a comprehensive, coordinated approach that involves individuals, communities, governments, and global organizations. A comprehensive approach is crucial, requiring the integration of legislative reforms, community engagement, education, and support services, as well as the active participation of everyone to promote gender equality and challenging harmful masculinities. By integrating educational initiatives, utilizing online platforms for awareness and support, and fostering community engagement, technology can create a safer environment for all individuals. These strategies not only empower individuals to recognize and address GBV but also promote a culture of respect and understanding that can lead to healthier relationships in the modern era. The efforts to end GBV must be sustained and inclusive, ensuring that all voices are heard and supported.

1.11 RECONSTRUCTING MARITAL BELIEFS FOR THE MODERN ERA

In today's world, the institution of marriage faces unprecedented challenges and transformations. The traditional marital beliefs that once served as a foundation for strong, healthy marriages are no longer sufficient on their own. The erosion of these beliefs, coupled with the challenges posed by modern influences, has left many couples struggling to find their footing. As traditional notions of marriage grapple with contemporary values, it is important to explore the intersection of long-held beliefs and modern realities. Factors such as technological advancements, shifting gender roles, and increasing cultural diversity influence people's understanding of partnership. The rapid evolution of technology, changing gender dynamics, and growing cultural diversity are collectively reshaping traditional notions of partnerships. This may lead to a more nuanced and inclusive understanding of relationships that values mutual respect, trust, and communication.

Rather than abandoning the wisdom of traditional marital beliefs altogether, it is crucial to propose a reconstruction of these beliefs for the modern era. Reconstructing marital beliefs for the modern era is essential for fostering healthy fulfilling relationships. By emphasizing equality, personal fulfillment, open communication, and inclusivity, couples can navigate the complexities of contemporary life while nurturing their connection. As societies continues to evolve, so to must our understanding of marriage, ensuring that it reflects the values and needs of individuals in the modern era. This reconstruction not only enhances the quality of marital relationships but also contributes to a more compassionate and understanding of the whole society. The collective efforts foster a ripple effect of empathy, tolerance, and understanding that extends beyond the family unit, ultimately contributing to a more harmonious, compassionate, and inclusive society.

1.12 CONCLUSION

The erosion of traditional marital beliefs among the Shona people of Zimbabwe has been exacerbated by modern influence, driven by the interplay of cultural exchange, and societal transformation. This is evident that the erosion is not merely a byproduct of modern influences but a complex interplay of education, evolving societal roles, and the pervasive impact of technology. The decline of traditional courtship practices, the commercialization of bride price, and the rise of online dating have all contributed to a shift away from traditional values that once underpinned Shona marital traditions. The transition towards modern marriages has brought both opportunities and risks, including the issue of gender-based violence. The increase in gender-based violence, to the perceived haunting of avenging spirits highlight the urgent need for strategies that promote prevention, support, and reconstruction of marital beliefs in the modern era. The journey towards redefining marital beliefs is multifaceted and requires a collective effort from individuals, communities, and institutions. To navigate this evolving landscape, it is essential to advocate for a reconstruction of marital beliefs that honours traditional values while embracing equity, consent, and mutual respect. By honoring traditional values while embracing modern perspectives, the chapter emphasizes the importance of creating a more inclusive and supportive environment that empowers individuals to building healthy, respectful, and fulfilling relationships. Through education, awareness, and proactive support, different stakeholders can transform the narrative around marriage and gender-based violence, fostering a culture of respect and understanding that benefits all. Ultimately, this requires a nuanced understanding of the complex interplay between tradition and modernity. By engaging in this critical reflection and collective effort, communities can work towards revitalizing the institution of marriage and promoting a brighter future for generations to come. Through this collaborative and introspective process, communities can transform the institution of marriage into a vibrant, inclusive, and supportive framework that nurtures healthy relationships, fosters personal growth, and contributes to a more loving, equitable, and thriving society.

COMPETING INTERESTS

The author of this publication declare there are no competeting interests

FUNDING

This research received no specific grant from any funding agency in the public, commercial, or not-for-profit sectors. Funding for this research was covered by the author of the article.

REFERENCES

Ainamon, A., Gbaguigi, C., & Agonglovi, M. K. (2022). Exploring the Issue of Marriage Rites in Selected African Novels. [IJSELL]. *International Journal on Studies in English Language and Literature, 10*(2), 1–14. DOI: 10.20431/2347-3134.1002001

Akram, M. S., & Albalawi, W. (2016). Youths' social media adoption: Theoretical model and empirical evidence. *International Journal of Business and Management, 11*(2), 22–30.

Baloyi, G. T. (2022). Marriage and Culture within the Context of African Indigenous Societies: A Need for African Cultural Hermeneutics. *Studia Historiae Ecclesiasticae, 48*(1), 1–12. DOI: 10.25159/2412-4265/10113

Cardno, C. (2018). Policy Document Analysis: A practical educational leadership tool and a qualitative research method. *Educational Administration: theory and Practice,* 24 (4), 623-640: doi: DOI: 10.14527/kuey.2018.016

Emelone, C. (2020). Marriage and Procreation in Africa: The Portrayal in Lola Shoneyn's the Secret Lives of Baba Segi's Wives. *Preorc Journal of Gender and Sexuality Studies, 1*(1), 31–51.

Fufunwa, A. B., & Aisiku, J. U. (2022). *Education in Africa: a comparative survey.* Routledge., DOI: 10.4324/9781003316114

Goredema, D. (2016). *Zimbabwe's Liberation Struggle Era Conflicts and the Pitfalls of Reconciliation after Independence: a case Study of Bikita District 1976-2013.* A thesis submitted to the Midlands State University in partial fulfilment of the requirements for the degree of Doctor of Philosophy in History.

Kakembo, A. A. (2024). Evaluating the effectiveness of educational policies. *IDOSR Journal of Communication and English., 9*(3), 6–10. DOI: 10.59298/IDOSR/JCE/93.610.202400

Kaylo, P. (2012). A Reflection on the African Traditional Values of Marriage and Sexuality. *International Journal of Academic Research in Progressive Education and Development, 1*(2), 84–92.

Mambure, D. (2024). Status Quo of Divorces in Zimbabwe. [IJRISS]. *International Journal of Research and Innovation in Social Science, 8*(8), 4452–4461.

Mawere, M., & Mawere, A. M. (2010). The Changing Philosophy of African Marriage: The Relevance of the Shona customary marriage practice of kukumbira. *Journal of African Studies and Development*, *2*(9), 224–233.

Montemayor, R. (2019). *Sexuality in adolescence and emerging adulthood.* Guiford Press.

Musanga, T. (2017). "Ngozi" (Avenging Spirit), Zimbabwean Transnational Migration, and Restorative Justice in Brian Chikwava's Harare North (2009). *Journal of Black Studies*, *48*(8), 1–17. DOI: 10.1177/0021934717720563

Ogoma, D. E. (2014). Reflection on an African Traditional Marriage System. *Journal of Social Sciences and Public Affairs*, *4*(1), 1–11.

Stewart, O., & Linden, M. (2022). *Mother of Abominations: Witchcraft Dreamscapes.* Black Moon Publishing.

United Nations. (2008). *An Introduction to Human Trafficking: Vulnerability, Impact and Action.* Office on Drugs and Crime.

KEY TERMS AND DEFINITIONS

Community involvement: refers to an active participation of individuals to the development and improvement of their communities.

Culture: refers to the traditional beliefs, values, customs, and practices that are unique to a certain group of people.

Education: is the process of facilitating holistic development in individuals

Gender roles: refer to societal beliefs associated with duties for men and women

Gender-based violence: refers to harmful acts directed to an individual based on their gender.

Marital beliefs: are cultural and social norms that influence various aspects of marriage.

Compilation of References

Abdulharis, R., Kusdiwanggo, S., Nurlinda, I., Iskandar, G. H., Dwiartama, A., Hernandi, A., & Sidiq, T. P. (2022). *Cultural Space Methodology on Assessment of Authenticity of Indigenous Community Forest and Conservation Area in Indonesia.* SOCIAL SCIENCES. DOI: 10.20944/preprints202212.0402.v1

Achebe, C. (2009). Spelling our proper name. In *The education of a British-protected child* (pp. 79–94). Penguin Books.

Adade, W. P., Sikutshwa, L., & Shackleton, S. (2020). Acknowledging indigenous and local knowledge to facilitate collaboration in landscape approaches—Lessons from a systematic review. *Land (Basel)*, *9*(9), 331. https://www.mdpi.com/2073 445X/9/9/331

Adam, A. A., Othman, N., & Halim, A. A. (2021). Indigenous knowledge transfer among Dusuns and Bajaus in Kota Belud, Sabah, Malaysia: Approaches and challenges. *International Journal of Academic Research in Business & Social Sciences*, *11*(11), 1722–1734. https://www.researchgate.net/profile/Anis-Amalina Adam/publication/

Adam, A. A., Othman, N., & Halim, A. A. (2022). Indigenous Knowledge Documentation: Perspectives of Dusun and Bajau Communities in Kota Belud, Sabah, Malaysia. *International Journal of Academic Research in Business & Social Sciences*, *12*(10), 162–175. DOI: 10.6007/IJARBSS/v12-i10/15205

Adams, M. S., Carpenter, J., Housty, J. A., Neasloss, D., Paquet, P. C., Service, C., Walkus, J., & Darimont, C. T. (2014). Toward increased engagement between academic and indigenous community partners in ecological research. *Ecology and Society*, *19*(3), art5. DOI: 10.5751/ES-06569-190305

Adefila, E., Oladokun, B. D., & Adewojo, A. A. (2024). Beginning of a new era: Leveraging digital twin technology in the preservation of indigenous knowledge systems in the disruptive era. *Library Hi Tech News*. Advance online publication. DOI: 10.1108/LHTN-05-2024-0088

Adeola, A., Olaniyi, E., & Ngare, I. (2024). *Gender equality, climate action, and technological innovation for sustainable development in Africa*. Palgrave Macmillan.

Adeyemo, A.A. and Adebayo, J.O. (2017) Documentation and Dissemination of Indigenous Knowledge by Library Personnel in Selected Research Institutes in Nigeria. *Library Philosophy and Practice* (e-journal)

Adu, K. K., & Ngulube, P. (2016). Key threats and challenges to the preservation of digital records of public institutions in Ghana. *Information Communication and Society*, *20*(8), 1127–1145. DOI: 10.1080/1369118X.2016.1218527

Adu, K. K., & Ngulube, P. (2017). Key Threats and Challenges to the Preservation of Digital Records of Public Institutions in Ghana. *Information Communication and Society*, *20*(8), 1127–1145.

Afful-Arthur, P., Kwafoa, P. N. Y., Ampah-Johnston, M., & Mensah, V. (2021). Managing and Accessing Indigenous Knowledge for National Development: The Role of Academic Libraries in Ghana. *Information Development*, *38*(4), 535–548. DOI: 10.1177/02666669211009916

Ainamon, A., Gbaguigi, C., & Agonglovi, M. K. (2022). Exploring the Issue of Marriage Rites in Selected African Novels. [IJSELL]. *International Journal on Studies in English Language and Literature*, *10*(2), 1–14. DOI: 10.20431/2347-3134.1002001

Aiseng, K. (2023). Challenges and opportunities of preserving African indigenous knowledge using digital technologies: The case of Bogwera. Digital Preservation and Documentation of Global Indigenous Knowledge Systems, 123–141.

Ajani, Y. A., Oladokun, B. D., Olarongbe, S. A., Amaechi, M. N., Rabiu, N., & Bashrun, M. T. (2024). Revitalizing indigenous knowledge systems via digital media technologies for sustainability of indigenous languages. *PDT&C*, *53*(1), 35–44. DOI: 10.1515/pdtc-2023-0051

Akinwale, A. A. (2012) Digitization of indigenous knowledge for natural resources management in Africa. *Proceedings of the AERN Summit*. Paper presented at the 20th Anniversary Summit of the African Educational Research Network at North Carolina State University, Raleigh, USA, 19 May, pp.1-18.

Akpabio, E. (2023). Toolal Communication. In *Indigenous Communication: A Global Perspective* (pp. 31–55). Springer International Publishing., https://books.google.com/books?hl=en&lr=&id=ArrmEA

Akram, M. S., & Albalawi, W. (2016). Youths' social media adoption: Theoretical model and empirical evidence. *International Journal of Business and Management, 11*(2), 22–30.

Aksa, F. I. (2020). Wisdom of indigenous and tacit knowledge for disaster risk reduction. *The Indonesian Journal of Geography, 52*(3), 418–426. https://www.researchgate.net/profile/Furqan-Ishak-Aksa/publication/3483

Albornoz, D., & Chan, L. (2018). Power and Inequality in Open Science Discourses. *IRIS - Revista de Informação. Memória e Tecnologia, 4*(1), 70. DOI: 10.51359/2318-4183.2018.238912

Alexander, S. M., Provencher, J. F., Henri, D. A., Taylor, J. J., & Cooke, S. J. (2019). Bridging Indigenous and science-based knowledge in coastal-marine research, monitoring, and management in Canada: A systematic map protocol. *Environmental Evidence, 8*(1), 15. DOI: 10.1186/s13750-019-0159-1

Ali, S. H., & Rose, J. R. (2022). The post-colonialist condition, suspicion and social resistance during the West African Ebola epidemic: The importance of Frantz Fanon for global health. *Social Science & Medicine, 305*, 115066. DOI: 10.1016/j.socscimed.2022.115066 PMID: 35636052

Allard, D., & Ferris, S. (2014). *The Digital Archives and Marginalized Communities Project: Building Anti-Violence Archives*. DOI: 10.9776/14358

Allard-Tremblay, Y., & Coburn, E. (2024). "Reviewer 2 must be stopped"; Indigenous Contributions and Epistemic Domination in Political Science. *Appartenances & Altérités, 5*. Advance online publication. DOI: 10.4000/12kpw

Al-Sakran, H. O., & Alsudairi, M. A. (2021). Usability and accessibility assessment of Saudi Arabia mobile e-government websites. *IEEE Access : Practical Innovations, Open Solutions, 9*, 48254–48275. DOI: 10.1109/access.2021.3068917

Alviz-Meza, A., Vásquez-Coronado, M. H., Delgado-Caramutti, J. G., & Blanco-Victorio, D. J. (2022). Bibliometric analysis of fourth industrial revolution applied to heritage studies based on web of science and scopus databases from 2016 to 2021. *Heritage Science, 10*, 189. PMID: 36467714

Amanze-Unagha, B. N., & Unagha, A. O. (2023). Building community partnerships and collaboration for effective. Library and Information Services Delivery. *International Journal of Applied Technologies in Library and Information Management, 9*(3), 31–39.

Amechi, E.P. (2015). Leveraging traditional knowledge on the medicinal uses of plants within the patent system: The digitisation and disclosure of knowledge in South Africa. Potchefstroom Electronic Law Journal/Potchefstroomse Elektroniese Regsblad 18(1): 3071–3101.

Amoah, J., & Yeboah, W. (2018). Ethical dilemmas among academic librarians: Evidence from Sam Jonah Library, University of Cape Coast, Ghana. *Ghana Library Journal, 30*(1), 14–27.

Amunkete, K. N. (2020). A model for the digital preservation of indigenous knowledge on medicinal plants in Namibia via an e-learning platform. https://uir.unisa.ac.za/items/78076e2a-b80e-46fa-a3b6-ee2a9333e874

Amunkete, K. N., Staden, C. V., & Schoeman, M. (2021). *Preserving Indigenous Knowledge Through E-Learning: A Conceptual Theoretical Model*. DOI: 10.33965/icwi_ac2021_2021091022

ANTARA News. (2021, November 6th). Rumah Rusak Akibat Banjir Bandang di Kota Batu Bertambah. Antara News. Retrieved from https://www.antaranews.com/berita/2506153/rumah-rusak-akibat-banjir-bandang-di-kota-batu-bertambah Access on October 11th 2024.

Anti. S.L., and Awanda, I. (2025). Preserving traditional knowledge in the digital era: challenges and strategies. *Journal for Social Science and Society Studies.*

Aremu, O. A., & Makunga, N.. (2022). Africa is a treasure of medicinal plants, here are the seven that are popular, In *The Conversation Africa, SAST*

Armeni, K., Brinkman, L., Carlsson, R., Eerland, A., Fijten, R., Fondberg, R., Heininga, V. E., Heunis, S., Koh, W. Q., Masselink, M., & Moran, N. (2021). ´ Baoill, A, O.M., Sarafoglou, A., Schettino, Schwamm, H., Sjoerds, Z., Teperek, M., van den Akker, O.R., van't Veer, A., and Zurita-Milla, R. (2021). Towards wide-scale adoption of open science practices: The role of open science communities. *Science & Public Policy, 48*(5), 605–611. DOI: 10.1093/scipol/scab039

Arsenault, R., Diver, S., McGregor, D., Witham, A., & Bourassa, C. (2018). Shifting the Framework of Canadian Water Governance through Indigenous Research Methods: Acknowledging the Past with an Eye on the Future. *Water (Basel), 10*(1), 49. DOI: 10.3390/w10010049

Asadi Ojaei, S. K., & Abbasnejad Seresti, R. (2023). Symbols for the Neolithization Process: Ritual Animals of the Eastern Fertile Crescent. *International Journal of the Humanities*, *30*(2), 72–97. https://eijh.modares.ac.ir/article-27-58950-fa.html

Asante, M. K. (2007). *An Afrocentric manifesto: Toward an African renaissance*. Polity Press.

Asante, M. K. (2017). The philosophy of Afrocentricity. In Afolayan, A., & Falola, T. (Eds.), *The Palgrave handbook of African philosophy* (pp. 231–245). Palgrave Macmillan. DOI: 10.1057/978-1-137-59291-0_16

Asante, M. K. (2020). Africology, Afrocentricity, and what remains to be done. *The Black Scholar*, *50*(3), 48–57. DOI: 10.1080/00064246.2020.1780859

Assou-Dodji, H., & Serwah Boateng, J. (2025). Voices of empowerment: women's role in safeguarding cultural heritage and advancing human rights in Africa. https://strathprints.strath.ac.uk/93364/

Atalay, S. (2020). Indigenous Science for a World in Crisis. *Public Archaeology*, *19*(1–4), 37–52.

Athayde, S., Silva-Lugo, J., Schmink, M., & Heckenberger, M. (2017). The Same, but Different: Indigenous Knowledge Retention, Erosion, and Innovation in the Brazilian Amazon. *Human Ecology*, *45*(4), 533–544. DOI: 10.1007/s10745-017-9919-0

Austin, P. K. (2019). Digital Technology and Endangered Indigenous Languages: Challenges and New Opportunities. *Language Documentation and Conservation*, *13*, 356–382.

Avsharova, I. (2020). The reflection of" Bull" cult at the Bronze Age material culture of Azerbaijan. *International Journal of Psychosocial Rehabilitation*, *24*(01). https://www.researchgate.net/profile/Imcra

Awori, K., Vetere, F., & Smith, W. (2015). *Transnationalism, Indigenous Knowledge and Technology*. 3759–3768. DOI: 10.1145/2702123.2702488

Awuor, A.S., Kamau, G.W. & Owano, A. (2023). The role of digital humanities in the preservation of indigenous knowledge at the National Museums of Kenya.

Ayeni, A. O., & Aborisade, A. G. (2022). African indigenous knowledge systems and the world. In Oloruntoba, S. O., & Falola, T. (Eds.), *The Palgrave handbook of Africa and the changing global order* (pp. 149–166). Palgrave Macmillan. DOI: 10.1007/978-3-030-77481-3_8

Babacan, E. Y., Polat, R., Güler, O., Moyan, A., Paksoy, M. Y., & Cakilcioglu, U. (2022). An ethno-veterinary study on plants used for the treatment of livestock diseases in Genç (Bingöl-Turkey). [IJTK]. *Indian Journal of Traditional Knowledge, 21*(1), 81–88. http://op.niscpr.res.in/index.php/IJTK/article/view/41752

Bacon, E., Williams, M. D., & Davies, G. (2020). Coopetition in innovation ecosystems: A comparative analysis of knowledge transfer configurations. *Journal of Business Research, 115*, 307–316. https://www.sciencedirect.com/science/article/pii/S0148296319306745

Bagley, M. A. (2019). The fallacy of defensive protection for traditional knowledge. *Washburn Law Journal, 58*, 323.

Baker, M. (2016). 1,500 scientists lift the lid on reproducibility. *NATNews, 533*, 452–454. DOI: 10.1038/533452a PMID: 27225100

Balogun, T. (2023). Data management of digitised indigenous knowledge systems in repositories. *Information Development, 39*(3), 425–438.

Balogun, T. (2023). Data Management of Digitized Indigenous Knowledge System in Repositories. *Information Development, 39*(3), 425–438. DOI: 10.1177/02666669231186777

Balogun, T. (2023). Digitization of Indigenous knowledge systems in Africa: The case of South Africa's National Recorded System (NRS). *Records Management Journal, 33*(1), 88–102.

Balogun, T., & Kalusopa, T. (2021). A framework for digital preservation of Indigenous knowledge system (IKS) in repositories in South Africa. *Records Management Journal, 31*(2), 176–196. DOI: 10.1108/RMJ-12-2020-0042

Balogun, T., & Kalusopa, T. (2021). Web Archiving of Indigenous Knowledge Systems in South Africa. *Information Development, 38*(4), 658–671. DOI: 10.1177/02666669211005522

Baloyi, G. T. (2022). Marriage and Culture within the Context of African Indigenous Societies: A Need for African Cultural Hermeneutics. *Studia Historiae Ecclesiasticae, 48*(1), 1–12. DOI: 10.25159/2412-4265/10113

Bandarin, F., & Van Oers, R. (2012). *The Historic Urban Landscape: Managing Heritage in an Urban Century*. John Wiley & Sons.

Bandi, S., Angadi, M., & Shivarama, J. (2015). Best practices in digitization: Planning and workflow processes. In *International Conference on 'Emerging technologies and future of libraries: Issues and challenges'*, Gulbarga, Karnataka, India, 30–31 January 2015, pp. 332–339.

Barrera, T. G. V., Quintero-Sánchez, G. A., & Pérez-Salazar, B. (2023). From terra Nullius to Indigenous Collective Land Rights: Cases Before the Inter-American Court of Human Rights. *Alternative an International Journal of Indigenous Peoples*, *19*(1), 101–112. DOI: 10.1177/11771801221148790

Bassani, M. (2017). International Cultural Heritage Law and World Heritage Listing: A Vehicle for "White Control of Indigenous Heritage"? *Santander Art and Culture Law Review*, *2*. Advance online publication. DOI: 10.4467/2450050XSNR.17.032.8433

Bastos Lima, M. G., & Persson, U. M. (2020). Commodity-centric landscape governance as a double-edged sword: The case of soy and the Cerrado Working Group in Brazil. *Frontiers in Forests and Global Change*, *3*, 27.

Bates, D. W., Ebell, M., Gotlieb, E., Zapp, J., & Mullins, H. C. (2003). A proposal for electronic medical records in US primary care. *Journal of the American Medical Informatics Association : JAMIA*, *10*(1), 1–10. DOI: 10.1197/jamia.m1097 PMID: 12509352

Bawack, R., Roderick, S., Badhrus, A. M., Dennehy, D., & Corbett, J. (2025). Indigenous knowledge and information technology for sustainable development. *Information Technology for Development*, *31*(2), 233–250. DOI: 10.1080/02681102.2025.2472495

BBC News. (2022). Kenya bans 'biopiracy' export of notorious baobabs. https://www.bbc.com/news/world-africa-63716286

Belarde-Lewis, M., Littletree, S., Braine, I. R., Srader, K., Guerrero, N., & Palmer, C. L. (2024). Centering Relationality and CARE for Stewardship of Indigenous Research Data. *Data Science Journal*, *23*. Advance online publication. DOI: 10.5334/dsj-2024-032

Bennis, W., & O'Toole, J. (2005). How Business Schools Lost Their Way. *Harvard Business Review*, *83*, 96–104. PMID: 15929407

Berio, L., & Moore, R. (2023). Great ape enculturation studies: A neglected resource in cognitive development research. *Biology & Philosophy*, *38*(2), 17. https://link.springer.com/article/10.1007/s10539-023-09908-y

Berkes, F. and Berkes, M. K. (2009) Ecological complexity, fuzzy logic, and holism in indigenous knowledge, *Futures 41: 6–12, https://doi.org/. futures.2008.07.003.* DOI: 10.1016/j

Besançon, L., Peiffer-Smadja, N., Segalas, C., Jiang, H., Masuzzo, P., Smout, C., Billy, E., Deforet, M., & Leyrat, C. (2021). Open science saves lives: Lessons from the COVID-19 pandemic. *BMC Medical Research Methodology, 21*(1), 117. DOI: 10.1186/s12874-021-01304-y PMID: 34090351

Bett, S. K and Makori, E.O. (2024). Opening gates to open science: Impact of digital technologies in fostering open science principles in research. *Current Journal of Library and Information Sciences. 9(3).*

Bigambo, R. (2020). *Safeguarding practices for intangible cultural heritage in Tanzania: national vs local perspectives* (Doctoral dissertation, University of Birmingham). https://etheses.bham.ac.uk/id/eprint/

Bihari, S. (2023). Cultural Heritage and Indigenous Knowledge: Reviving Traditions for Future Generations. *Sustainable Development Goals in SAARC Countries: Key Issues, Opportunities and Challenges, 1*, 24-32. https://www.researchgate.net/profile/Santosh-Mane-2/publication/

Bihari, S. (2023). Cultural heritage and indigenous knowledge: Reviving traditions for future generations. Sustainable Development Goals in SAARC Countries: Key Issues, Opportunities and Challenges, 1, 24-32.102.

Bingham, J. A., Milne, S., Murray, G., & Dorward, T. (2021). Knowledge Pluralism in First Nations' Salmon Management. *Frontiers in Marine Science, 8*. Advance online publication. DOI: 10.3389/fmars.2021.671112

Bird, S. K., Flick, S. D., Carroll, S., Austin, A., & Lim, A. (2019). Technology for Indigenous Language Revitalization: The State of the Field. *Language, 95*(S1), e1–e29.

Bishoff, L., & Smith, C. (2015). Managing digital collections survey results. *D-Lib Magazine : the Magazine of the Digital Library Forum, 21*(3), 1–11.

Bishop, M. (2022). Elders' Conversations: Perspectives on Leveraging Digital Technology in Language Revival. *The Open/Technology in Education Society and Scholarship Association Journal, 2*(2), 1–13. DOI: 10.18357/otessaj.2022.2.2.21

Biyela, N., Oyelude, A., & Haumba, E. (2016) Digital preservation of indigenous knowledge (IK) by cultural heritage institutions: a comparative study of Nigeria, South Africa and Uganda. (Paper presented at the SCESCAL Conference, Ezulwini, Swaziland, 25 - 29 April). http://hdl.handle.net/20.500.11910/10196 http://hdl.handle.net/20.500.11910/10196

Biyela, N., Oyelude, A., & Haumba, E. (2016) Digital preservation of Indigenous Knowledge (IK) by cultural heritage institutions: a comparative study of Nigeria, South Africa and Uganda. *Paper presented at the SCESCAL Conference*, Ezulwini, Swaziland, 25 - 29 April, 2016.

Biyela, N., Oyelude, A., & Haumba, E. (2016, April). Digital preservation of indigenous knowledge (IK) by cultural heritage institutions: a comparative study of Nigeria, South Africa and Uganda. In *The proceedings of the digital transformation and the changing role of libraries and information centres in the sustainable development of Africa at the 22nd Standing Conference of Eastern, Central and Southern Africa Library and Information Associations*.

Blank, D. A. (2021). Artiodactyl vocalization. *Neuroendocrine Regulation of Animal Vocalization*, 159-188. https://www.sciencedirect.com/science/article/pii/B9780128151600000062

Boamah, E., & Liew, C. L. (2017). Conceptualising the digitisation and preservation of Indigenous knowledge: The importance of attitudes. In *Digital Libraries: Data, Information, and Knowledge for Digital Lives: 19th International Conference on Asia-Pacific Digital Libraries, ICADL 2017, Bangkok, Thailand, November 13-15, 2017, Proceedings* (pp. 65-80). Springer International Publishing.

Boamah, E., & Liew, C. L. (2017). Conceptualising the Digitisation and Preservation of Indigenous Knowledge: The Importance of Attitudes. Lecture Notes in Computer Science, *International Conference on Asian Digital Libraries*. DOI: 10.1007/978-3-319-70232-2_6

Bodo, B. (2016). Pirates in the library – an inquiry into the guerilla open access movement. Paper prepared for the 8th Annual Workshop of the International Society for the History and Theory of Intellectual Property, CREATe, University of Glasgow, UK, July 6-8, 2016.

Bohensky, E. L., & Maru, Y. (2011). Indigenous Knowledge, Science, and Resilience: What Have We Learned from a Decade of International Literature on 'Integration'? *Ecology and Society*, *16*(4), art6. DOI: 10.5751/ES-04342-160406

Bontsa, N. V., Mushunje, A., & Ngarava, S. (2023). Factors Influencing the Perceptions of Smallholder Farmers Towards Adoption of Digital Technologies in Eastern Cape Province, South Africa. *Agriculture*, *13*(8), 1471. DOI: 10.3390/agriculture13081471

Borda, A., & Bowen, J. P. (2020, July). *Turing's Sunflowers: Public research and the role of museums. Proceedings of EVA London 2020*. DOI: 10.14236/ewic/EVA2020.5

Borrero, R. M. (2013). *Innovation and technology for Indigenous peoples. United Nations*. Retrieved from UN Website.

Breidlid, A. (2009). Culture, indigenous knowledge systems and sustainable development: A critical view of education in an African context. *International Journal of Educational Development*, *29*, 140–148.

Breidlid, A. (2009). Culture, Indigenous Knowledge Systems and Sustainable Development: A Critical View of Education in An African Context. *International Journal of Educational Development*, *29*, 140–148.

Brewster, C., Nouwt, B., Raaijmakers, S., & Verhoosel, J. (2020). Ontology-based access control for FAIR data. *Data Intelligence*, *2*(1-2), 66–77. DOI: 10.1162/dint_a_00029

Brinklow, N. T. (2021). Indigenous Language Technologies: Anti-Colonial Oases in a Colonizing (Digital) World. *Winhec International Journal of Indigenous Education Scholarship*, *1*. Advance online publication. DOI: 10.18357/wj1202120288

Bruchac, M. M. (2014). Indigenous Knowledge and Traditional Knowledge. In Smith, C. (Ed.), *Encyclopedia of Global Archaeology*. Springer.

Burgess, C., Bishop, M., & Lowe, K. (2020). Decolonising Indigenous Education: The Case for Cultural Mentoring in Supporting Indigenous Knowledge Reproduction. *Discourse (Abingdon)*, *43*(1), 1–14. DOI: 10.1080/01596306.2020.1774513

Burrows, T. (2011). Sharing humanities data for e-research: Conceptual and technical issues. In Thieberger, N. (Ed.), *Sustainable Data from Digital Research* (pp. 177–192). PARADISEC.

Buthelezi, S. C., Ocholla, D., & Dlamini, P. (2024). Strategies for documenting and Disseminating indigenous knowledge at a South African university. *South African Journal of Information Management*, *26*(1), a1648. DOI: 10.4102/ sajim.v26i1.1648

Buthelezi, S. C., Ocholla, D., & Dlamini, P. (2024). Strategies for documenting and disseminating indigenous knowledge at a South African university'. *South African Journal of Information Management*, *26*(1), a1648.

Cakata, Z., & Ramose, M. B. (2023). When ukucelwa ukuzalwa becomes bride price: Spiritual meaning lost in translation. *African Identities*, *21*(3), 478–490. https://doi/abs/10.1080/14725843.2021.1940091

Callison, C., Ludbrook, A., Owen, V., & Nayyer, K. (2021). Engaging Respectfully With Indigenous Knowledges. *Kula Knowledge Creation Dissemination and Preservation Studies*, *5*(1). Advance online publication. DOI: 10.18357/kula.146

Cardno, C. (2018). Policy Document Analysis: A practical educational leadership tool and a qualitative research method. *Educational Administration: theory and Practice,* 24 (4), 623-640: doi: DOI: 10.14527/kuey.2018.016

Carroll, S. R., Garba, I., Figueroa-Rodríguez, O. L., Holbrook, J., Lovett, R., Materechera, S., Parsons, M., Raseroka, K., Rodriguez-Lonebear, D., & Rowe, R.. (2020). The CARE Principles for Indigenous Data Governance. *Data Science Journal, 19,* 1–12. DOI: 10.5334/DSJ-2020-043

Carroll, S. R., Herczog, E., Hudson, M., Russell, K., & Stall, S. (2021). Operationalizing the CARE and FAIR Principles for Indigenous data futures. *Scientific Data, 8,* 108. DOI: 10.1038/s41597-021-00892-0 PMID: 33863927

Chai, A., & Gichuhi, Z. W. (2023). Use of Information Communication Technologies for Documenting Indigenous Farming Knowledge for Improved Preservation, Access and Use in Kilifi County, Kenya. *International Journal of Current Aspects, 7*(3), 99–133. DOI: 10.35942/c2szz470

Chakrabarty, S., & Hossain, K. (2024). Recognising indigenous traditional knowledge with medicinal value within a legal framework: An overview of the issues and challenges with special focus on India. *Man in India, 104*(1&2).

Chambers, R. (2014). *Rural development: Putting the last first.* Routledge.

Chanda, A., & Alfred, K. (2025). Ethical Integration of Indigenous and Scientific Knowledge Systems for a Sustainable Mitigation of Environmental and Climate Change Challenges. *International Journal of Global Community, VIII*(1).

Chan, E., Bourgeois-Doyle, D., Donaldson, M., & Haine-Bennett, E. (2020). *Toward a UNESCO Recommendation on Open Science: Canadian Perspectives.* UNESCO.

Chapman, J. M., & Schott, S. (2020). Knowledge coevolution: Generating new understanding through bridging and strengthening distinct knowledge systems and empowering local knowledge holders. *Sustainability Science, 15*(3), 931–943. DOI: 10.1007/s11625-020-00781-2

Chatwood, S., Paulette, F., Baker, R., Eriksen, A., Hansen, K. L., Eriksen, H., Hiratsuka, V., Lavoie, J., Lou, W., Mauro, I., Orbinski, J., Pabrum, N., Retallack, H., & Brown, A. (2015). Approaching *Etuaptmumk* – introducing a consensus-based mixed method for health services research. *International Journal of Circumpolar Health, 74*(1), 27438. DOI: 10.3402/ijch.v74.27438 PMID: 26004427

Chaudhuri, B. (2015). Science in society: Challenges and opportunities for indigenous knowledge in the present-day context. *Global Bioethics = Problemi di Bioetica, 26*(2), 78–85. DOI: 10.1080/11287462.2015.1037140

Chesbrough, H. (2015). *From Open Science to Open Innovation*. ESADE.

Chigora, P., Masocha, R., & Mutenheri, F. (2007). The role of indigenous medicinal knowledge in the treatment of ailments in rural Zimbabwe: The case of Mutirikwi Communal lands. *Journal of Sustainable Development in Africa*, *9*(2).

Chigwada, J. (2024). Harnessing indigenous knowledge in the digital era. https://www.unisa.ac.za/sites/myunisa/default/News/Articles/Harnessing-indigenous-knowledge-in-the-digital-era (10 April 2025).

Chigwada, J. P., & Ngulube, P. (2023). Librarians' Role in the Preservation and Dissemination of Indigenous Knowledge. *IFLA Journal*, *50*(2), 242–256. DOI: 10.1177/03400352231217270

Chigwada, J., & Ngulube, P. (2023). Librarians' role in the preservation and dissemination of indigenous knowledge. *IFLA Journal*, *50*(2), 242–256.

Chigwada, J., & Ngulube, P. (2024). Librarians' role in the preservation and dissemination of indigenous knowledge. *International Federation of Library Associations and Institutions.*, *50*(2), 242–256.

Chikozho, C., & Mabhena, R. (2020). Harnessing Traditional Ecological Knowledge for Sustainable Environmental Management: Digital Preservation in Southern Africa. *The International Journal of Environmental Studies*, *35*(4), 280–294.

Chiteyi, L. I., Lusambili, K. M., & Ochieng, A. (2023). Assessing The Origins and Development of Bullfighting Culture among the Abakakamega of Western Kenya. *East African Journal of Arts and Social Sciences*, *6*(2), 80–100.

Chiwanza, K., Musingafi, M. C. C., & Mupa, P. (2012). Challenges in preserving Indigenous knowledge systems: Learning from past experiences. [Retrieved from Scientific Research Publishing.]. *Information and Knowledge Management*, *3*(2), 19–25.

Chng, K. S., & Narayanan, S. (2017). Culture and social identity in preserving cultural heritage: An experimental study. *International Journal of Social Economics*, *44*, 1078–1091.

Chowdhury, G., McLeod, J., Lihoma, P., Teferra, S., & Wato, R. (2021). Promoting Access to Indigenous Information in Africa: Challenges and Requirements. *Information Development*, *39*(3), 611–623. DOI: 10.1177/02666669211048488

Christensen, G., Wang, Z., Levy Paluck, E., Swanson, N., Birke, D., Miguel, E., and Littman, R. (2020). Open science practices are on the rise: The state of social science (3S) survey.

Christensen, G., Wang, Z., Paluck, E., Swanson, N., Birke, D., Miguel, E., & Littman, R. (2019). Open Science Practices Are On The Rise: The State Of Social Science (3s) Survey. https://doi.org/DOI: 10.31222/osf.io/5rksu

Christian, G. E. (2009). Digitization, intellectual property rights and access to traditional medicine knowledge in developing countries: the Nigerian experience, *A research papers prepared for the International Development Research Centre (IDRC)* Ottawa, Canada. Available at http://idl-bnc.idrc.ca/dspace/handle/10625/41341

Christian, G. E. (2009). *Digitisation, intellectual property, rights and access to traditional medicine knowledge in developing countries-The Nigerian Experience. A research paper prepared for the International Development Research Centre*. IDRC.

Chukwuma, A., & Mpofu, B. (2019). The relevance of Indigenous knowledge systems in social work education and practice in Zimbabwe. *Journal of Social Development in Africa.*

Church, R. M. (2002). The effective use of secondary data. *Learning and Motivation*, *33*(1), 32–45.

Cindi, L. (2021). *Incorporating African Indigenous Knowledge Systems into the Basic Education curriculum: Experiences from two schools in the Gauteng and KwaZulu-Natal Provinces, South Africa. Master of Philosophy (MPHIL) in African Studies. Bloemfontein: University of the Free State Cohen, S. (1972). Folk Devils and Moral Panics: The Creation of the Mods and Rockers*. Routledge.

Cobey, K., Alayche, M., Saba, S., Barnes, N., Ebrahimzadeh, S., Alarcon, E., & Moher, D. (2024). Cardiology Researchers' Practices and Perceived Barriers to Open Science: An International Survey. *Open Heart*, *11*(1), e002433. DOI: 10.1136/openhrt-2023-002433 PMID: 38233041

Cocq, C. (2023). Open science in Sámi research: Researchers' dilemmas. *Frontiers in Research Metrics and Analytics*, *8*, 1095169. DOI: 10.3389/frma.2023.1095169 PMID: 37124717

Cole, T. (2022). Developing and Maintaining Practical Archives: A How-To-Do-It Manual. 3rd ed. By Gregory S. Hunter. *Archival Issues, 41*(2). https://doi.org/DOI: 10.31274/archivalissues.15645

Cole, N. L., Kormann, E., Klebel, T., Apartis, S., & Ross-Hellauer, T. (2024). The Societal impact of Open Science: A scoping review. *Royal Society Open Science*, *11*, 240286. DOI: 10.1098/rsos.240286 PMID: 39100167

Conroy, A. B. (2022). Oxen: Status, Uses and Practices in the USA, Encouraging a Historic Tradition to Thrive. https://scholars.unh.edu/faculty_pubs/1523/

Conway, P. (2011). Archival quality and long-term preservation: A research framework for validating the usefulness of digital surrogates. *Archival Science*, *11*, 293–309. DOI: 10.1007/s10502-011-9155-0

Cook, B. G., Johnson, A. H., Maggin, D. M., Therrien, W. J., Barton, E. E., Lloyd, J. W., Reichow, B., Talbott, E., & Travers, J. C. (2021). Open Science and Single-Case Design Research. *Remedial and Special Education*, *43*(5), 359–369. DOI: 10.1177/0741932521996452

Cosijn, E., Pirkola, A., & Bothma, T.. (2002). Information access in indigenous languages: A case study in Zulu. *South African Journal of Library and Information Science*, *8*(2), 94–103.

Creswell, J. W. (2009). *Research Design: Qualitative, quantitative, and mixed method approaches* (3rd ed.). Sage.

Cronje, A. (2015). Epistemological border-crossing between Western science and indigenous knowledge and its implications for teacher professional development. University of Johannesburg: Unpublished PhD thesis.

Crüwell, S., van Doorn, J., Etz, A., Makel, M. C., Moshontz, H., Niebaum, J. C., Orben, A., Parsons, S., & Schulte-Mecklenbeck, M. (2019). Seven Easy Steps to Open Science An Annotated Reading List. *Zeitschrift für Psychologie mit Zeitschrift für Angewandte Psychologie*, *227*(4), 237–248.

Cucchiara, R., Grana, C., Borghesani, D., Agosti, M., & Bagdanov, A. D. (2011). *Multimedia for Cultural Heritage: Key Issues*. Springer.

Cugueró-Escofet, N., & Rosanas, J. M. (2017). The Ethics of Metrics: Overcoming the Dysfunctional Effects of Performance Measurements Through Justice. *Journal of Business Ethics*, *140*, 615–631.

Cullen-Unsworth, L. C., Hill, R., Butler, J. R. A., & Wallace, M. (2011). A research process for integrating Indigenous and scientific knowledge in cultural landscapes: Principles and determinants of success in the Wet Tropics World Heritage Area, Australia. *The Geographical Journal*, *178*(4), 351–365. DOI: 10.1111/j.1475-4959.2011.00451.x

Culturally. (2025). What is cultural competence training? https://www.cultureally.com/blog/what-is-cultural-competence-training. (Accessed 30 April 2025).

Cunningham, H., & Kim, C. (2024, March 5). Weaving Open Dialogue Using Canada's Open Science Roadmap Framework. *Toronto Metropolitan University*. DOI: 10.32920/25343236.v1 da Silva, C., Pereira, F., & Amorim, J. P. (2023). The integration of indigenous knowledge in school: A systematic review. *Compare: A Journal of Comparative Education, 54*(7), 1210–1228. DOI: 10.1080/03057925.2023.2184200

Dalvit, L., Murray, S., & Terzoli, A. (2008). The role of indigenous knowledge in computer education in Africa. In IFIP International Federation for Information Processing. 281, pp. 287–294). DOI: 10.1007/978-0-387-09729-9_43

David-Chavez, D., & Gavin, M. C. (2018). A Global Assessment of Indigenous Community Engagement in Climate Research. *Environmental Research Letters, 13*(12), 123005. DOI: 10.1088/1748-9326/aaf300

David-Chavez, D., Gavin, M. C., Ortiz, N., Valdez, S., & Carroll, S. R. (2024). A Values-Centered Relational Science Model: Supporting Indigenous Rights and Reconciliation in Research. *Ecology and Society, 29*(2). Advance online publication. DOI: 10.5751/es-14768-290211

De la Porte, B., & Higgs, R. (2019). Challenges in digitization of cultural heritage material in the de la PorteI, B and Higgs, R. (2019). Challenges in digitisation of cultural heritage material in the Western Cape, South Africa. https://scielo.org.za/scielo.php?script=sci_arttext&pid=S1560-683X2019000100030 (Accessed 07 April 2025).

De Souza, P., Edmonds, F., McQuire, S., Evans, M., & Chenhall, R. (2016). Aboriginal Knowledge, Digital Technologies and Cultural Collections Policy, Protocols, Practice. Melbourne Networked Society Institute Research Paper 4.

De Souza, P., Edmonds, F., McQuire, S., Evans, M., & Chenhall, R. (2016). Aboriginal Knowledge, Digital Technologies and Cultural Collections Policy, Protocols, Practice. Melbourne Networked Society Institute Research Paper 4. Dienlin, T., Johannes, N., Bowman, N. D., Masur, P. K., Engesser, S., Kümpel, A.

Dei, D. G. J. (2024). Strategies for capturing, managing, and sharing indigenous knowledge. *Information Development, 02666669241248832*, •••. https://doi/abs/10.1177/02666669241248832

Dei, G. J. S. (2000). Rethinking the role of Indigenous knowledges in the academy. *International Journal of Inclusive Education, 4*(2), 111–132. DOI: 10.1080/136031100284849

Demchenko, Y., De Laat, C., Los, W., & Gommans, L. (2022). Defining Platform Research Infrastructure as a Service (PRIaaS) for Future Scientific Data Infrastructure. In Otto, B., Ten Hompel, M., & Wrobel, S. (Eds.), *Designing Data Spaces* (pp. 241–260). Springer International Publishing., DOI: 10.1007/978-3-030-93975-5_15

Department of Science. Technology and Innovation (DSTI). (2019). Government Establishes a New Digital Centre to Promote Indigenous Languages. https://www.dsti.gov.za/index.php/media-room/latest-news/2885-government-establishes-a-new-digital-centre-to-promote-indigenous-languages

Dias, Á., Silva, G. M., Patuleia, M., & González-Rodríguez, M. R. (2023). Developing sustainable business models: Local knowledge acquisition and tourism lifestyle entrepreneurship. In *Knowledge Management, Organisational Learning and Sustainability in Tourism* (pp. 37-56). Routledge. https://www.taylorfrancis.com/chapters/edit/10.4324/9781032626512

Digital Preservation Coalition. (2025). What is digital preservation? https://www.dpconline.org/digipres/what-is-digipres

Dikopoulou, A., & Mihiotis, A. (2012). The contribution of records management to good governance. *The TQM Journal*, *24*(2), 123–141. DOI: 10.1108/17542731211215071

Diop, C. A. (1974). *The African Origin of Civilization: Myth or Reality*. Lawrence Hill Books.

Dlamini, P. N. (2020). Use of Information and Communication Technologies Tools to Capture, Store, and Disseminate Indigenous Knowledge. 325–347. DOI: 10.4018/978-1-7998-0423-9.ch018

Dlamini, P., & Khanyile, N.C. (2021). Preservation of traditional medicinal knowledge: Initiatives and techniques in rural communities in KwaZulu-Natal. *Library Philosophy and Practice (e-journal)*. 4824.

Donaldson, D. R. (2020). Certification information on trustworthy digital repository websites: A content analysis. *PLoS One*, *15*(12), e0242525. PMID: 33296371

Drucker, P. (1967). *The Effective Executive*. Harper & Collins.

Duarte, C. (2021). Africa Goes Digital. Finance & Development. Washington DC. https://www.imf.org/en/Publications/fandd/issues/2021/03/africas-digital-future-after-COVID19-duarte

Dunning, A., de Smaele, M., & Boehmer, J. (2017). Are the FAIR Data Principles fair? *International Journal of Digital Curation*, *12*, 177–195.

Duranti, L., Eastwood, T., & MacNeil, H. (2013). *Preservation of the integrity of electronic records* (Vol. 2). Springer Science & Business Media., DOI: 10.1007/978-94-015-9892-7

Durie, M. (2004). Understanding health and illness: Research at the interface between science and indigenous knowledge. *International Journal of Epidemiology, 33*(5), 1138–1143. DOI: 10.1093/ije/dyh250 PMID: 15218015

Dutta, U. (2019). Digital preservation of indigenous culture and narratives from the global south: In search of an approach. *Humanities (Washington), 8*(2), 68.

Dvurechenskaya, N. D. (2022). Ceramic collections from the citadel of the Uzundara fortress have been dated as the first third of the 3rd century BCE. *Archaeological Research in Asia, 31*, 100394. https://www.sciencedirect.com/science/article/pii/S2352226722000459

Dzulkifli, S. (2020). *Regional Report: Open Science & the Decolonization of Knowledge International Webinar Series. EUROPE AND NORTH AMERICA (ENGLISH)*. UNESCO.

EcoLiterateLaw. (2024). The impact of digital technology on Indigenous peoples. Retrieved from EcoLiterateLaw.

Egeruoh-Adindu, I. (2022). Leveraging Indigenous Knowledge for Effective Environmental Governance in West Africa. *Beijing Law Review, 13*(04), 931–947. DOI: 10.4236/blr.2022.134060

Elijah, I. S., & Mary Taiye, D. M. (2025). The Significance of The Digitalization Of Ifa in the Committee of Knowledge. *Cognizance Journal of Multidisciplinary Studies, 5*(2), 208–216.

Emelone, C. (2020). Marriage and Procreation in Africa: The Portrayal in Lola Shoneyn's the Secret Lives of Baba Segi's Wives. *Preorc Journal of Gender and Sexuality Studies, 1*(1), 31–51.

Essoungou, A. M. (2011). A social media boom begins in Africa: Using mobile phones, Africans join the global conversation. *Africa Renewal, 3*, •••. https://www.un.org/africarenewal/magazine/december-2010/social-media-boom-begins-Africa

European Commission. (2015). Impact of Open Science methods and practices on the economics of research and science. Field. *Language, 95*(S1), e1–e29.

Ezema, I. J., & Ugwu, C. I. (2013). Electronic theses and dissertations in Nigeria university libraries: Status, challenges and strategies. *The Electronic Library, 31*(4), 493–507.

Fafunwa, A. B., & Aisiku, J. U. (2022). *Education in Africa: a comparative survey*. Routledge., DOI: 10.4324/9781003316114

Fairclough, G., Rippon, S., Bull, D., & de l'Europe, C. (2002). *Europe's Cultural Landscape: Archaeologists and the Management of Change: Europae Archaeologiae Consilium, Mars 2000*. Europae Archaeologiae Consilium.

Fecher, B., & Friesike, F. (2013). Open Science: One Term, Five Schools of Thought, RatSWD Working Paper, No. 218, Rat für Sozial- und Wirtschaftsdaten (RatSWD), Berlin.

Ferguson, M., Brown, C., Georga, C., Miles, E., Wilson, A., & Brimblecombe, J. (2017). Traditional food availability and consumption in remote Aboriginal communities in the Northern Territory, Australia. *Australian and New Zealand Journal of Public Health, 41*(3), 294–298. PMID: 28370899

Fernández-Llamazares, Á., Lepofsky, D., Lertzman, K., Armstrong, C. G., Brondizio, E. S., Gavin, M. C., Lyver, P. O., Nicholas, G. P., Pascua, P., Reo, N. J., Reyes-García, V., Turner, N. J., Yletyinen, J., Anderson, E. N., Balée, W., Cariño, J., David-Chavez, D. M., Dunn, C. P., Garnett, S. C., & Vaughan, M. B. (2021). Scientists' Warning to Humanity on Threats to Indigenous and Local Knowledge Systems. *Journal of Ethnobiology, 41*(2), 144–169. DOI: 10.2993/0278-0771-41.2.144

Finlay, K., & Barnabas, S. (2012). Shifting representations of the Bushmen. In *Cultural Tourism and Identity* (pp. 71–83). Brill.

First Nations Information Governance Centre. (2014). Ownership, Control, Access, and Possession (OCAP): The Path to First Nations Information Governance. Retrieved from https://fnigc.ca/

Flor, A. (2013). Exploring the Downside of Open Knowledge Resources: The Case of Indigenous Knowledge Systems and Practices in the Philippines. *Open Praxis, 5*(1), 75. DOI: 10.5944/openpraxis.5.1.15

Forutnani, S., Nowkarizi, M., & Kiani, M. R.. (2018). The role of rural libraries in preserving the indigenous knowledge of rural residents. *World Journal of Science Technology and Sustainable Development, 15*(3), 245–256.

Franks, P. C. (2013). *Records and information management*. American Library Association.

Friesike, S., Widenmayer, B., Gassmann, O., & Schildhauer, T. (2015). Opening science: Towards an agenda of open science in academia and industry. *The Journal of Technology Transfer, 40*(4), 581–601.

Funk, J., & Guthadjaka, K. (2020). Indigenous Authorship on Open and Digital Platforms: Social Justice Processes and Potential. *Journal of Interactive Media in Education, 2020*(1). DOI: 10.5334/jime.560

Galla, C. K. (2018). Digital Realities of Indigenous Language Revitalization: A Look at Hawaiian Language Technology in the Modern World. *Language and Literature*, *20*(3), 100–120. DOI: 10.20360/langandlit29412

Gallert, P, Winschiers-Theophilus, H, Kapuire, G & Stanley, C. (2016). Indigenous Knowledge for Wikipedia: A Case Study with an OvaHerero Community in Eastern Namibia.

Gardner, H., & Davis, K. (2017). *The app generation: How today's youth navigate identity, intimacy, and imagination in a digital world.* Yale University Press., DOI: 10.12987/9780300199185

Garrett-Ruffin, S., Hindash, A. C., Kaczkurkin, A. N., Mears, R. P., Morales, S., Paul, K., Pavlov, Y. G., & Keil, A. (2021). Open science in psychophysiology: An overview of challenges and emerging solutions. *International Journal of Psychophysiology*, *162*, 69–78. DOI: 10.1016/j.ijpsycho.2021.02.005 PMID: 33556468

Garwood-Houng, A. (2005). Protocols: Meeting the challenges of indigenous information needs. *Australian Academic and Research Libraries*, *36*(2), 143–151.

Geia, L. K., Hayes, B., & Usher, K. (2013). Yarning/Aboriginal Storytelling: Towards an understanding of an Indigenous perspective and its implications for research practice. *Contemporary Nurse*, *46*(1), 13–17. DOI: 10.5172/conu.2013.46.1.13 PMID: 24716757

Ghimire, P. (2021). Digitalization of Indigenous Knowledge in Nepal – Review Article. *Acta Informatica Malaysia*, *5*(2), 42–47. DOI: 10.26480/aim.02.2021.42.47

Ghoshal, S. (2005). Bad Management Theories are Destroying Good Management Practices. *Academy of Management Learning & Education*, *4*, 75–91.

Gilliland, A. J., & Caswell, M. (2015). Records and their imaginaries: Imagining the impossible, making possible the imagined. *Archival Science*, *16*, 53–75.

Gitonga, G. M., Ngeranwa, J., Machocho, A. K., & Kiruki, S. (2020). Qualitative Phytochemical Screening and Antipyretic Activity of Methanolic Stem Bark Extract Of Mangifera Indica in Wistar Rats. *The Journal of Phytopharmacology*, *9*(4), 218–223.

Global Indigenous Data Alliance. (2019). CARE principles for Indigenous Data Governance.

Gomez, R. (2012). *Libraries, Telecentres, Cybercafés and Public Access to ICT: International Comparisons*. IGI Global., DOI: 10.4018/978-1-60960-771-5

Goredema, D. (2016). *Zimbabwe's Liberation Struggle Era Conflicts and the Pitfalls of Reconciliation after Independence: a case Study of Bikita District 1976-2013.* A thesis submitted to the Midlands State University in partial fulfilment of the requirements for the degree of Doctor of Philosophy in History.

Gorjestani, N. (2000). Indigenous knowledge for development. In *UNCTAD conference on traditional knowledge* (pp. 1–8). UNCTAD., DOI: 10.1111/an.1999.40.7.20

Gownaris, N., Vermeir, K., Bittner, M., Gunawardena, L., Kaur-Ghumaan, S., Lepenies, R., & Zakari, I. (2022). Barriers to full participation in the open science life cycle among early career researchers. *Data Science Journal, 21*. Advance online publication. DOI: 10.5334/dsj-2022-002

Graham, B., & Howard, P. (2016). Heritage and identity. In *The Routledge Research Companion to Heritage and Identity* (pp. 1–15). Routledge.

Grandin, T., & Deesing, M. J. (2022). Genetics and behavior during handling, restraint, and herding. In *Genetics and the behavior of domestic animals* (pp. 131–181). Academic press., https://www.sciencedirect.com/science/article/pii/B9780323857529000032

Grenier, L. (1998). *Working with indigenous knowledge: a guide for researchers.* International Development Research Centre. https://idrc-crdi.ca/en/books/working-indigenous-knowledge-guide-researchers

Greyling, E., & McNulty, N. (2011). *How to build an indigenous digital library through community participation: The case of the Ulwazi Programme.* Available at http://www.academia.edu/837039/How_to_build_an_Indigenous_Digital_Library_through_Community_Participation_the_case_of_the_Ulwazi_Programme

Greyling, E., & McNulty, N. (2012), June). "How to build an indigenous digital library through community participation: the case of the ulwazi programme", Information for Sustainable Development in a Digital Environment. *XXth Standing Conference of Eastern, Central and South Africa Library and Information Associations (SCECSAL).*

Gruber, S. (1972). *Convention Concerning the Protection of the World Cultural and Natural Heritage 1972.* Edward Elgar Publishing.

Gruen, L. (2024). Tangled Manes, Tangled Chains. *The Georgic Mode in Twentieth-century American Literature: The Satisfactions of Soil and Sweat*, 137. https://books.google.com/books?hl=en&lr=&id=

Gundakanal, S. S., & Kaddipujar, M. (2024). Cultural custodians: the role of libraries in preserving and promoting language, literature, and heritage. http://dspace.rri.res.in/handle/2289/8297?mode=full

Guo, L., Yan, F., Li, T., Yang, T., & Lu, Y. (2022). An automatic method for constructing machining process knowledge base from knowledge graph. *Robotics and Computer-integrated Manufacturing*, *73*, 102222. https://www.sciencedirect.com/science/article/pii/S0736584521001058

Gupta, J., Pouw, N. R., & Ros-Tonen, M. A. (2015). Towards an elaborated theory of inclusive development. *European Journal of Development Research*, *27*(4), 541–559.

Gurib-Fakim, A. (2006). Medicinal plants: Traditions of yesterday and drugs of tomorrow. *Molecular Aspects of Medicine*, *27*(1), 1–93. DOI: 10.1016/j.mam.2005.07.008 PMID: 16105678

Gutam, S. (2024). Empowering Research Communities: Open Access as a Catalyst for a Sustainable Knowledge Society. Researchgate.

Guzzo, R. A., Schneider, B., & Nalbantian, H. R. (2022). Open science, closed doors: The perils and potential of open science for research in practice. *Industrial and Organizational Psychology: Perspectives on Science and Practice*, *15*(4), 495–515. DOI: 10.1017/iop.2022.61

Haddad, A., Habaebi, M. H., Islam, M. R., Hasbullah, N. F., & Zabidi, S. A. (2022). Systematic review on ai-blockchain based e-healthcare records management systems. *IEEE Access : Practical Innovations, Open Solutions*, *10*, 94583–94615. DOI: 10.1109/access.2022.3201878

Hadjimichael, D., & Tsoukas, H. (2019). Toward a better understanding of tacit knowledge in organizations: Taking stock and moving forward. *The Academy of Management Annals*, *13*(2), 672–703. https://doi/abs/10.5465/annals.2017.0084

Hagger, M. S. (2021). Developing an open science 'mindset'. *Health Psychology and Behavioral Medicine*, *10*(1), 1–21. DOI: 10.1080/21642850.2021.2012474 PMID: 34993003

Hamwaalwa, N. (2021). Indigenous knowledge management: roles of public libraries and archives in preserving and providing access to indigenous knowledge (IK) in Namibia: Case study of "Preserve Namibia Indigenous Knowledge" project. In C. Mnubi-Mchombu, C. Nengomasha, & W. Uutoni (Eds.), *Inclusive libraries and information services towards achieving prosperity for sustainable development in Africa: Proceedings of the 24th Standing Conference of Eastern, Central and Southern African Library and Information Associations* (pp. 289-292). National Information Workers Association of Namibia. https://www.scecsal.org/publications/papers2021/scecsal_papers2021_hamwaalwa.pdf

Handique, P., & Dubey, A. (2022). Legal mechanisms for protecting Indigenous knowledge. *Interdisciplinary Studies in Society, Law, and Politics*, *1*(1), 4–13.

Handique, P., & Dubey, A. (2022). Legal Mechanisms for Protecting Indigenous Knowledge. *Isslp*, *1*(1), 4–13. DOI: 10.61838/kman.isslp.1.1.2

Harisanty, D., Variant Anna, N. E., Putri, T. E., Firdaus, A. A., & Azizi, N. A. A. (2023). Is adopting artificial intelligence in libraries urgency or a buzzword? A systematic literature review. *Journal of Information Science*, •••, 1–12.

Hautemo, A. M., & Mushaandja, T. (2024). Technauriture: The Intersection Between Orature and Technology. In Oojorah, V. A., & Udhin, W. (Eds.), *Imagining the Futures of Higher Education in Southern Africa. Future Education and Learning Spaces*. Springer. DOI: 10.1007/978-981-97-6473-0_10

Heinle, M., Ross, N., & Sauma, R. (2014). A Theory of Participative Budgeting. *The Accounting Review*, *89*, 1025–1050.

Hendrawan, M. R., Isa, A. M., & Samsudin, A. Z. H. (2024). Initiating Memory Institutions Convergence Through Digital Convergence in Indonesian World Heritage Sites. In *Multidisciplinary Approach to Information Technology in Library and Information Science* (pp. 231–251). IGI Global., DOI: 10.4018/979-8-3693-2841-5.ch013

Hernandez, A. (2025). Preserving and transferring Indigenous knowledge in a digital world. https://www.giga-hamburg.de/en/events/conferences-and-workshops/co-la-borate-2-preserving-and-transferring-indigenous

Higginson, A. D., & Munafò, M. R. (2016). Current incentives for Scientists lead to Underpowered studies with erroneous conclusions. *PLoS Biology*, *14*, e2000995. DOI: 10.1371/journal.pbio.2000995 PMID: 27832072

Hilbolling, S., Berends, H., Deken, F., & Tuertscher, P. (2020). Complementors as connectors: Managing open innovation around digital product platforms. *R & D Management*, *50*(1), 18–30. DOI: 10.1111/radm.12371

Hildebrand, J., & Magnan, P. (2025). How Indigenous Communities Advance Climate Action at COP Conferences. https://www.weforum.org/stories/2025/01/indigenous-communities-advance-climate-action-cop-conferences/

Hinton, L. (2019). Language Revitalization and New Technologies: Cultures of Electronic Mediation and the Refiguring of Communities. In Jaffe, J., Morita, D., Vogel, N. D., & Spolsky, G. (Eds.), *Language Ideologies and the Globalization of 'Standard' Languages* (pp. 244–264). Routledge.

Hipolito, M. F. G. (2024). A Systematic Literature Review of the Significance of the School of Living Tradition in the Preservation of Cultural Heritage of the Indigenous Peoples. *International Journal of Multidisciplinary: Applied Business and Education Research*, *5*(2), 605–614. https://www.ijmaberjournal.org/index.php/ijmaber/article/view/1401

Hippolyte, N., & Namaseb, L. (2011). Can IK be Communicated through Foreign Languages without Losing its Authenticity? *IKTC2011: Embracing Indigenous Knowledge Systems in a New Technology Design Paradigm.* ndigenous Knowledge Technology Conference. 2011. Windhoek, Namibia, 2 - 4 November 2011, pp.116–117.

Hoffman, A. (2005). Ethical challenges in librarianship. *Journal of Academic Librarianship*, *31*(4), 271–274.

Hollaway, M. J., Dean, G., Blair, G. S., Brown, M., Henrys, P. A., & Watkins, J. (2020). Tackling the Challenges of 21st-Century Open Science and Beyond: A Data Science Lab Approach. *Patterns (New York, N.Y.)*, *1*(7), 100103. DOI: 10.1016/j.patter.2020.100103 PMID: 33205137

Hombalmath, M. M., & Mathad, M. A. (2021, November). Design Optimization of Bullock Cart Yoke. []. IOP Publishing. https://iopscience.iop.org/article/10.1088/1742-6596/2070/1/012154/meta]. *Journal of Physics: Conference Series*, *2070*(1), 012154.

Hoppers, C. O. (2021). Research on Indigenous knowledge systems: The search for cognitive justice. *International Journal of Lifelong Education*, *40*(4), 310–327. DOI: 10.1080/02601370.2021.1966109

Horowitz, H. (1981). The UNESCO framework for cultural statistics and a cultural data bank for Europe. *Journal of Cultural Economics*, *5*, 1–17.

Horsthemke, K. (2008). The Idea of Indigenous Knowledge. *Arch, 4*, 129–143. DOI: 10.1007/s11759-008-9058-8

Horsthemke, K. (2008). The idea of indigenous knowledge. *Archaeologies, 4*, 129–143.

Horsthemke, K. (2017). Indigenous (African) Knowledge Systems, Science, and Technology. In Afolayan, A., & Falola, T. (Eds.), *The Palgrave Handbook of African Philosophy*., DOI: 10.1057/978-1-137-59291-0_38

Houtkoop, B. L., Chambers, C., Macleod, M., Bishop, D. V., Nichols, T. E., & Wagenmakers, E. J. (2018). Data sharing in psychology: A survey on barriers and preconditions. *Advances in Methods and Practices in Psychological Science*, (1), 70–85. DOI: 10.1177/2515245917751886

Huang, T.-L., & Tseng, C.-H. (2015). Using augmented reality to reinforce vivid memories and produce a digital interactive experience. *Journal of Electronic Commerce Research, 16*, 307.

Hunter, A. (2021). *Co-developing Openness: Indigenous Knowledge and Data Governance and Open Science in Canada* [Master of Arts, Carleton University]. DOI: 10.22215/etd/2021-14620

Hunter, J. (2005). The Role of Information Technologies in Indigenous Knowledge Management. *Australian Academic and Research Libraries, 36*(2), 109–124. DOI: 10.1080/00048623.2005.10721252

Hutchinson, P. J., McIlduff, C. D., Legare, M., Keewatin, M., Hagel, M., Chapados, M., & Acharibasam, J. B. (2023). Indigenous knowledge mobilization: Reflection on context, content, and relationship. *AlterNative: An International Journal of Indigenous Peoples, 19*(4), 902–913. DOI: 10.1177/11771801231198082

IFLA. (2002). *Guidelines for digitization projects for collections and holdings in the public domain, particularly those held in libraries and archives.* Available at https://www.ifla.org/VII/s19/sconsv.htm

IFLA. (2002). IFLA Statement on Indigenous Traditional Knowledge. https://www.ifla.org/publications/ifla-statement-on-indigenous-traditional-knowledge/

Igbafe, E. C. (2023). Researching African Indigenous Population in Contemporary Times: What Scholars and Practitioners Need to Know. In Gbendazhi Barnabas, S. (Ed.), *Indigenous and Minority Populations—Perspectives From Scholars and Writers across the World*. IntechOpen., DOI: 10.5772/intechopen.107591

Ijatuyi, E. J., Lamm, A., Yessoufou, K., Suinyuy, T., & Patrick, H. O. (2025). Integration of indigenous knowledge with scientific knowledge: A systematic review. *Environmental Science & Policy*, *170*, 104119.

Imenda, S. (2014). Is there a conceptual difference between theoretical and conceptual frameworks? *Journal of Social Sciences*, *38*(2), 185–195.

International Federation of Library Associations and Institutions (IFLA). (2002). *IFLA Statement on Indigenous Traditional Knowledge.* The Hague: IFLA. https://www.ifla.org/publications/ifla-statement-on-indigenous-traditional-knowledge

Intezari, A., Namvar, M., & Taghinejad, R. (2022). Knowledge identity (KI): A determining factor in the effective use of analytics. *Knowledge Management Research and Practice*, *20*(1), 14–33. https://doi/abs/10.1080/14778238.2021.1967213

Isa, W. M. W., Zin, N. A. M., Rosdi, F., & Sarim, H. M. (2018). Digital Preservation of Intangible Cultural Heritage. *Indonesian Journal of Electrical Engineering and Computer Science*, *12*(3), 1373. DOI: 10.11591/ijeecs.v12.i3.pp1373-1379

Islam, R. (2023). The Gods Must Be Crazy (1980 & 1989): A Humorous Reflection on Modernity, Nature, and Contrasting Human Cruelty on Nature and Simplicity. https://www.probinism.com/the-gods-must-be-crazy-human-cruelty-on-nature/

Jacobsen, A., Azevedo, R. M., Juty, N., Batista, D., Coles, S., Cornet, R., Courtot, M., Crosas, M., Dumontier, M., & Evelo, C. T.. (2020). FAIR Principles: Interpre tations and Implementation Considerations. *Data Intelligence*, *2*, 10–29.

Jaya, E. (2006). The role of university in promoting indigenous knowledge systems in Zimbabwe with reference to traditional practices in rural areas", *2nd International Appropriate Technology Conference*, Bulawayo, Zimbabwe, 12–15 July 2006, pp. 1-7.

Jennings, D., Johnson-Jennings, M., & Little, M. (2020). Utilizing webs to share ancestral and intergenerational teachings: The process of co-building an online digital repository in partnership with indigenous communities. *Genealogy*, *4*(70). Advance online publication. DOI: 10.3390/genealogy4030070

Jia, J., Zhang, Y., & Saad, M. (2022). An approach to capturing and reusing tacit design knowledge using relational learning for knowledge graphs. *Advanced Engineering Informatics*, *51*, 101505. https://www.sciencedirect.com/science/article/pii/S1474034621002548

Jin, X., Peng, B., Wu, Y., Liu, Y., Liu, J., Liang, D., . . . Hu, X. (2019). Knowledge distillation via route constrained optimization. In *Proceedings of the IEEE/CVF International Conference on Computer Vision* (pp. 1345-1354). https://openaccess.thecvf.com/content_ICCV_2019/html/Jin_Knowledge_

Jiri, O., Mafongoya, P. L., Mubaya, C., & Mafongoya, O. (2016). Seasonal Climate Prediction and Adaptation Using Indigenous Knowledge Systems in Agriculture Systems in Southern Africa: A Review. *Journal of Agricultural Science*, *8*(5), 156. DOI: 10.5539/jas.v8n5p156

Jones, S., Pergl, R., Hooft, R., Miksa, T., Samors, R., Ungvari, J., Davis, R. I., & Lee, T. (2020). Data Management Planning: How Requirements and Solutions are Beginning to Converge. *Data Intelligence*, *2*, 208–219.

Jonnaert, P., Ndinga, P., Ettayebi, M., Barry, A., Rabinovitch, L., & Malu, R. (2021). *Towards indigenous curricula. In-Progress Reflection. 41. On Current and Critical Issuesin Curriculum, Learningand Assessment*. UNESCO.

Ka'ai-Mahuta, R. (2019). Preserving Indigenous voices: Web archiving in Aotearoa/ New Zealand. *Interaction Design and Architecture(s) Journal - IxD&A, 41, (2019): 24 – 30.*

Ka'ai-Mahuta, R. (2012). Digital Technology: Contemporary methods for the Dissemination of Ancient Knowledge. The Use of Digital Technology in the Preservation of Māori Song. *Te Kaharoa. 5(2012), Special Edition, 2012, ISSN 1178-6035.*

Ka'ai-Mahuta, R. (2019). Preserving Indigenous Voices: Web Archiving in Aotearoa/ New Zealand. [s]. *Interaction Design and Architecture*, *41*, 24–30. DOI: 10.55612/s-5002-041-002

Kahakalau, K. (2004). Indigenous Heuristic Action Research: Bridging Western and Indigenous Research Methodologies. *Multidisciplinary Research on Hawaiian Well-Being. 1 (1).*

Kakembo, A. A. (2024). Evaluating the effectiveness of educational policies. *IDOSR Journal of Communication and English.*, *9*(3), 6–10. DOI: 10.59298/IDOSR/ JCE/93.610.202400

Kalle, R., & Kass, M. (2020). Local practice of cattle farming and ethnoveterinary medicine in Estonia: Case study of Saaremaa and Muhumaa. *Ethnoveterinary medicine: Present and future concepts*, 345-374. https://link.springer.com/chapter/10 .1007/978-3-030-32270-0_15

Kamal, S. & Amani M. (2020). Digitization and Indigenous Knowledge: Challenges and Opportunities in Africa

Kamboya, U. (2018). Indigenous knowledge and sustainable development: Role of libraries. *International Journal of Library and Information Studies*, *8*(1), 495–499.

Karega, P., Mwaura, D. K., Mwangi, K. W., Wanjiku, M., Landi, M., & Kibet, C. K. (2023). Building awareness and capacity of bioinformatics and open science skills in Kenya: A sensitize, train, hack, and collaborate model. *Frontiers in Research Metrics and Analytics*, *8*, 1070390. DOI: 10.3389/frma.2023.1070390 PMID: 37324282

Karenga, M. (2008). *Kawaida and Questions of Life and Struggle: African American, Pan-African, and Global Issues*. University of Sankore Press.

Karjalainen, J., & Okere, I. (2018). Local institutions and Indigenous knowledge in climate-smart agricultural innovations. *International Journal of Climate Change Strategies and Management*.

Kaschula (2013).Reviewed Work(s): Oral Literature in Africa by Ruth Finnegan. Review by: Russell H. Kaschula. Source: Journal of African Cultural Studies, March 2013, Vol. 25, No. 1, Special Issue: New Media Entrepreneurs and Changing Styles of Public Communication in Africa (March 2013), pp. 141-144 .Published by: Taylor & Francis, Ltd

Kaschula, R. H. (2004). Imbongi to slam: The emergence of a technologised Auriture. *Southern African Journal of Folklore Studies*, *14*(2), 45–58.

Kaschula, R. H. (2017). Technauriture as a platform to create an inclusive environment for the sharing of research. *Searching for Sharing: Heritage and Multimedia in Africa*, *7*(2), 41–60. DOI: 10.11647/OBP.0111

Kaylo, P. (2012). A Reflection on the African Traditional Values of Marriage and Sexuality. *International Journal of Academic Research in Progressive Education and Development*, *1*(2), 84–92.

Keane, M., Khupe, C., & Muza, B. (2016). It matters who you are: Indigenous knowledge research and researchers. *Education as Change*, *20*(2), 163–183.

Kennedy, M., Ninomiya, M. M., & Ninomiya, M. M., Brascoupé (Anishinabeg/ Haudenausanee), S., Smylie (M tis), J., Calma (Kungarakan, Iwaidja), T., Mohamed (Narrunga Kaurna), J., Stewart (Taungurung), P. J., & Maddox (Bagumani, Modewa), R. (2024). Knowledge translation in Indigenous health research: Voices from the field. *The Medical Journal of Australia*, *221*(1), 61–67. DOI: 10.5694/mja2.52357 PMID: 38946651

Kessi, S., Marks, Z., & Ramugondo, E. (2020). Decolonizing African Studies. *Critical African Studies*, *12*(3), 271–282. DOI: 10.1080/21681392.2020.1813413

Ketelaar, E. (2017). Archival turns and returns: Studies of the archive. In *Research in the Archival Multiverse* (pp. 228–268). Monash University Publishing.

Khalafzai, M. A. K. (2023). Climate Change-Related Natural Hazards and Risk Communication: Incorporating Traditional Indigenous Knowledge. In Mokhtari, M. (Ed.), *Natural Hazards—New Insights*. IntechOpen., DOI: 10.5772/intechopen.108302

Khalil, A. T., Shinwari, Z. K., & Islam, A. (2022). Fostering openness in open science: An ethical discussion of risks and benefits. *Frontiers in Political Science*, *4*, 930574. DOI: 10.3389/fpos.2022.930574

Khan, M. and Gope, L. (2025). Aligning Indigenous Knowledge with Digital Technology for Achieving Sustainable Development Goals.

Kiely, T. J. (2024). What Is Desk Research? Meaning, Methodology, Examples https://www.meltwater.com/en/blog/desk-research (Accessed 18 November 2024).

Kimberly, C., Davis, L., Griffith, Z., & Neely, J. (2018). Traditional Knowledge and Digital Archives: An Interview with Kim Christen. *Discl. A J. Sociological Theory*, *27*, 5.

Kinyua, J. (2020). Challenges in Implementing Digital Technologies in Rural Kenya. https://www.engineeringforchange.org/news/challenges-implementing-digital-technologies-rural-kenya/

Kiplagat, J., Chepken, C. and Kiptoo, C. (2024). Digital Repository for African History Culture and Heritage: A Case of Keiyo Community of Kenya. *iPRES 2024 Proceedings. iPRES 2024.*

Klump, J., Bertelmann, R., Brase, J., Diepenbroek, M., Grobe, H., Höck, H., Lautenschlager, M., Schindler, U., Sens, I., & Wächter, J. (2006). Data publication in the open access initiative. *Data Science Journal*, *5*, 79–83.

Kolbe, V. (2022). Open Science versus Data Protection—Challenges and Solutions in Sign Language Acquisition Studies. *Hrvatska Revija za Rehabilitacijska Istraživanja*, *58*, 109–120.

Koley, M. (2022). Analysis of Open Science Policy Recommendations Proposed in India's 5th Science, Technology & Innovation Policy Draft. *Journal of Science Policy & Governance.*, *21*(02). Advance online publication. DOI: 10.38126/JSPG210208

Koster, R., Baccar, K., & Lemelin, R. H. (2012). Moving from research On, to Research WITH and FOR Indigenous communities: A critical reflection on community-based participatory research. *The Canadian Geographer. Geographe Canadien*, *56*(2), 195–210.

Kovach, M. (2021). *Indigenous methodologies: Characteristics, conversations, and contexts*. University of Toronto press. https://books.google.com/books?hl=en&lr=&id=0V47EAA

Kovach, M. (2005). Emerging from the margins: indigenous methodologies. In Brown, L., & Strega, S. (Eds.), *Research as resistance: Critical, Indigenous and anti-oppressive approaches* (pp. 19–36). Canadian Scholars' Press.

Kugara, S. L. and Mokgoatsana, S.(2022). Challenges presented by the digitization of the Vha Venda oral tradition: An African Indigenous Knowledge perspective: HTS Teologies Studies.

Kuhn, R., Davis, F., Désilets, A., Joanis, E., Kazantseva, A., Knowles, R., Littell, P., Lothian, D., Pine, A., Running Wolf, C., Santos, E., Stewart, D., Boulianne, G., Gupta, V., Maracle Owennatékha, B., Martin, A., Cox, C., Junker, M.-O., Sammons, O., & Souter, H. (2020). The Indigenous Languages Technology project at NRC Canada: An empowerment-oriented approach to developing language software. *Proceedings of the 28th International Conference on Computational Linguistics*, 5866–5878. DOI: 10.18653/v1/2020.coling-main.516

Kumar, A. (2022). Protection of Traditional Knowledge. Issue 2 Indian JL & Legal Rsch., 4, 1.

Kumar, P., Madaan, S., & Bhargava, G. (2023, January). An Insight About Bell Metal Craft—Scopes of Improvement and Promotion. In *International Conference on Research into Design* (pp. 965–975). Springer Nature Singapore., https://link.springer.com/chapter/10.1007/978-981-99-0264-4_79

Kunduru, A. R., & Kandepu, R. (2023). Data archival methodology in enterprise resource planning applications (Oracle ERP, People soft). *Journal of Advances in Mathematics and Computer Science, 38*(9), 115–127. http://public.paper4promo.com/id/eprint/1106/

Kwiatkowski, R. E. (2011). Indigenous community-based participatory research and health impact assessment: A Canadian example. *Environmental Impact Assessment Review, 31*(4), 445–450. DOI: 10.1016/j.eiar.2010.02.003

Landborn, A. (2019). Chapter Sixteen La Malagueña Y El Torero: Flamenco Dance and Bullfighting. *Transatlantic Malagueñas and Zapateados in Music, Song and Dance: Spaniards, Natives, Africans, Roma*, 294. https://books.google.com/books?hl=en&lr=&id=i0OeDwAAQBAJ&oi=

Landi, A., Thompson, M., Giannuzzi, V., Bonifazi, F., Labastida, I., Bonino da Silva Santos, L. O., & Roos, M. (2020). The "A" of FAIR—As open as possible, as closed as necessary. *Data Intelligence*, *2*, 47–55.

Langmia, K., & Sani, M. H. (2023). Decolonizing the African Mind in the Digital Space. In Akpan, U. S. (Ed.), *African Media Space and Globalization*. Palgrave Macmillan. DOI: 10.1007/978-3-031-35060-3_15

Larrière, S., & Maurice, S. (2021). Orality in the Digital Age. https://leblob.fr/en/visions-africa/orality-digital-age

Lauter, O. (2020). *Challenges in Combining Indigenous and Scientific Knowledge in the Arctic. SocArXiv*. DOI: 10.31235/osf.io/kvn2c

Lauter, O. (2023). Challenges in combining Indigenous and scientific knowledge in the Arctic. *Polar Geography*, *46*(1), 62–74. DOI: 10.1080/1088937X.2023.2233578

Lawton, B. (2023). *Various and Ingenious Machines. Volume Two: Manufacturing and Weapons Technology* (Vol. 6). Brill. https://books.google.com/books?hl=en&lr=&id=

le Roux, A. (2017). *Indigenous knowledge in a virtual context: Sustainable digital preservation. Unisa Library*. Retrieved from WiredSpace.

Lee, D. (2011). Indigenous knowledge organization: a study of concepts, terminology, structure and (Mostly) indigenous voices. *Partnership the Canadian Journal of Library and Information Practice and Research*, 6(1).

Lee, P. C. (2024). Unveiling community resilience: The integral role of public libraries. *Journal of Library Administration*, *64*(2), 194–216. DOI: 10.1080/01930826.2024.2305072

Levin, N., Leonelli, S., Weckowska, D., Castle, D., & Dupré, J. (2016). How Do Scientists Define Openness? Exploring the Relationship Between Open Science Policies and Research Practice. *Bulletin of Science, Technology & Society*, *36*(2), 128–141. DOI: 10.1177/0270467616668760 PMID: 27807390

Lindemann, T., & Häberlein, L. (2023). Contours of a research ethics and integrity perspective on open science. *Frontiers in Research Metrics and Analytics*, *8*, 1052353. DOI: 10.3389/frma.2023.1052353 PMID: 37235074

Lopez, J. M. C., Canalejo, A. M. C., & Márquez, C. M. (2022). Perception of and motivation for attending bullfights: A comparative analysis of tourists and residents. *Cuadernos de Turismo*, (50), 333–353. https://revistas.um.es/turismo/article/view/542001

Louis, R. P. (2007). Can You Hear us Now? Voices from the Margin: Using Indigenous Methodologies in Geographic Research. *Geographical Research*, *45*(2), 130–139. DOI: 10.1111/j.1745-5871.2007.00443.x

Loupis, N., & Saffy, L. (2020). Indigenous Knowledge and Digital Repositories: Ethical and Legal Considerations. *International Journal of Cultural Heritage*, *15*(2), 134–146.

Louw, E., & King, G. (2021). Digital Preservation and the Revitalization of Endangered Indigenous Languages in Southern Africa. *Journal of Language and Cultural Revitalization*, *8*(1), 67–82.

Lugmayr, A., Sutinen, E., Suhonen, J., Sedano, C. I., Hlavacs, H., & Montero, C. S. (2017). Serious storytelling–a first definition and review. *Multimedia Tools and Applications*, *76*, 15707–15733. https://link.springer.com/article/

Lui, P.O., Skewes, M.C., Gobrial, S., and Rollock, D. (2021). Advancing Transparency and Impact of Research and Impact of Research: Initiating ch: Initiating Crosstalk between Indigenous Research and Mainstream "Open Science". *Journal of Indigenous. .9(2021).*

Lui, P. P., Skewes, M. C., Gobrial, S., & Rollock, D. (2021). Advancing transparency and impact of research: Initiating crosstalk between Indigenous research and mainstream "open science". *Journal of Indigenous Research*, *9*, 10.

Luo, L. (2016). Professional ethics in librarianship: A case study of ethical dilemmas faced by reference librarians. *The Library Quarterly*, *86*(2), 174–189.

Lwoga, E. T., Ngulube, P., & Stilwell, C. (2010). Managing indigenous knowledge for sustainable agricultural development in developing countries: Knowledge management approaches in the social context. *The International Information & Library Review*, *42*(3), 174–185.

Ma, J., & Jing, E. (2017). *Mission Alignment as a Substitute or Incentive: How Nonprofits Utilize the Alignment Between Mission Statement and Daily Operation.* Rochester, NY, USA, 2017.

Maasz, D., Winschiers-Theophilus, H., Stanley, C., Rodil, K., & Mbinge, U. (2018). *A digital Indigenous knowledge preservation framework: The 7C model—Repositioning IK holders in the Digitisation of IK. Digitisation of Culture: Namibian and International Perspectives*. Retrieved from SpringerLink.

Macdonald, S. (2004). A people's story: Heritage, identity and authenticity. In *Heritage, Museums and Galleries* (pp. 299–319). Routledge.

Madonsela, B. S., Semenya, K., & Shale, K. (2024). A Review of Indigenous Knowledge Systems and Their Application in Sustainable Solid Waste Management. *WORLD (Oakland, Calif.)*, *5*(2), 219–239. DOI: 10.3390/world5020012

Mahdi, O. R., Nassar, I. A., & Almsafir, M. K. (2019). Knowledge management processes and sustainable competitive advantage: An empirical examination in private universities. *Journal of Business Research*, *94*, 320–334. https://www.sciencedirect.com/science/article/pii/S0148296318300882

Mahomoodally, M. F. (2013). Traditional Medicines in Africa: An Appraisal of Ten Potent African Medicinal Plants. *Evidence-Based Complemetary and Altenative Medicine*, *2013*(1), 1–14. DOI: 10.1155/2013/617459 PMID: 24367388

Mahuika, N. (2019). *Rethinking oral history and tradition: An Indigenous perspective*. Oxford University Press., https://books.google.com/books?hl=en&lr=&id=z5KwDwAAQBAJ&oi=

Mahwasane, N. P. (2017). Library support and the innovative application of indigenous knowledge systems. *Journal of Sociology and Social Anthropology*, *8*(2), 77–81.

Makgabo, C., & Quintero, G. J. (2024). Philippine and South African Experiences on Folk Literature Research: Relevance, Gains, and Challenges. *Journal of Languages and Language Teaching*, *12*(2), 985–995. https://e-journal.undikma.ac.id/index.php/jollt/article/view/10667

Makonya, L., (2024). Safeguarding traditional Knowledge in Zimbabwe. *Manica Post*, 08 November 2024.

Malapane, O. L., Chanza, N., & Musakwa, W. (2024). Transmission of indigenous knowledge systems under changing landscapes within the Vhavenda community, South Africa. *Environmental Science & Policy*, *161*, 103861.

Malapane, O. L., Musakwa, W., Chanza, N., & Radinger-Peer, V. (2022). Bibliometric Analysis and Systematic Review of Indigenous Knowledge from a Comparative African Perspective: 1990–2020. *Land (Basel)*, *11*, 1167. DOI: 10.3390/land11081167

Maldonado, J., & Lazrus, H. (2019). A Story of "Rising Voices" and Intercultural Collaboration. *Practical Anthropology*, *41*(3), 34–37. DOI: 10.17730/0888-4552.41.3.34

Mambure, D. (2024). Status Quo of Divorces in Zimbabwe. [IJRISS]. *International Journal of Research and Innovation in Social Science*, *8*(8), 4452–4461.

Manco, A. (2023). Open science policies as regarded by the communities of researchers from the basic sciences in the scientific periphery. *Online Information Review*, *48*(6), 1065–1087. DOI: 10.1108/OIR-03-2023-0135

Manesh, M. F., Pellegrini, M. M., Marzi, G., & Dabic, M. (2020). Knowledge management in the fourth industrial revolution: Mapping the literature and scoping future avenues. *IEEE Transactions on Engineering Management*, *68*(1), 289–300. https://ieeexplore.ieee.org/abstract/document/8964410/

Markowitz, D. M., Song, H., & Taylor, S. H. (2021). *Tracing the Adoption and Effects of Open Science in Communication Research. PsyArXiv*. DOI: 10.31234/osf.io/dsf67

Maroyi, A. (2013). Traditional Use of Medicinal Plants in South-Central Zimbabwe: Review and Perspectives. *Journal of Ethnobiology and Ethnomedicine*, *9*(31). Advance online publication. DOI: 10.1186/1746-4269-9-31 PMID: 23642285

Martin, K. L. (2016). Vision led = vision fed: NIRAKN's role in progressing scholarship in Aboriginal early childhood education, raising the bar in Aboriginal knowledges in higher education and holding ground in Aboriginal research capacity building. *International Journal of Critical Indigenous Studies*, *9*(2), 1–14. DOI: 10.5204/ijcis.v9i2.137

Masango, C. A. (2010). Indigenous Traditional Knowledge Protection:prospects in South Africa's Intellectual Property Framework? *South African Journal of Library and Information Science*, *76*(1). Advance online publication. DOI: 10.7553/76-1-88

Masenya, T. M. (2022). *Digital Preservation of Indigenous Knowledge in South African Rural Communities*. 317–340. DOI: 10.4018/978-1-7998-7492-8.ch017

Masenya, T. M. (2024). Revitalization and Digital Preservation of Indigenous Knowledge Systems for Sustainable Development of Indigenous Communities in South Africa. https://www.tandfonline.com/doi/abs/10.1080/0361526X.2023.2277962 (Accessed 21 November 2024).

Masenya, T. M. (2022). Digital preservation of indigenous knowledge in South African rural communities. In Tshifhumulo, R., & Makhanikhe, T. (Eds.), *Handbook of Research on Protecting and Managing Global Indigenous Knowledge Systems* (pp. 317–340). IGI Global.

Masenya, T. M. (2023). Revitalization and digital preservation of Indigenous Knowledge Systems for sustainable development of Indigenous communities in South Africa. *The Serials Librarian*, *84*(5-8), 86–102.

Masenya, T. M. (2023). Revitalization and Digital Preservation of Indigenous Knowledge Systems for Sustainable Development of Indigenous Communities in South Africa. *The Serials Librarian*, *84*(5–8), 86–102. DOI: 10.1080/0361526X.2023.2277962

Masenya, T. M. (Ed.). (2023). *Digital Preservation and Documentation of Global Indigenous Knowledge Systems*. IGI Global.

Masenya, T. M., & Ngulube, P. (2019). Digital preservation practices in academic libraries in South Africa in the wake of the digital revolution. *South African Journal of Information Management, 21*(1), 1–9.

Masenya, T., & Ngulube, P. (2020). Factors that influence digital preservation sustainability in academic libraries in South Africa. *South African Journal of Library and Information Science, 86*. Advance online publication. DOI: 10.7553/86-1-1860

Masocha, M. (2002). *Identification of medicinal plants used in the treatment of abdominal diseases*. Prince Press.

Mawere, M. (2011). Ethical quandaries in spiritual healing and herbal medicine: A critical analysis of the morality of traditional medicine advertising in southern African urban societies. *The Pan African Medical Journal, 10*, 1–10. DOI: 10.4314/pamj.v10i0.72212 PMID: 22187588

Mawere, M., & Mawere, A. M. (2010). The Changing Philosophy of African Marriage: The Relevance of the Shona customary marriage practice of kukumbira. *Journal of African Studies and Development*, 2(9), 224–233.

Mazama, A. (2003). The Afrocentric Paradigm. Africa World Press. melisallela.com. About Melisa Allela. https://melisallela.com/about

Mazama, A. (1998). The Eurocentric Discourse on Writing: An Exercise in Self-Glorification. *Journal of Black Studies, 29*(1), 3–16. https://www.jstor.org/stable/2668053

Mazama, A. (2001). The Afrocentric Paradigm: Contours and Definitions. *Journal of Black Studies, 31*(4), 387–405. DOI: 10.1177/002193470103100401

Mazzocchi, F. (2020). A deeper meaning of sustainability: Insights from indigenous knowledge. *The Anthropocene Review, 7*(1), 77–93. DOI: 10.1177/2053019619898888

Mbigi, L. (2019). Digital Tools for Safeguarding Endangered Indigenous Knowledge in Southern Africa. *International Journal of Cultural Heritage Studies, 14*(3), 142–156.

McGregor, D. (2021). Indigenous Knowledge Systems in Environmental Governance in Canada. *KULA: Knowledge Creation, Dissemination, and Preservation Studies, 5*(1). Advance online publication. DOI: 10.18357/kula.148

McKemey, M. B., Rangers, B., Rangers, Y. M., Costello, O., Hunter, J. T., & Ens, E. J. (2022). 'Right-way' science: Reflections on co-developing Indigenous and Western cross-cultural knowledge to support Indigenous cultural fire management. *Ecological Management & Restoration, 23*(S1), 75–82. DOI: 10.1111/emr.12532

McKemey, M., Ens, E., Rangers, Y. M., Costello, O., & Reid, N. (2020). Indigenous Knowledge and Seasonal Calendar Inform Adaptive Savanna Burning in Northern Australia. *Sustainability, 12*(3), 995. DOI: 10.3390/su12030995

Mdhluli, T. D., Mokgoatšana, S., & Kugara, S. L.. (2021). Knowledge management: Preserving, managing and sharing indigenous knowledge through digital library. *Hervormde Teologiese Studies*, *77*(2), a6795.

Mdhluli, T. D., Mokgoatšana, S., Lee, K. S., & Vuma, L. (2021). Knowledge Management: Preserving, Managing and Sharing Indigenous Knowledge Through Digital Library. *Hervormde Teologiese Studies*, *77*(2). Advance online publication. DOI: 10.4102/hts.v77i2.6795

Meertens, D. (2018). Respecting protocols: Cultural preservation and digitisation in Indigenous communities. *Journal of Cultural Heritage Management.*

Meertens, D. (2018). Conceptualising the digitisation and preservation of Indigenous knowledge. In *The Digital Humanities and Indigenous Knowledge* (pp. 83–95). Springer., https://link.springer.com/chapter/10.1007/978-3-319-70232-2_6

Meighan, P. J. (2021). Decolonizing the Digital Landscape: The Role of Technology in Indigenous Language Revitalization. *Alternative an International Journal of Indigenous Peoples*, *17*(3), 397–405. DOI: 10.1177/11771801211037672

Melka, T. L., Warkineh, T. Z., Gizaw, A. M., Bedada, Y. M., & Birhanu, E. T. (2022). How Can We Achieve Lifelong Learning Opportunities for All in Ethiopia? Making a Case for Intergenerational Approaches to Literacy and Learning. Policy Brief. *Online Submission.* https://eric.ed.gov/?id=ED647423

Mendez, D., Graziotin, D., Wagner, S., & Seibold, H. (2020). Open Science in Software Engineering. In Felderer, M., & Travassos, G. H. (Eds.), *Contemporary Empirical Methods in Software Engineering* (pp. 477–501). Springer International Publishing., DOI: 10.1007/978-3-030-32489-6_17

Mercer, J., Kelman, I., Taranis, L., & Suchet-Pearson, S. (2009). Framework for integrating indigenous and scientific knowledge for disaster risk reduction. *Disasters*, *34*(1), 214–239. DOI: 10.1111/j.1467-7717.2009.01126.x PMID: 19793324

Merolla, D. (2014). Introduction. Orality and techauriture of African literatures. In. Dossier "Oraliteit en tegno-ouditUur van Afrikaliterature / Orality and Techauriture of African Literatures / Oralité et technoauriture de littératures africaines, D.Merolla (ed.), *Tydskrif vir Letterkunde – A Journal for African Literature,* 51, 1.

Meyer, A., & Fourie, I. (2021). Ethical Challenges and Protection of Privacy. In *Autoethnography for Librarians and Information Scientists* (pp. 92–111). Routledge., DOI: 10.4324/9781003014775-7

Mhlongo, M., & Ngulube, P. (2018). Public libraries as conduits for indigenous knowledge in South Africa. DigitalCommons@University of Nebraska - Lincoln. https://digitalcommons.unl.edu/libphilprac/1757/

Mhlongo, M. (2020). Harnessing Indigenous Knowledge Through Community Involvement in Public Libraries in South Africa. *Libri*, *70*(4), 291–303. DOI: 10.1515/libri-2019-0194

Mhlongo, M., & Ngulube, P. (2020). Resource provision and access to indigenous knowledge in public libraries in South Africa. *Information Development*, *36*(2), 2. Advance online publication. DOI: 10.1177/0266666919841095

Mielkov, Y. (2023). The paradigm of open science and democratization of the research activity of universities. *Paradigm of knowledge*, *3*(57). Advance online publication. DOI: 10.26886/2520-7474.3(57)2023.4

Miller, B. G. (2024). *Oral history on trial: Recognizing Aboriginal narratives in the courts*. UBC Press. https://books.google.com/books?hl=en&lr=&id=SjOvU2iJ6JwC&oi=fnd&pg=PP2

Miri, S. J., & Aryamanesh, S. (2022). Ritual as a Social Phenomenon in Ancient Near Eastern Societies. *International Journal of the Humanities*, *29*(4), 60–74. https://eijh.modares.ac.ir/article-27-59542-en.html

Mirza, N. (2020). Nursing, Indigenous health, water, and climate change. *Witness: The Canadian Journal of Critical Nursing Discourse*.

Mirzadeh, S. I., Farajtabar, M., Li, A., Levine, N., Matsukawa, A., & Ghasemzadeh, H. (2020, April). Improved knowledge distillation via teacher assistant.). https://aaai.org/ojs/index.php/AAAI/article/view/5963. *Proceedings of the AAAI Conference on Artificial Intelligence*, *34*(04), 5191–5198.

Mlilo, T., & Bandama, F. (2024). Collective Memory and Memory-Making within the Mapungubwe and Makgabeng Heritage Landscapes: Indigenous Voices Amongst the Sotho- and Venda- Speaking Groups of Southern Africa. *Arts and Humanities*. DOI: 10.20944/preprints202403.0850.v1

Mohan, D. (2024). Wisdom of the ancestors: Harnessing indigenous knowledge for contemporary challenges. *International Journal of Applied Research*, *10*(3), 112–119. DOI: 10.22271/allresearch.2024.v10.i3b.11591

Mons, B., Neylon, C., Velterop, J., Dumontier, M., Santos, L. O. B. S., & Wilkinson, M. D. (2017). Cloudy, increasingly FAIR; revisiting the FAIR Data guiding principles for the European Open Science Cloud. *Information Services & Use*, *37*, 49–56.

Mons, B., Schultes, E., Liu, F. H., & Jacobsen, A. (2020). The FAIR principles: First generation implementation choices and challenges. *Data Intelligence*, *2*, 1–9.

Montemayor, R. (2019). *Sexuality in adolescence and emerging adulthood*. Guiford Press.

Monti, L., Delnevo, G., Mirri, S., Salomoni, P., & Callegati, F. (2018). *Digital Invasions within Cultural Heritage: Social Media and Crowdsourcing*. Springer.

Morgan, H. (2022). Conducting a qualitative document analysis. *The qualitative report*, *27*(1), 64–77.

Morgan, J. A., Lindsay, B. L., & Moran, C. (2022). Opening a "closed door": A call for nuance in discussions of open science. *Industrial and Organizational Psychology: Perspectives on Science and Practice*, *15*(4), 537–541. DOI: 10.1017/iop.2022.72

Mostert, A., Lisney, B.,MarokoG.M., & Kaschula., R.H. (2021). From Technauriture to Cultauriture: Developing a Coherent Digitisation Paradigm for Enhancing Cultural Impact. International Journal of Society, Culture and Language.

MSN. (2025). *Indigenous knowledge and climate adaptation in Vanuatu*. MSN Reports.

Mthembu, P. (2021). Challenges in Digital Archiving and Preservation of Indigenous Knowledge in Southern Africa. *Journal of Digital Preservation*, *12*(4), 45–59.

Mudau, P. P. (2024). Exploring the Importance of Traditional Schools in Maintaining the Heritage Language: A Study of Tshiven a From Tshimbupfhe, South Africa. *Forum for Linguistic Studies*, *6*(5), 213–225. DOI: 10.30564/fls.v6i5.7156

Muir, C., Rose, D., & Sullivan, P. (2010). From the other side of the knowledge frontier: Indigenous knowledge, social–ecological relationships and new perspectives. *The Rangeland Journal*, *32*(3), 259. DOI: 10.1071/RJ10014

Mukasa, G., & Kamusiime, N. (2012). The value of managing library information resources in digital form in Uganda. *African Research & Documentation*, *118*, 69–78.

Mulalap, C. Y., Frere, T., Huffer, E., Hviding, E., Paul, K., Smith, A., & Vierros, M. K. (2020). Traditional knowledge and the BBNJ instrument. *Marine Policy*, *122*, 104103. DOI: 10.1016/j.marpol.2020.104103

Muli, K. K., Nyambura, E., & Onyango, J. (2023). *Preservation of Knowledge Systems through Integration of Indigenous Knowledge System: A Case Study of Ogiek Community in Mau Forest.*

Murphy, L. (2021). The Ethics of Digitizing Indigenous Knowledge: An African Perspective, Mwanza, A.J. (2022) Can digital technology be used for addressing digital literacy in the dissemination of indigenous knowledge in rural areas? *South African Journal of Library and Information Science*, *88*(1), 1–11.

Murphy, M. C., Mejia, A. F., Mejia, J., Yan, X., Cheryan, S., Dasgupta, N., Destin, M., Fryberg, S. A., Garcia, J. A., Haines, E. L., Harackiewicz, J. M., Ledgerwood, A., Moss-Racusin, C. A., Park, L. E., Perry, S. P., Ratliff, K. A., Rattan, A., Sanchez, D. T., Savani, K., & Pestilli, F. (2020). Open science, communal culture, and women's participation in the movement to improve science. *Proceedings of the National Academy of Sciences of the United States of America*, *117*(39), 24154–24164. DOI: 10.1073/pnas.1921320117 PMID: 32929006

Musanga, T. (2017). "Ngozi" (Avenging Spirit), Zimbabwean Transnational Migration, and Restorative Justice in Brian Chikwava's Harare North (2009). *Journal of Black Studies*, *48*(8), 1–17. DOI: 10.1177/0021934717720563

Musili, T. K. (2022). Ndwae ngone mwaitu: A Postmodern Cultural Phenomenon of Dowry among the Akamba and its influence on Spousal Violence. *Journal of International Women's Studies*, *24*(4), 11. https://vc.bridgew.edu/jiws/vol24/iss4/11/

Mutula, M. S. (2014). Status of Digital Heritage Preservation Management in Eastern Africa. *The Electronic Library*, *32*(3), 363–374.

Mvula, D. (2023). *Role of Traditional Healers (Asing'anga) in the Preservation of Indigenous Medical Knowledge (IMK).* DOI: 10.20944/preprints202311.1802.v1

Mwangi, K., Mainye, N., Ouso, D., Esoh, K., Muraya, A., Mwangi, C., & Kibet, C. (2021). Open Science In Kenya: Where Are We? *Frontiers in Research Metrics and Analytics*, *6*. Advance online publication. DOI: 10.3389/frma.2021.669675 PMID: 34056516

Mwelwa, J., Boulton, G., Wafula, J., & Loucobar, C. (2020). Developing open Science In Africa: Bariers, scholar and opportunities. *Data Science Journal*, *19*(31), 1–17.

Nagoya Protocol. (2010). The Nagoya Protocol on Access and Benefit-sharing. Retrieved from https://www.cbd.int/abs/

Nagoya Protocol. (2010). The Nagoya Protocol on Access to Genetic Resources and the Fair and Equitable Sharing of Benefits Arising from Their Utilization. *Convention on Biological Diversity. (Open Science Framework (OSF).* Available from: https://www.researchgate.net/publication/315794339_Open_Science_Framework_OSF [accessed Mar 02 2025].

Nakashima, D. (Ed.). (2010). Indigenous Knowledge in Global Policies and Practice for Education, Science and Culture. https://unesdoc.unesco.org/ark:/48223/pf0000265855

Nakata, M., & Langton, M. (2009). Australian indigenous knowledge and libraries. In *UTS ePRESS eBooks*.

Nakata, M. (2002). Indigenous Knowledge and the Cultural Interface: Underlying issues at the intersection of knowledge and information systems. *IFLA Journal, 28*(5–6), 281–291. DOI: 10.1177/034003520202800513

Nakata, M., Hamacher, D. W., Warren, J., Byrne, A., Pagnucco, M., Harley, R., Venugopal, S., Thorpe, K., Neville, R., & Bolt, R. (2014). Using Modern Technologies to Capture and Share Indigenous Astronomical Knowledge. *Australian Academic and Research Libraries, 45*(2), 101–110. DOI: 10.1080/00048623.2014.917786

Nappi, M. L., Buono, M., Chivăran, C., & Giusto, R. M. (2024). Models and tools for the digital organisation of knowledge: Accessible and adaptive narratives for cultural heritage. *Heritage Science, 2024*(12), 112. DOI: 10.1186/s40494-024-01219-z

National Indigenous Knowledge Systems Office. (2019). Porttfolio on Science and Technology. Cape Town: Department of Science and Technology Cape Town 29 August 2006. Decade of Traditional Medicine [2001 – 2010] Lusaka Summit of Heads of States and Government.https://www.slideserve.com/zalman/role-of-nikso-in-research-and-development-powerpoint-ppt-presentation

National Library of Australia. (2003). Guidelines for the Preservation of Digital Heritage.

Natural Justice. (2023). Rooibos robbery: a story of bioprospecting in South Africa", [video], available at: https://naturaljustice.org/video/rooibosrobbery-a-story-of-bioprospecting-in-south-africa/

Nawroth, C., & Krause, E. T. (2021). The academic, societal and animal welfare benefits of Open Science for animal science. *Open Science Framework.* DOI: 10.31219/osf.io/dnfvs

Ncube, B. (2022). Indigenous knowledge perspectives on water management and its challenges in South Africa. In *Current Directions in Water Scarcity Research (4)* (pp. 227–242). Elsevier.

Ndung'u, J. K., Nguta, J. M., Mapenay, I. M., & Moriasi, G. A. (2024). A Comprehensive Review of Ethnomedicinal Uses, Phytochemistry, Pharmacology, and Toxicity of Prunus africana (Hook. F.) Kalkman from Africa. Scientifica (Cairo). DOI: 10.1155/2024/8862996

Neale, T., Carter, R., Nelson, T., & Bourke, M. (2019). Walking together: A decolonising experiment in bushfire management on Dja Dja Wurrung country. *Cultural Geographies*, *26*(3), 341–359. DOI: 10.1177/1474474018821419

Neitzel, A. L., Smalls, B. L., Walker, R. J., Dawson, A. Z., Campbell, J. A., & Egede, L. E. (2019). Examination of dietary habits among the indigenous Kuna Indians of Panama. *Nutrition Journal*, *18*(1), 44. DOI: 10.1186/s12937-019-0469-8 PMID: 31370836

Nesterova, Y. (2019). Teaching Indigenous Children in Taiwan: Tensions, Complexities and Opportunities. *Global Studies of Childhood*, *9*(2), 156–166. DOI: 10.1177/2043610619846349

Nettleton, K. (2020). The Role of Digital Repositories in Enhancing the Accessibility of Indigenous Knowledge. *Journal of Information Science*, *46*(2), 231–245.

Ng, J. Y., Wieland, L. S., Lee, M. S., Liu, J., Witt, C. M., Moher, D., & Cramer, H. (2024). Open science practices in traditional, complementary, and integrative medicine research: A path to enhanced transparency and collaboration. *Integrative Medicine Research*, *13*, 101047. PMID: 38799120

Ng, J., Lin, B., Kreuder, L., Cramer, H., & Moher, D. (2024). Open science practices among authors published in complementary, alternative, and integrative medicine journals: an international, cross-sectional survey. DOI: 10.1101/2024.02.26.24303402

Ngugi, W. T. (1986). *Decolonising the Mind: The Politics of Language in African Literature. London: James Currey*. Heinemann Kenya.

Ngugi, W. T. (2009). *Something Torn and New: An African Renaissance*. Basic Civitas Books.

Ngulube, P. (2012). Ghost in our machines: Preserving public digital information for the sustenance of electronic government in Sub-Sahara Africa. *Mousaion*, *30*(2), 128–136.

Ngulube, P. (2024). From Orality to Digital Assets: Managing Indigenous Knowledge in Africa in the Wake of the Open Science Movement. *The Serials Librarian, 84*(5–8), 71–85. DOI: 10.1080/0361526X.2023.2296880

Ngulube, P. (2025). Leveraging information and communication technologies for sustainable agriculture and environmental protection among smallholder farmers in tropical Africa. *Discover Environment, 3*(1), 1–17.

Ngulube, P., & Chigwada, J. (2023). Stakeholders in the acquisition, preservation and dissemination of indigenous knowledge projects. *Information Development*, •••, 1–18.

Nherera, C., & Mukora, F. N. (2024). Digitalisation of Higher Education in Zimbabwe: A Challenging Necessity and Emerging Solutions. *Journal of Comparative & International Higher Education, 16*(2).

Nijdam, E. (2021). Sami-Digital Storytelling: Survivance and Revitalization in Indigenous Digital Games. *New Media & Society, 25*(11), 3093–3116. DOI: 10.1177/14614448211038902

Njogu, H. W. (2022). Effects of droughts on the delivery of infrastructure services in Kenya. *In Natural Resources Forum* (Vol. 46, No. 2, pp. 221-244). UK: Blackwell Publishing Ltd. https://doi.org/DOI: 10.1111/1477-8947.12251

Nobel Prize Outreach. 2025. Wangari Maathai-Speed Read. nobelprize.org. https://www.nobelprize.org/prizes/peace/2004/maathai/speedread

Ntsebeza, L. (2022). Traditional Knowledge and Intellectual Property Law in Southern Africa: The Case for Digital Protection. *Journal of African Law, 66*(1), 58–79.

O'Boyle, E. H., Banks, C. B., & Gonzalez-Mulé, E. (2014). The Chrysalis effect: How ugly initial results metamorphosize into beautiful articles. *Journal of Management, 43*, 376–399. DOI: 10.1177/0149206314527133

Ocholla, D. N. (2025). The complex marriage: interfacing open science and open indigenous knowledge. *Library Management. 46(3/4): 252-267. Emerald Publishing Limited e-ISSN: 1758-7921 p-ISSN: 0143-5124 DOI*DOI: 10.1108/LM-11-2024-0125

Ocholla, D. (2025). The complex marriage: Interfacing open science and open indigenous knowledge. *Library Management, 46*(3/4), 252–267. DOI: 10.1108/lm-11-2024-0125

Odulwa, K. K. K., Wafula, P., & Choti, C. (2021). The Pre-Colonial Socio-Economic Practices of the Khayo Community of Busia County in Western Kenya. https://www.academia.edu/download/90892691/ijsrp-p11725.pdf

Odunlade, R. O., & Okiki, O. C. (2019). Exploring the Role of Information Managers in Knowledge Transfer and Preservation Among the "Elewe Omos" (Indigenous Herbal Medicine Practitioners) of the Yorubas in Nigeria. *Information Impact Journal of Information and Knowledge Management*, *9*(4), 142. DOI: 10.4314/iijikm.v9i4.14

Ogoma, D. E. (2014). Reflection on an African Traditional Marriage System. *Journal of Social Sciences and Public Affairs*, *4*(1), 1–11.

Okeke, C. (2023). Reclaiming Digital Spaces: Indigenous Knowledge and Technology in Africa.

Oliveira, E. J., Chavez, C. V. F. G., Cordeiro, A. F. R., & Feitosa, D. (2021). How do Brazilian Software Engineering Researchers Perceive and Practice Open Science? *Anais Do I Workshop de Práticas de Ciência Aberta Para Engenharia de Software (OpenScienSE 2021)*, 28–33. DOI: 10.5753/opensciense.2021.17142

Ondari-Okemwa, E. (2014). Ethical Issues and Indigenous Knowledge Production and Use in Sub-Saharan Africa in the 21st Century. *Mediterranean Journal of Social Sciences*, *5*(23).

Onwuatuegwu, I. N., & Paul-Mgbeafulike, V. S. (2023). African traditional educational framework: Unveiling the wisdom beyond western education. *Nnadiebube Journal of Philosophy, 6*(2). https://www.acjol.org/index.php/NJP/article/view/3675

Onyancha, O. B. (2024). Indigenous knowledge, traditional knowledge and local knowledge: What is the difference? An informetrics perspective. *Global Knowledge, Memory and Communication*, *73*(3), 237–257. DOI: 10.1108/GKMC-01-2022-0011

Opertti, R. (2021). Ten clues for rethinking curriculum. In. On Current and Critical Issues in Curriculum, Learning and Assessment. In-Progress Reflection No. 42 BE/2021/WP/CD/42

Osagie, F. O., Fagbemi, V. Y., Agubata, N. M., & Dogara, P. D. (2023). Public libraries as pivotal agents in preservation of indigenous knowledge for sustainable development. [RIST]. *Review of Information Science and Technology*, *9*, 27–34.

Osedo, O. A. (2013). Challenges of digital preservation in Africa. Available at SSRN 3496885. Paris: UNESCO.

Osman, A. (2009). Indigenous Knowledge in Africa: Challenges and Opportunities. An Inaugural Lecture. Centre for Africa Studies. https://www.ufs.ac.za/docs/librariesprovider20/centre-for-africa-studies-documents/all-documents/osman-lecture-1788-eng.pdf?sfvrsn=e436fb21_0

Osypinska, M., Żurawski, B., Belka, Z., Osypinski, P., & Łopaciuk, R. (2022). "Oh, My Beloved, Great Bull!" An Interdisciplinary Approach to Cattle as a Marker of Economic and Political Status in Medieval Sub-Saharan Societies. *Journal of African Archaeology, 21*(1), 46–62. https://brill.com/view/journals/jaa/21/1/article-p46_4.xml

Othman, M. K., Petrie, H., & Power, C. (2011). *Engaging Visitors in Museums with Technology: Scales for the Measurement of Visitor and Multimedia Guide Experience*. Springer.

Owiny, S., Mehta, K., & Maretzki, A. N. (2014). The use of social media technologies to create, preserve, and disseminate indigenous knowledge and skills to communities in East Africa. *International Journal of Communication, 8*, 234–247.

Owolabi, K. A., Ovwasa, D. E., Ajayi, T. B., & Odewale, M. O. (2021). Preservation and Use of Indigenous Knowledge Practices in Public Libraries in Nigeria. *Public Library Quarterly, 41*(5), 485–502. DOI: 10.1080/01616846.2021.1938897

Oyelude, A. A. (2023). Indigenous knowledge preservation as a sign of respect for culture: concerns of libraries, archives and museums', *Insights: the UKSG journal*, 36(1), 1-21.

Oyelude, A. A. (2023). Indigenous Knowledge Preservation as a Sign of Respect for Culture: Concerns of Libraries, Archives and Museums. *Insights the Uksg Journal, 36*. Advance online publication. DOI: 10.1629/uksg.628

Pace, G. (2021). *Heritage Conservation and Community Empowerment. Tools for Living Labs. In book: Underground Built Heritage Valorisation: A Handbook*. CNR Edizioni.

Pallathadka, L. K., Pallathadka, H., & Devi, M. S. (2022). A Review of marriage rituals in different cultures. *Integrated Journal for Research in Arts and Humanities, 2*(5), 152-160. https://www.academia.edu/download/102036846/234.pdf

Pampel, H. (2023). Promoting Open Access in Research-Performing Organisations: Spheres of Activity, Challenges, and Future Action Areas. *Publications, 2023*(11), 44. DOI: 10.3390/ publications11030044

Pandey, R., & Kumar, V. (2020). Exploring the Impediments to Digitization and Digital Preservation of Cultural Heritage Resources: A Selective Review. *Preservation Digital Technology & Culture, 49*(1), 26–37. DOI: 10.1515/pdtc-2020-0006

Panopoulos, P., & Chatzopoulou, V. (2015). Homeland as Sound and Sound as Homeland: Cultural and Personal Soundscapes in Christos Christovasilis's Short Stories. By Panayotis Panopoulos. Translated by Vasiliki Chatzopoulou. *Ethnomusicology Translations*, (1), 1-19. https://scholarworks.iu.edu/journals/index.php/emt/article/view/20266

Parekh, H. (2000). *Digitization: An overview of issues*. Available at: https://ir.inflibnet.ac.in/server/api/core/bitstreams/81bdcf00-f3b9-4f63-b3d1-752d33869d4e/content

Parsons, W. (2004). Not just steering but weaving: Relevant knowledge and the craft of building policy capacity and coherence. *Australian Journal of Public Administration*, *63*(1), 43–57. https://onlinelibrary.wiley.com/doi/abs/10.1111/j.

Pathak, K., Marwaha, J., Chen, H., Krumholz, H., & Matthews, J. (2024). Use of open science practices in surgical journals. *JAMA Surgery*, *159*(2), 228. DOI: 10.1001/jamasurg.2023.5389 PMID: 38117492

Paul, J. (2023). Traditional Knowledge Protection and Digitisation: A Critical Decolonial Discourse Analysis. *International Journal for the Semiotics of Law*, *2023*(36), 2133–2215. DOI: 10.1007/s11196-023-09989-8

Pedota, L. (2011). *Master of Arts Program in Cultural and Educational Policy Studies*. Loyola University Chicago.

Penna, K. (2023). Culturally sensitive practices and intercultural communication competence for the protection of indigenous peoples' cultures. ENCATC Congress - Artificial Intelligence embraced: the future of the culture and creative sector. ISBN: 978-92-990088-5-0 At: Helsinki, Finland. https://www.encatc.org/media/7193-2023-book-of-proceedings.pdf

Pepple, D. G., & Dore-Okuiomse, M. M. (2023). An enabler framework for developing knowledge management practices: Perspectives from Nigeria. *South African Journal of Information Management*, *25*(1). Advance online publication. DOI: 10.4102/sajim.v25i1.1647

Perry, S. R. (2014) Digitization and digital preservation: a review of the literature. http://scholarworks.sjsu.edu/cgi/viewcontent.cgi?article=1170&context=slissrj

Petrelli, D., Dulake, N., Marshall, M. T., Pisetti, A., & Not, E. (2016). Voices from the war: Design as a means of understanding the experience of visiting heritage. In *Proceedings of the 2016 CHI Conference on Human Factors in Computing Systems*, San Jose, CA, USA, 7–12 May, pp. 1033–1044.

Pfeffer, J., & Fong, C. (2002). The end of business schools? Less success than meets the eye. *Academy of Management Learning & Education*, *1*, 78–95.

Pickover, M. (2008). The DISA project. Packaging South African heritage as a continuing resource: Content, access, ownership and ideology. *IFLA Journal*, *34*(2), 192–197.

Pirson, M. (2016). Editorial: Welcome to the Humanistic Management Journal. *Humanistic Management Journal*, *1*, 1–7.

Pivec, F. (2003). *The Ethics of Librarianship: An International Survey*. Academia.edu.

Piwowar, H., Priem, J., Larivière, V., Alperin, J. P., Matthias, L., Norlander, B., & Haustein, S. (2018). The state of Open Access: A large-scale analysis of the prevalence and impact of Open Access articles. *PeerJ*, *6*, e4375. DOI: 10.7717/peerj.4375 PMID: 29456894

Pizzi, G. (2020). Open-Science Platform for Computational Materials Science: AiiDA and the Materials Cloud. In Andreoni, W., & Yip, S. (Eds.), *Handbook of Materials Modeling* (pp. 1813–1835). Springer International Publishing., DOI: 10.1007/978-3-319-44677-6_64

Plockey, F. (2014). The role of Ghana public libraries in the digitization of indigenous knowledge: Issues and prospects. *The Journal of Pan African Studies*, *6*(10), 20–36.

Powell, T. B. (2016). Digital Knowledge Sharing: Forging Partnerships between Scholars, Archives, and Indigenous Communities. *Museum Anthropology*.

Pretorius, R. I. Ë. T. T. E., & Bezuidenhout, H. E. N. N. I. E. (2011). National Recordal System IK holder catalogue process.

Priest, N., Mackean, T., Waters, E., Davis, E., & Riggs, E. (2009). Indigenous child health research: A critical analysis of Australian studies. *Australian and New Zealand Journal of Public Health*, *33*(1), 55–63. DOI: 10.1111/j.1753-6405.2009.00339.x PMID: 19236360

Putra, D., Sofwany, B., Zulfinanda, H., & Kamaruddin, I. (2022). Flash Flood (Δ) Risk and Damage Assessment in Batu, East Java. *Jurnal Teknosains*, *12*(1), 72–84. DOI: 10.22146/teknosains.78483

Pyett, P. (2002). Towards reconciliation in Indigenous health research: The responsibilities of the non-Indigenous researcher. *Contemporary Nurse*, *14*(1), 56–65. DOI: 10.5172/conu.14.1.56 PMID: 16114194

Qobilova, E. (2024). The Place of Folk Songs in Isajan Sultan's Stories. *American Journal of Social Sciences and Humanity Research*, *4*(02), 44–49. https://theusajournals.com/index.php/ajsshr/article/view/2450

Radcliffe, C., & Parissi, C. (2022). Understanding and Preserving Indigenous Agricultural Knowledge: A Suggested Way Forward for Future Research. *Global Knowledge, Memory and Communication, 73*(8/9), 1005–1024. DOI: 10.1108/gkmc-03-2022-0058

Rafi, N., Ahmed, A., Shafique, I., & Kalyar, M. N. (2022). Knowledge management capabilities and organizational agility as liaisons of business performance. *South Asian Journal of Business Studies, 11*(4), 397–417. https://doi/10.1108/SAJBS-05-2020-0145/full/html

Ragnier, M., & Dacos, M. (2023). *Report of the G7 Open Science—Research on Research Sub-Working Group*. Ministère de l'enseignement supérieur et de la recherche. DOI: 10.52949/32

Ranasinghe, P. (2008). Preservation and provision of access to indigenous knowledge in Sri Lanka. World Library and Information Congress: 74th IFLA General Conference and Council, 10-14 August 2008, Quebec, Canada. https://www.ifla.org/iu/ifla7u/index

Raven, E., & Theuns-de Boer, G. (Eds.). (2021). *Reading Śiva: An Illustrated Selection from the ABIA Online Bibliography on the Arts and Material Culture of South and Southeast Asia* (Vol. 38). Brill., https://books.google.com/books?hl=en&lr=&id=CAxWEAAAQBAJ&oi=

Ray, D. (2022). *Ethics in Librarianship*. Quest Journals. *Journal of Research in Humanities and Social Science, 10*(5), 53–56.

Recht, J. (2009). Hearing Indigenous Voices, Protecting Indigenous Knowledge. *International Journal of Cultural Property, 16*(3), 233–254. DOI: 10.1017/S0940739109990166

Reed, I. (1972). *Mumbo Jumbo*. Doubleday.

Reichmann, S., & Wieser, B. (2022). Open science at the science–policy interface: Bringing in the evidence? *Health Research Policy and Systems, 20*(1), 70. DOI: 10.1186/s12961-022-00867-6 PMID: 35725491

Reid, A. J., Eckert, L. E., Lane, J., Young, N., Hinch, S. G., Darimont, C. T., Cooke, S. J., Ban, N. C., & Marshall, A. (2020). "Two-Eyed Seeing": An Indigenous framework to transform fisheries research and management. *Fish and Fisheries, 22*(2), 243–261. DOI: 10.1111/faf.12516

Revi, A. T., Millard, D. E., & Middleton, S. E. (2020). A systematic analysis of user experience dimensions for interactive digital narratives. In *Proceedings of the Interactive Storytelling—13th International Conference on Interactive Digital Storytelling*, ICIDS 2020, Bournemouth, UK, 3–6 November, pp. 58–74.

Ribeiro, R. G. T., Kaminski, M. R., Lübeck, M., & Boscarioli, C. (2019). Pedagogical Usage of Digital Technologies in an Indigenous Village in Brazil: Reflections and Challenges. [s]. *Interaction Design and Architecture*, *41*, 50–61. DOI: 10.55612/s-5002-041-004

RickShaw Travel Group. (2024). Bull fighting in western Kenya. https://rickshawtravels.com/about/articledetails/Bull-Fighting-in-Western-

Ritchie, D. A. (2014). *Doing oral history*. Oxford University Press.

Rokka, J. (2021). Consumer Culture Theory's future in marketing. *Journal of Marketing Theory and Practice*, *29*(1), 114–124.

Romualdo, A., Arroyo, J., Suriaga, E. G., & Fuentes, R. (2022). Traditional Knowledge Management (TKM) Through Biosprospecting: Mainstreaming Scientific Research Trends and Techniques for Inclusive and Sustainable Education. *East Asian Journal of Multidisciplinary Research*, *1*(7), 1497–1506. DOI: 10.55927/eajmr.v1i7.837

Roshetko, J. M., Pingault, N., Quang, T. A., Meybeck, R., & Gitz, M. V. (2022). Asia-Pacific Roadmap for Innovative Technologies in the Forest Sector. CIFOR. 15, Rome and Bogor.

Rosner, D., Roccetti, M., & Marfia, G. (2014). The digitization of cultural practices. *Communications of the ACM*, *57*, 82–87.

Rössler, M. (2006). World Heritage cultural landscapes: A UNESCO flagship programme 1992–2006. *Landscape Research*, *31*, 333–353.

Rothenberg, N. R. (2015). Communication and Information Sharing in Teams. *The Accounting Review*, *90*, 761–784.

Roy, L. 2013. The Role of Tribal Libraries and Archives in the Preservation of Indigenous Cultural Identity Through Supporting Native Language Revitalization. *International Preservation News* No. 61.

Ryan, J., & Silvanto, S. (2009). The World Heritage List: The making and management of a brand. *Place Brand. Publi*, *5*, 290–300.

Sabbagh, K., Roman, F., Bahjat, E., Milind, S., & Sandeep, G. (2012). Maximizing the Impact of Digitization." *The Global Information Technology Report*, pp. 121–33. https://www3.weforum.org/docs/GITR/2012/GITR_Chapter1.11_2012.pdf

SadanandaNayak. G. (2021). Historical Analysis on traditional Tools and Technology for Farming and Food Processing: A Study of South Odisha. *PalArch's Journal of Archaeology of Egypt/Egyptology, 18*(4), 1627-1635. https://archives.palarch.nl/index.php/jae/article/view/6540

Saeed, S., Bajwa, I. S., & Mahmood, Z. (2015). *Human Factors in Software Development and Design*. DOI: 10.4018/978-1-4666-6485-2

Saffady, W. (2021). *Records and information management: fundamentals of professional practice*. Rowman & Littlefield.

Samarasinghe, K., & Nethsinghe, R. (2023). Social processes that shaped Sri Lankan Sinhala folk music and its characteristics based on the purposes of usage. http://repository.lib.vpa.ac.lk/handle/123456789/1836

Sam, J., Schmeisser, C., & Hare, J. L. (2021). Grease Trail Storytelling Project. *Kula Knowledge Creation Dissemination and Preservation Studies*, *5*(1). Advance online publication. DOI: 10.18357/kula.149

Sanka, C. G., Adu-Gyamfi, S., & Marfo, C. O. (2020). Tales in the Paasaali Dirge: Structure and Moral Lessons from the Past. *Manusya: Journal of Humanities, 23*(1), 19-39. https://brill.com/view/journals/mnya/23/1/article-p19_19.xml

Sarkhel, J. K. (2017). Strategies of indigenous knowledge management in libraries. *Qualitative and Quantitative Methods in Libraries*, *5*(2), 427 439.

Scazzosi, L. (2004). Reading and assessing the landscape as cultural and historical heritage. *Landscape Research*, *29*, 335–355.

Scherer, F. M. (2019). *Managerial Control and Executive Compensation. In Faculty Research Working Paper Series*. Harvard Kennedy School of Government: Cambridge, MA, USA.

Schöpfel, J., Ferrant, C., André, F., & Fabre, R. (2016). Ready for the future? A survey on open access with scientists from the French National Research Center (CNRS). *Interlending & Document Supply*, *44*(4), 141–149. DOI: 10.1108/ILDS-06-2016-0023

Science and Technology Department. (nd). Indigenous Knowledge Systems. Republic of South Africa.

Senanayake, S. G. J. N. (2006). Indigenous Knowledge as a key to sustainable development. Journal of Agricultural Sciences – Sri Lanka 2(1). https://www.researchgate.net/publication/265197993_Indigenous_knowledge_as_a_key_to_sustainable_development

Senanayake, S. G. J. N. (2006). Indigenous knowledge as a key to sustainable development. *Journal of Agricultural Sciences – Sri Lanka, 2*(1), 87–94. DOI: 10.4038/jas.v2i1.8117

Sen, B. (2005). Indigenous knowledge for development: Bringing research and practice together. *The International Information & Library Review, 2005*(37), 375–382.

Seretse, M., Chukwuere, J., Lubbe, S., & Klopper, R. (2018). Problems around accessing information in rural communities. *Alternation Journal, 25*(1), 214–244.

Settee, P. (2008). *Native Languages Supporting Indigenous Knowledge*. United Nations, Department of Economic and Social Affairs.

Sharma, P., Martin, M., Swanlund, D., & Latham, C. (2024). Anderson, D and Waitangi Wood, W. (2023). A cloud-based solution for trustless indigenous data sovereignty: Protecting Māori biodiversity management data in Aotearoa New Zealand. *Transactions in GIS, 28*, 836–857.

Shepherd, E., & Yeo, G. (2003). *Managing records: A handbook of principles and practice*. Facet Publishing.

Shiferaw, D., Takele, S., & Hailemichael, A. (2017). Indigenous Practices of Ganta Community in Protecting Natural Resources, Gamo Gofa Zone, Ethiopia. *International Journal of Environmental Protection and Policy, 5*(5), 74. DOI: 10.11648/j.ijepp.20170505.12

Shiri, A., Howard, D., & Farnel, S. (2021). Indigenous Digital Storytelling for Cultural Heritage Access and Preservation. *Proceedings of the Annual Conference of CAIS Actes Du Congrès Annuel De L'ACSI*. DOI: DOI: 10.29173/cais1202

Shobaruddin, M., Hendrawan, M. R., & Bafadhal, A. S. (2024). Multi-Disaster Risk Mitigation through Augmented Reality Conservation Map for Enhancing Cultural Heritage Attitude. *Environment-Behaviour Proceedings Journal, 9*(SI18), 205–213. DOI: 10.21834/e-bpj.v9isi18.5484

Siliutina, I., Tytar, O., Barbash, M., Petrenko, N., & Yepyk, L. (2024). Cultural preservation and digital heritage: Challenges and opportunities. *Amazonia Investiga, 13*(75), 262–273. DOI: 10.34069/AI/2024.75.03.22

Silverman, D. (2013). *Interpreting Qualitative Data*. 5th Ed. Sage, London The United Nations Educational, Scientific and Cultural Organization (UNESCO) (2025). Local and Indigenous Knowledge Systems (LINKS). https://www.unesco.org/en/links

Singh, B., & Kumar, A. (2024). Examining the Impact of Digitalization on Indigenous Cultural Heritage Preservation in India. *Cultural, 2*(1), 43–57. DOI: 10.61996/cultural.v2i1.59

Sites, O. C. H. (2008). The ICOMOS charter for the interpretation and presentation of cultural heritage sites. *International Journal of Cultural Property, 15*, 377–383.

Sithole, M. (2016.) Opportunities of Incorporating African Indigenous Knowledge Systems (AIKS) in the Physics Curriculum. Alternation Special Edition 18. 255 - 294 255

Small, H. (1973). Co-citation in the scientific literature: A new measure of the relationship between two documents. *Journal of the American Society for Information Science, 24*, 265–269.

Smallwood, R. F. (2019). *Information governance: Concepts, strategies and best practices*. John Wiley & Sons., DOI: 10.1002/9781119491422

Smith, L. T. (1999). *Decolonizing Methodologies: Research and Indigenous Peoples*. Bloomsbury Academic. https://books.google.co.zw/books?id=Nad7afStdr8C

Smith, K. (2016). The importance of records in the right to information. In *Integrity in Government through Records Management* (pp. 149–159). Routledge.

Smith, L. (2006). *The Uses of Heritage*. Routledge.

Smith, L. T. (2012). *Decolonizing methodologies: Research and Indigenous peoples* (2nd ed.). Zed Books.

Smylie, J., Martin, C. M., Kaplan-Myrth, N., Steele, L., Tait, C., & Hogg, W. (2004). Knowledge translation and indigenous knowledge. *International Journal of Circumpolar Health, 63*(sup2), 139–143. DOI: 10.3402/ijch.v63i0.17877

Snyder, H. (2019). Literature review as a research methodology: An overview and guidelines. *Journal of Business Research, 104*, 333–339.

Soini, K., & Birkeland, I. (2014). Exploring the scientific discourse on cultural sustainability. *Geoforum, 51*, 213–223.

Song, H., Markowitz, D., & Taylor, S. (2021). Trusting on the shoulders of open giants? open science increases trust in science for the public and academics. DOI: 10.31219/osf.io/g328c

South African Centre for Digital Language Resources (SADiLaR). About SADiLaR https://sadilar.org/en/about/

Sperveslage, G., & Heagy, T. C. (2023). A Tail's Tale: Narmer, the Catfish, and Bovine Symbolism. *The Journal of Egyptian Archaeology*, *109*(1-2), 65–75. https://doi/abs/10.1177/03075133231188144

Spitschan, M., Schmidt, M. H., & Blume, C. (2021). Principles of open, transparent and reproducible science in author guidelines of sleep research and chronobiology journals. *Wellcome Open Research*, *5*, 172. DOI: 10.12688/wellcomeopenres.16111.2 PMID: 33748432

Spoor & Fisher. (2019). Kenya: Amendments to Copyright Law. https://spoor.com/kenya-amendments-to-copyright-law/

Sraku-Lartey, M., Acquah S.B., Samar, S.B., & Djagbletey, G.D. (2017). Digitization of indigenous knowledge on forest foods and medicines. *International Federation of Library Associations and Institutions*, 1(11). .

Sraku-Lartey, M., Acquah, S. B., Samar, S. B., & Djagbletey, G. D. (2017). Digitization of indigenous knowledge on forest foods and medicines. IFLA Journal, 43(2), 187-197. (Original work published 2017).DOI: 10.1177/0340035216681326

Stephenson, J. (2023). *Culture and sustainability: Exploring stability and transformation with the cultures framework*. Springer Nature.

Stevens, A. and Stevens, A. (2008). A Different Way of Knowing: Tools and Strategies for Managing Indigenous Knowledge. 58(2008): 25–33, 2008.

Stewart, O., & Linden, M. (2022). *Mother of Abominations: Witchcraft Dreamscapes*. Black Moon Publishing.

Stewart, S. L. (2020). One Indigenous Academic's Evolution: A Personal Narrative of Native Health Research and Competing Ways of Knowing. *First Peoples Child & Family Review*, *4*(1), 57–65. DOI: 10.7202/1069350ar

Stojanovski, J. (2019). Promicanje ideja otvorene znanosti u Hrvatskoj. Promoting the ideas of open science in Croatia. *Soc. Teme*, *1*, 166–170.

Stracke, C. M. (2020). Open Science and Radical Solutions for Diversity, Equity and Quality in Research: A Literature Review of Different Research Schools, Philosophies and Frameworks and Their Potential Impact on Science and Education. In D. Burgos (Ed.), *Radical Solutions and Open Science: An Open Approach to Boost Higher Education* (pp. 17–37). Springer Singapore. DOI: 10.1007/978-981-15-4276-3_2

Stranger-Johannessen, E., Doherty, L., & Norton, B. (2018). The African Storybook and Storybooks Canada: Digital stories for linguistically diverse children.

Suárez-Inclán Ducassi, M. R. (2005). A new category of heritage for understanding, cooperation and sustainable development. Their significance within the macrostructure of cultural heritage. The role of the CIIC of ICOMOS: Principles and methodology. In *Proceedings of the 15th ICOMOS General Assembly and International Symposium: 'Monuments and Sites in Their Setting—Conserving Cultural Heritage in Changing Townscapes and Landscapes'*, Xi'an, China, 17–21 October 2005.

Suber, P. (2012). *Open Access*. MIT Press., DOI: 10.7551/mitpress/9286.001.0001

Sullivan (2016). The John Marshall review of intellectual property law. https://repository.law.uic.edu/cgi/viewcontent.cgi?article=1392&context=ripl

Sundararajan, S. A. (2023). CONCEPTUAL STUDY ON INDIGENOUS KNOWLEDGE MANAGEMENT.https://isarpublisher.com/backend/public/assets/articles/

Su, X., Li, X., & Kang, Y. (2019). A bibliometric analysis of research on intangible cultural heritage using CiteSpace. *SAGE Open*, *9*, 2158244019840119.

Tahil, S. K., & Alibasa, J. T., Shamier – Rahman K. Tahil, Marsin, J., & Shanaia – Shedri K. Tahil. (2023). Preserving and Nurturing Tausug Language: The Bahasa Sug Mobile Learning Application Tool for Enhancing Mother Tongue Development for Toddlers. *International Journal of Learning Teaching and Educational Research*, *22*(11), 18–35. DOI: 10.26803/ijlter.22.11.2

Tang, R., & Gavin, M. C. (2016). A Classification of Threats to Traditional Ecological Knowledge and Conservation Responses. *Conservation & Society*, *14*(1), 57–70.

Tapfuma, M. M., & Hoskins, R. (2017). *Visibility and Accessibility of Indigenous Knowledge on Open Access Institutional Repositories at Universities in Africa*. 248–266. DOI: 10.4018/978-1-5225-0833-5.ch011

Taringa, N. T., & Museka, G. (2021). No to bride price/bride wealth, yes to Roora: Investigating the meaning, function and purpose of Roora as a ritual. *Lobola (Bridewealth) in Contemporary Southern Africa: Implications for Gender Equality*, 13-27. https://link.springer.com/chapter/10.1007/978-3-030-59523-4_2

Tata, E. F., Yongabi, K. A., & Dzelamonyuy, E., Forbang, & Geh, E. A. (2023). Decolonizing African science: Efforts, challenges, and future directions. *International Journal of Science and Research Archive*, *10*(1), 823–833. DOI: 10.30574/ijsra.2023.10.1.0840

Taylor, D. L., & Cameron, A. (2016). Valuing IKS in successive South African Physical Sciences Curricula. *African Journal of Research in Mathematics. Science and Technology Education*, *20*(1), 35–44. DOI: 10.1080/10288457.2016.1147800

Teffera, T. (2021). The Nägarit in Past and Present Day Ethiopia. Drums and Drum Ensembles of the Silk Road. Edited by Xiao Mai & Gisa Jähnichen. 28–30. December 2020, Shanghai Conservatory of Music. Logos, Berlin, pp. 39-56. https://www.academia.edu/45644565/Timkehet_Teffera_2021_The_

Tembe, M.A.,& Nsibirwa, Z. (2022). Archiving the voices of the once voiceless: strategies for digital preservation of oral history at the KwaZulu-Natal Provincial Archives. *ESARBICA Journal: Journal of the Eastern and Southern Africa* Regional Branch of the International Council on Archives: African Journals Online (AJOL). DOI: 10.4314/ESARJO.V41I.10

Tembe, M. A. (2019). *Preservation of, and access to oral history records at Pietermaritzburg Archives Repository. Master of Information Studies. Durban.* University of KwaZulu-Natal., https://researchspace.ukzn.ac.za/handle/10413/18140

Teng¨o, M., Hill, R., Malmer, P., Raymond, C. M., Spierenburg, M., Danielsen, F., Elmqvist, T., & Folke, C. (2017). Weaving knowledge systems in IPBES, CBD and beyond—Lessons learned for sustainability. *Current Opinion in Environmental Sustainability, 26–27*, 17–25. DOI: 10.1016/j.cosust.2016.12.005

Terzes, C. (2020). Musical tools of Greek and Roman antiquity. *A Companion to Ancient Greek and Roman Music*, 213-227. https://doi/abs/10.1002/9781119275510.ch16

Tesfa, S. (2024). *Preservation of Indigenous knowledge and appropriate technologies. ResearchGate. The Department of Science and Technology. (nd).* Indigenous Knowledge Systems. Republic of South Africa.

Thairu, W. (2007). A critical analysis of the impact of digitizing indigenous knowledge. *AAJIS*, 1(001).

Thaldar, D., Gooden, A., & Steytler, M. (2023). Open science and human genetic data: Recommendations on South Africa's Draft National Open Science Policy. *Frontiers in Genetics, 14*, 1248747. DOI: 10.3389/fgene.2023.1248747 PMID: 37849503

Tharakan, J. (2017). *Indigenous Knowledge Systems for Appropriate Technology Development*. DOI: 10.5772/intechopen.69889

The South African National Biodiversity Institute (SANBI). 2025. Annual Report 2023/24. https://www.sanbi.org/wp-content/uploads/2024/10/SANBI-Annual-Report-2023-24.pdf

The South African National Biodiversity Institute (SANBI). 2025. What SANBI does. https://www.sanbi.org/about/about-sanbi

The Strait Times. (2019). Activist Greta Thunberg turns spotlight on indigenous struggle at climate summit. https://www.straitstimes.com/world/europe/activist-thunberg-turns-spotlight-on-indigenous-struggle-at-climate-summit

The United Nations Educational, Thorpe K and Galassi M. (2014). Rediscovering Indigenous Languages: The role and impact of libraries and Archives in cultural revitalisation. *Australian Academic and Research Libraries*, *45*(2), 81–100.

Thompson, K.-L., Reece, N., Robinson, N., Fisher, H.-J., Ban, N. C., & Picard, C. R. (2019). "We monitor by living here": Community-driven actualization of a social-ecological monitoring program based in the knowledge of Indigenous harvesters. *Facets*, *4*(1), 293–314. DOI: 10.1139/facets-2019-0006

Thornton, R. J. (2020). Therianthropes as human-animal relations: Contextualising rock art in anthropological perspective. *Anthropology Southern Africa*, *43*(3), 181–196. https://www.tandfonline.com/doi/abs/10.1080/23323256.2020.1773290?

Thorpe, K., Christen, K., Booker, L., & Galassi, M. (2021). Designing Archival Information Systems Through Partnerships With Indigenous Communities. *AJIS. Australasian Journal of Information Systems*, *25*. Advance online publication. DOI: 10.3127/ajis.v25i0.2917

Thursby, M., Thursby, J., Haeussler, C., and Jiang, L. (2009). Do scientists freely share information? Not necessarily. Vox News 29 Nov. 2009.

Tora, A., & Heliso. (2017). Assessment of the indigenous knowledge and use of traditional medicinal plants in Wolaita Zone, Southern Ethiopia. *International Journal of Medicinal Plants and Natural Products*, *3*(1). Advance online publication. DOI: 10.20431/2454-7999.0301003

Torri, M. C. (2010). Increasing knowledge and traditional use of medicinal plants by local communities in Tamil Nadu: Promoting self-reliance at the grassroots level through a community-based entrepreneurship initiative. *Complementary Health Practice Review*, *15*(1).

Traynor, C. (2017). Data Management Plan: Empowering Indigenous Peoples and Knowledge Systems Related to Climate Change and Intellectual Property Rights. *Research Ideas and Outcomes*, *3*, e15111. DOI: 10.3897/rio.3.e15111

Tredinnick, L. (2017). Artificial intelligence and professional roles. *Business Information Review*, *34*(1), 37–41.

Tsekea, S. (2016). The position of intellectual property systems in the protection of indigenous knowledge. *Paper presented at SCECSAL XXII*, Ezulwini, Eswatini.

Tsosie, K. S., Yracheta, J. M., Kolopenuk, J. A., & Geary, J. (2021). We Have "Gifted" Enough: Indigenous Genomic Data Sovereignty in Precision Medicine. *The American Journal of Bioethics, 21*(4), 72–75. DOI: 10.1080/15265161.2021.1891347 PMID: 33825628

Tulius, J. (2020). Lesson from the past, knowledge for the future: roles of human memories in earthquake and tsunami narratives in Mentawai, Indonesia. *Paradigma, Jurnal Kajian Budaya.* https://dr.ntu.edu.sg/handle/10356/165267

urcan, N., Cuciureanu, G., Cujba, R., Cojocanu, I., & Cojocaru, I. (2023). Readiness Toward the Implementation of Open Science Initiatives In the Republic of Moldova. *Central European Journal of Public Policy, 17*(2), 97–118. DOI: 10.2478/cejpp-2023-0009

Turner, N. J., Cuerrier, A., & Joseph, L. (2022). Well grounded: Indigenous Peoples' knowledge, ethnobiology and sustainability. *People and Nature, 4*(3), 627–651. https://doi/abs/10.1002/PAN3.10321

Twarog, S. (2004). Preserving, Protecting and Promoting Traditional Knowledge: National Actions and International Dimensions. In *Protecting And Promoting Traditional Knowledge: Systems, National Experiences And International Dimensions*, Twarog S, and Kapoor, P. United Nations Conference on Trade and Development: United Nations

Tzanova, S. (2020). Changes in academic libraries in the era of Open Science. *Education for Information, 36*(3), 281–299. DOI: 10.3233/EFI-190259

UNESCO Institute for Statistics. (2009). *UNESCO Framework for Cultural Statistics*. UNESCO.

UNESCO. (2016). *The role of Indigenous knowledge in sustainable development.* United Nations Educational, Scientific and Cultural Organization.

UNESCO. (2017). *Safeguarding Indigenous knowledge through digitisation: Challenges and opportunities*. United Nations Educational, Scientific and Cultural Organization.

UNESCO. (2021) UNESCO Recommendation on Open Science. The General Conference of the United Nations Educational, Scientific and Cultural Organization (UNESCO), meeting in Paris, from 9 to 24 November 2021, at its 41st session.

UNESCO. (2021). UNESCO Recommendation on Open Science. The General Conference of the United Nations Educational, Scientific and Cultural Organization (UNESCO), meeting in Paris, from 9 to 24 November 2021, at its 41st session. UNESCO.

United Nations Department of Economic and Social Affairs. (2019). 'Preserve, Revitalize and Promote' Indigenous Languages, or Lose A 'Wealth of Traditional Knowledge', UN Chief says. https://www.un.org/development/desa/en/news/social/indigenous-peoples-day-2019.html

United Nations General Assembly. (2015). *A/RES/70/1. Transforming Our World: The 2030 Agenda for Sustainable Development*. Resolution Adopted by the General Assembly. DOI: 10.1007/s13398-014-0173-7.2

United Nations. (2008). *An Introduction to Human Trafficking: Vulnerability, Impact and Action*. Office on Drugs and Crime.

United Nations. (2021). Indigenous peoples and connectivity: Bridging the digital divide. Available at: https://www.un.org/development/desa/dspd/2021/04/indigenous-peoples-digital-divide/

Vaarzon-Morel, P., Barwick, L., & Green, J. (2021). Sharing and storing digital cultural records in Central Australian Indigenous communities. *New Media & Society*, *23*(4), 692–714. DOI: 10.1177/1461444820954201

Valkonen, J., & Valkonen, S. (2014). Contesting the Nature Relations of Sámi Culture. *Acta Borealia*, *31*(1), 25–40. DOI: 10.1080/08003831.2014.905010

Van Laar, D. M. (2023). *Knowledge Management Implementation in US Army Headquarters: A Case Study* (Doctoral dissertation, Walden University). https://search.proquest.com/openview/1d5dddccf01d7e287b8a9913c

Van Laar, E., Van Deursen, A. J., Van Dijk, J. A., & De Haan, J. (2017). The relation between 21st-century skills and digital skills: A systematic literature review. *Computers in Human Behavior*, *72*, 577–588. DOI: 10.1016/j.chb.2017.03.010

Vazire, S. (2018). Implications of the credibility revolution for productivity, creativity, and progress. *Perspectives on Psychological Science*, (13), 411–417. DOI: 10.1177/1745691617751884 PMID: 29961410

Vicente-Saez, R., Gustafsson, R., & Van den Brande, L. (2020). The Dawn of an Open exploration era: Emergent principles and practices of open science and innovation of university research teams in a digital world. *Technological Forecasting and Social Change*, *156*, 120037.

Vicente-Saez, R., & Martinez-Fuentes, C. (2018). Open Science now: A systematic literature review for an integrated definition. *Journal of Business Research*, *88*, 428–436.

Vignoli, M., & Rörden, J. (2019). Why We Need Open Science Communication Experts. *Mitteilungen Der Vereinigung Österreichischer Bibliothekarinnen Und Bibliothekare*, *72*(2), 284–296. DOI: 10.31263/voebm.v72i2.3049

Viscogliosi, C., Asselin, H., Basile, S., Borwick, K., Couturier, Y., Drolet, M. J., & Levasseur, M. (2020). Importance of Indigenous elders' contributions to individual and community wellness: Results from a scoping review on social participation and intergenerational solidarity. *Canadian Journal of Public Health = Revue Canadienne de Santé Publique*, *111*, 667–681. https://link.springer.com/article/10.17269/s41997 -019- PMID: 32109314

Vos, M. G. D., Hazeleger, W., Bari, D., Behrens, J., Bendoukha, S., Garcia-Marti, I., Van Haren, R., Haupt, S. E., Hut, R., Jansson, F., Mueller, A., Neilley, P., Van Den Oord, G., Pelupessy, I., Ruti, P., Schultz, M. G., & Walton, J. (2020). Open weather and climate science in the digital era. *Geoscience Communication*, *3*(2), 191–201. DOI: 10.5194/gc-3-191-2020

Walker, C. D. B., & Burbanks, S. (2010). Afrocentricity Published in the Encyclopedia of Identity. In. Encyclopedia of Identity. Afrocentricity. Jackson, R.L II (Eds) Sage Publications

Walker, C. E. (2001). *We Can't Go Home Again: An Argument about Afrocentrism*. Oxford University Press.

Walker, W. H., & Berryman, J. (2023). Ritual closure: Rites de passage and apotropaic magic in an animate world. *Journal of Archaeological Method and Theory*, *30*(2), 449–494. https://link.springer.com/article/10.1007/s10816-022-09565-7 PMID: 35757009

Walsh, J. P. (2020). Social media and moral panics: Assessing the effects of technological change on societal reaction. *International Journal of Cultural Studies*, *23*(6), 840–859. DOI: 10.1177/1367877920912257 PMID: 40479194

Wamukoya, J., & Mutula, S. M. (2005). Capacity-building Requirements for E-records Western Cape, South Africa. *South African Journal of Information Management*, *21*(1).

Wang, Y., Li, H., Chau, L. P., & Kot, A. C. (2021, October). Embracing the dark knowledge: Domain generalization using regularized knowledge distillation. In *Proceedings of the 29th ACM International Conference on Multimedia* (pp. 2595-2604). https://doi/abs/10.1145/3474085.3475434

Wang, L., & Yoon, K. J. (2021). Knowledge distillation and student-teacher learning for visual intelligence: A review and new outlooks. *IEEE Transactions on Pattern Analysis and Machine Intelligence*, *44*(6), 3048–3068. https://ieeexplore.ieee.org/abstract/document/9340578/ PMID: 33513099

Wang, Y. (2024). Better Ways to Protect Indigenous Knowledge and Cultures Through Intellectual Property. *Lecture Notes in Education Psychology and Public Media*, *51*(1), 52–57. DOI: 10.54254/2753-7048/51/20240902

Waterton, E., & Watson, S. (2015). Methods in motion: Affecting heritage research. In *Affective Methodologies: Developing Cultural Research Strategies for the Study of Affect* (pp. 97–118). Springer.

Wepukhulu, A. N., Matisi, M. N., & Ngome, S. (2023). Evolution of Beliefs and Practices of Bride Wealth and Their Role on Marriage among the Bukusu. https://www.topflightvalet.com/formatedpaper/1704025.pdf

WHO. (2008), *Fact sheet N°134.* Available at: https://www.who.int/mediacentre/factsheets/2003/fs134/en/

WHO. (2023), *Traditional Medicine,* Available at https://www.who.int/news-room/questions-and-answers/item/traditional-medicine

Wiegers, K. (2021). *Software Development Pearls: Lessons from Fifty Years of Software Experience.* Addison-Wesley Professional. https://books.google.com/books?hl=en&lr=&id=

Wilkinson, M., Dumontier, M., Aalbersberg, I., Appleton, G., Axton, M., Baak, A., Blomberg, N., Boiten, J.-W., da Silva Santos, L. B., & Bourne, P. E.. (2016). The FAIR Guiding Principles for scientific data management and stewardship. *Scientific Data*, *3*, 160018. PMID: 26978244

Williams, D. H., & Shipley, G. P. (2023). Indigenous Research Methodologies: Challenges and Opportunities for Broader Recognition and Acceptance. *Open Journal of Social Sciences*, *11*(05), 467–500. DOI: 10.4236/jss.2023.115030

Wilson, D. (1987). African Council on Communication Education Traditional Systems of Communication in Modern African Development an Analytical Viewpoint 89

Wilson, S. (2008). Progressing toward an Indigenous research paradigm in Canada and Australia. *Canadian Journal of Native Education*, *27*(2), 161–178.

Wise, M., & Kostelecky, S. R. (2018). Respecting the language: Digitising Native American language materials. *Digital Library Perspectives*, *34*(3).

Woldeyohannes, T. (2020). Assessment of husbandry practices, production and reproductive performance of indigenous cattle in Hadiya zone, southern Ethiopia. *International Research Journal of Science and Technology, 1*(3), 177-198. https://www.researchgate.net/profile/Tariku-Woldeyohannes/

World Bank. (2025). Executive Summary. 2024. https://thedocs.worldbank.org/en/doc/4cafc3a906669ba0d34d4a39dc903472-0090012024/original/IndigenousPeoplesResilience-ExecutiveSummary-FINAL-Aug-8-2024.pdf

Wright, A. L., Gabel, C., Ballantyne, M., Jack, S. M., & Wahoush, O. (2019). Using Two-Eyed Seeing in Research with Indigenous People: An Integrative Review. *International Journal of Qualitative Methods, 18*, 1609406919869695. DOI: 10.1177/1609406919869695

Wright, K., Tapera, R. M., Stott, N. S., Sorhage, A., Mackey, A., & Williams, S. A. (2022). Indigenous health equity in health register ascertainment and data quality: A narrative review. *International Journal for Equity in Health, 21*(1), 34. DOI: 10.1186/s12939-022-01635-2 PMID: 35279132

Wynberg, R. (2022). Biopiracy: Crying wolf or a lever for equity and conservation? *Research Policy, 52*(2), 104674. DOI: 10.1016/j.respol.2022.104674

Wyrzykowski, R. (2023). Rural-urban digital divide still poses a challenge in South Africa. https://www.opensignal.com/2023/12/13/rural-urban-digital-divide-still-poses-a-challenge-in-south-africa

Yadav, R. (2025). The Intergenerational Transfer of Environmental Knowledge and Leadership. In *Generation Alpha and Next Generation Leadership for Greener Futures* (pp. 237–254). IGI Global Scientific Publishing., https://www.igi-global.com/chapter/the-intergenerational-

Yaga, D., Mell, P., Roby, N., & Scarfone, K. (2019). Blockchain technology overview. Crytography and Securitity. Available at: https://doi.org//arXiv.1906.11078 .DOI: 10.48550

Yanou, M. P., Ros-Tonen, M. A., Reed, J., Moombe, K., & Sunderland, T. (2023). Integrating local and scientific knowledge: The need for decolonising knowledge for conservation and natural resource management. *Heliyon, 9*(11).

Yeo, G. (2018). *Records, information, and data: exploring the role of record-keeping in an information culture*. Facet Publishing., DOI: 10.29085/9781783302284

Yeoh, B. S. (2020). The global cultural city? Spatial imagineering and politics in the (multi) cultural marketplaces of South-east Asia. *Culture-Led Urban Regeneration*, 102-115. https://www.taylorfrancis.com/chapters/edit/10.4324/9781315878768-6/

Yin, R. K. (2018). *Case study research and applications: design and methods* (6th ed.). Sage.

Yirga, G. (2010). Assessment of indigenous knowledge of medicinal plants in central zone of Tigray, North Ethiopia. *African Journal of Plant Science*, *4*(1).

Yiyun, Z. B. A. (2024). Museum digital repatriation and case studies: exploring Guidelines for the future practice and digital bridge to cultural continuity. Master of Arts in Communication, Culture, and Technology Thesis. Graduate School of Arts and Sciences of Georgetown University. Washington, D.C.

Yu, W., Chen, J., & Deng, S. (2024). Open Science Under Debate: Disentangling the Interest on Twitter and Scholarly Research. *SAGE Open*, *14*(3). DOI: 10.1177/21582440241271300

Zahidi, Z., Lim, Y. P., & Woods, P. C. (2013). User Experience for Digitization and Preservation of Cultural Heritage. In *Proceedings of the 2013 International Conference on Informatics and Creative Multimedia*, Kuala Lumpur, Malaysia, 4–6 September 2013, pp. 13–16.

Zanotti, L. (2024). Transdisciplinary Research Practice by, with, and for Indigenous Knowledge Holders. *Bulletin of the Ecological Society of America*, *106*(2), e2212. DOI: 10.1002/bes2.2212

Zečević, K., Houghton, C., Noone, C., Lee, H., Matvienko-Sikar, K., & Toomey, E. (2020). Exploring factors that influence the practice of Open Science by early career health researchers: A mixed methods study. *HRB Open Research*, *3*, 56. DOI: 10.12688/hrbopenres.13119.1 PMID: 33537552

Zellner, P. J., Claus, M., Dolezalova, T., Balogun, R. O., Eberle, J., Hodam, H., Eckardt, R., Meißl, S., Jacob, A., & Anghelea, A. (2024). MOOC Cubes and Clouds—Cloud Native Open Data Sciences for Earth Observation. *The International Archives of the Photogrammetry, Remote Sensing and Spatial Information Sciences, XLVIII-4/W12-2024*, 157–162. DOI: 10.5194/isprs-archives-XLVIII-4-W12-2024-157-2024

About the Contributors

Josiline Chigwada is a postdoctoral research fellow at the University of South Africa (UNISA). She is a librarian with 18 years of experience in academic librarianship. She holds an Information Science doctorate from UNISA and a Postgraduate Diploma in Higher Education. She is a member of the Library and Information Association of South Africa (LIASA), SPARC Africa management committee, Zimbabwe Library Association (ZIMLA), Zimbabwe Young Academy of Science, Zimbabwe Rural Schools Library Initiative (ZRSLI), Association for Information Science and Technology (ASIS&T), International Association of Social Science Information Service and Technology (IASSIST), and the Rotary Club of Msasa, Harare. She has authored on indigenous knowledge, open science, research data management, information and digital literacy, academic librarianship, and contemporary library and information science issues. She is an editorial board member of the IFLA Journal and the Journal of Web Librarianship.

Patrick Ngulube (PhD) is a professor of information science and interdisciplinary research at the University of South Africa. His research interests as in records and archives, e-government, application of information and communication technologies, research methods and knowledge management, including indigenous knowledge systems

***

Aniesa Samira Bafadhal is an assistant professor in the Department of Business Administration, Faculty of Administrative Science, Universitas Brawijaya, Malang City, East Java, Republic Indonesia. Since 2018, she has served as vice head of the Tourism Research Laboratory and has been involved in the Tourism Research Center at the Universitas Brawijaya. Her scholarly contributions are also evident in his capacity as a board member of the Association of the Indonesian Tourism

Education Institutes (HILDIKTIPARI). She is the author of ten books in the Tourism and Business Administration field, published in Indonesia and globally, consisting of encyclopedias, book chapters, and monographs. She is also certified in Drone Pilot by the Indonesian Drone Pilot Federation, augmented reality (AR) educator by Assemblr, and virtual reality (VR) creator by Millealab. Her research interests currently focus on several areas, including the psychology of tourism, virtual tourism, heritage tourism, and visual culture.

Muhammad Rosyihan Hendrawan is an assistant professor in the Department of Library and Information Science (LIS) at the Faculty of Administrative Science, University of Brawijaya, Indonesia. In his administrative capacity, previously, he served as the vice head of university archives at the University of Brawijaya in 2016–2020 and memory institutions (library, archives, and museums/LAMs) for the Faculty of Administrative Science at Brawijaya University in 2020–2022. His scholarly contributions are also evident in his capacity as a Chair of Research Administration and Innovation in the Indonesian Library and Information Science Scholars Association (ISIPII) 2024-2029 and a member of the International Council on Archives (ICA). His research interests include knowledge management, knowledge organization, cultural heritage information management, and archival and records management. He is the author of a number of books published in Indonesia and globally, which consist of encyclopedias, book chapters, and monographs. His multifaceted role reflects his commitment to academic excellence and significant contribution to the LIS fields.

Alfred Mapolisa is a Climate Change Scientist specializing in impacts of climate change mitigation and resilience on vulnerable communities in the Global South. He also has passion for Indigenous Knowledge Systems and their applications. Born in rural Birchenough Bridge, in 1982, his early life instilled in him a deep love for the natural world and Indigenous knowledge Systems. This passion led him to pursue a degree in Geography and Environmental Science at the University of Zimbabwe where he graduated among the top of the class. After graduating he taught High School A Level Geography for several years. He later pursued a Master of Science in Natural Resource Management and Environmental Sustainability at Bindura University of Science Education, after which he joined lectureship at Zimbabwe Open University in the Geography Department. He is also an Environmental Impact Assessment consultant.

Methembe Melusi Mhlope is a Research Support Librarian at Gwanda State University with expertise in metadata creation and management, digital curation, and FAIR-aligned research data management. In his current role, he leads digital scholarship initiatives, manages research repositories, and trains faculty and students in ethical data and information practices. Previously, he oversaw metadata standards, acquisitions, and support for open-source software implementations, and taught library systems as a teaching assistant. His academic qualifications include an MSc and BSc (Hons) in Library and Information Science from the National University of Science and Technology. His research interests include bibliometrics and research assessment, open science, indigenous knowledge, and research data management.

Artwell Moyo is a dedicated librarian with a Master of Science degree in Library Science. With experience at Bulawayo Public Library and Edward Ndlovu Memorial Library, he is passionate about ensuring effective information use within communities to support informed decision-making. His aspiration is to pursue a PhD in Library and Information Science, further advancing research and knowledge accessibility.

Kwanisai Mukwerete serves as the school librarian at Christian Brothers' College in Bulawayo. He has previously held the position of tutor at Zimbabwe Open University. His research interests encompass Indigenous Knowledge Systems, knowledge management, social media utilisation, the application of Artificial Intelligence in research, and the management of school libraries. Dr. Mukwerete has co-authored one scholarly article and is actively pursuing the publication of additional articles and book chapters in the future. He is committed and dedicated to advancing knowledge in the information sciences through continued scholarly contributions.

Shadreck Ndinde is a National University of Science and Technology Lecturer and an Advisor to the Zimbabwe Library Association (ZimLA). He previously held the position of Vice Chairperson for the African Library and Information Associations & Institutions (AFLIA), demonstrating his impactful contribution to the library and information sciences field at both national and continental levels. With a wealth of experience across various organizations, Dr. Ndinde has also co-authored scholarly articles emphasizing the significance of information in fostering community development. His professional passions are deeply rooted in Indigenous Knowledge Systems and the management of community development information projects. Dr. Ndinde's research pursuits are centered on critical areas such as the performance evaluation of information services, enhancing user experience, assessing service quality, and measuring user satisfaction. These interests underscore his commitment to advancing the standards of information services and their relevance to societal growth.

Elphas Muruli Ngaira Born on 25th August 1971 in Kakamega Current location: Nairobi City Education: PhD in Information and Knowledge Management (Candidate) Professional experience: Senior Management, Integrity Assurance, QMS-Auditing, ADR Civil and Commercial Mediator, E-learning and Automation of records Achievement and contribution: Over 20 years' work experience in strategic management of library information services

Ashah Owano holds a PhD and a MA Degree in Library and Information Studies from Moi University and the University of London (University College London) respectively. She is a librarian by profession, and is currently Chairperson and lecturer in the Department of Information, Language and Communication Studies at the Technical University of Kenya. Dr. Owano previously worked at the National Museums of Kenya as the Resource Centre Manager, and an Information Designer at Egerton University. She has vast experience in information industry; was member of the editorial committee for the Journal of East Africa Natural History and Steering Committee member of the Biodiversity Heritage Library (Africa). Her interests are in Knowledge management and the use ICTs to access information and knowledge. Dr. Owano has worked with researchers in the documentation of Kenya's indigenous knowledge, including the Suba community; and has collaborated with researchers in various multidisciplinary research projects and initiatives such as Endangered Archives Programmes (EAP) of the British Library in the digitization of historical botanical collection at the National Museums of Kenya Herbarium (formerly East African Herbarium), Goethe Institute/National Museums of Kenya Elimu Asilia initiative on digitization of Indigenous Knowledge, Digitization of the Journal of East Africa Natural History, Tanzania Notes and Records among others.

Lilian Ingutia Oyieke is a Senior Lecturer in the Department of Information Science, Languages and Communication Studies, The Technical University of Kenya. She is an accomplished academic and practitioner in Library and Information Science. She has a strong background in theory, practice, research and management of information services and knowledge utilization. As global institutions seek to build a strengthened framework for information and knowledge, her skills are applicable in various environments with a special focus on libraries and knowledge management. She is currently engaged in teaching, research and supervision of PhD and master's students. Her technical skills include development of information policy and strategy, service quality in libraries, knowledge management, personal information management, information organization, representation and retrieval, information dissemination platforms, critical information literacy, proposal writing, qualitative and mixed methods research designs. She has attended and presented papers at various conferences. She has published several articles and book chapters.

Notice Pasipamire is a Senior Lecturer and the current Chair of the Department of Library and Information Science at the National University of Science and Technology (NUST). He holds a Doctorate in Information Studies from University of KwaZulu Natal (UKZN) in South Africa, a Master of Philosophy Degree in Library and Information, a Postgraduate Diploma in Higher Education (NUST), and a Bachelor of Science Honours Degree in Library and Information Science (NUST). Dr. Pasipamire has a specialty in teaching Project Management, and Research Methods. He has written book chapters and articles on artificial intelligence, indigenous knowledge, research methods, research support services and academic libraries, and LIS education. Identifier:

Witness Roya is a postdoctoral fellow in the Department of Digital Transformation and Supply Chain Management at University of South Africa (Unisa) Graduate School of Business Leadership. Witness holds a PhD in Communication from Fort Hare University, a Master of Science in Media and Society Studies from Midlands State University (Zimbabwe) among other qualifications and has lectured English, Media and Communication in Zimbabwe and South Africa. He has authored and co-authored several research papers and attended a few conferences. Witness has few papers that are in press and manuscripts which have been accepted for publication.

Stephen Tsekea lectures in the Department of information Science and Records Management. He has over 19 years of experience in university libraries. He holds a Bachelor and Masters Degrees in Library and Information Science. Stephen also read for a Masters in Business Leadership and a Postgraduate Diploma in Tertiary Education. He has worked in different units of academic libraries especially information services and cataloguing and classification. He is a researcher and his subject areas include digital libraries, artificial intelligence, research data management, school library media centres, indigenous knowledge and intellectual property, libraries and ICT. He has published in several renowned journals and presented papers at different national and international conferences.

Index

A

B

C

D

E

F

G

I

K

M

N

www.ingramcontent.com/pod-product-compliance
Lightning Source LLC
LaVergne TN
LVHW080911300926
849162LV00003B/641